Scientists and Scholars in the Field

Scientists and Scholars in the Field

Studies in the History of Fieldwork and Expeditions

Edited by Kristian H. Nielsen, Michael Harbsmeier & Christopher J. Ries

Aarhus University Press

Scientists and Scholars in the Field
© The authors and Aarhus University Press 2012
Cover design: Harvey Macauley, Imperiet
Cover image: Galatheaekspeditionen på de Nikobariske øer 1948
© Hans Scherfig/billedkunst.dk

Typesetting and print: PJ Schmidt Grafisk
Printed in Denmark 2012

ISBN 978 87 7124 014 6

AARHUS UNIVERSITY PRESS

Aarhus
Langelandsgade 177
8200 Århus N

Copenhagen
Tuborgvej 164
2400 København NV

www.unipress.dk

INTERNATIONAL DISTRIBUTORS:

Gazelle Book Services Ltd.
White Cross Mills
Hightown, Lancaster, LA1 4XS
United Kingdom

ISD
70 Enterprise Drive
Bristol, CT 06010
USA

Contents

Kristian H. Nielsen, Michael Harbsmeier & Christopher J. Ries
    Studying Scientists and Scholars in the Field
    An Introduction   9

Michael Harbsmeier
    Fieldwork avant la lettre
    Practicing Instructions in the Eighteenth Century   29

Jeppe Strandsbjerg
    Surveying the Field of State Territory   51

Rengenier Rittersma
    Subterranean Fieldwork
    Marsili's Survey on the Biogeography and Ecobiology of
    Truffles in Eighteenth Century North and Central Italy   77

Kasper Risbjerg Eskildsen
    Exploring the Republic of Letters
    German Travellers in the Dutch Underground, 1690-1720   101

Daniel E. Clinkman
    The Civil-Military Enlightenment in Britain
    Links between the Royal Society of London and the
    British Military, 1761-1790   123

Anke Fischer-Kattner
    Entangled Experiences, Disentangling Disciplines
    Antoine and Arnauld d'Abbadie's Voyages in Ethiopia   147

Casper Andersen
    Explorer-Engineers Take the Field
    Imperial Engineers, Africa, and the Late Victorian Public   169

Palle O. Christiansen
    From Collection to Fieldwork
    The Field Research of Danish Folklorist
    Evald Tang Kristensen, 1870-1890   191

Jeremy Vetter
    Field Life in the American West
    Surveys, Networks, Stations, and Quarries   225

Neha Gupta
    Before Creation
    Competing Excavators, Imperial Interests, and the Making
    of the Indus Civilization in 1920s India   259

Esther Fihl
    The Rolling Field Station
    Danish Explorations of Central Asia in the
    Late Nineteenth Century   283

Serge Reubi
    Exploring the Disciplinary Significance of Fieldwork Methods
    A Case from the History of Swiss Anthropology   309

Christopher J. Ries
    Armchairs, Dogsleds, Ships, and Airplanes
    Field Access, Scientific Credibility, and Geological Mapping
    in Northern and North-Eastern Greenland 1900-1939   329

Kristian H. Nielsen
    Expedition "Live"
    Science, Media, and Politics on the Galathea 3
    Expedition 2006-2007   363

Jenny Beckmann
    The Swedish Taxonomy Initiative
        Managing the Boundaries of "Sweden" and "Taxonomy"   395

Mikkel Bunkenborg & Morten Axel Pedersen
    The Ethnographic Expedition 2.0
        Resurrecting the Expedition as a Social Scientific Research Method   415

Matthew Edney
    Field/Map
        A Historiographic Review and Reconsideration   431

Author information   457

Index   461

Kristian H. Nielsen, Michael Harbsmeier & Christopher J. Ries
Studying Scientists and Scholars in the Field
An Introduction

Fieldwork and expeditions have been – and still are – central to a wide range of disciplines from anthropology to zoology. Both fieldwork and the expedition signify specific types of practice that have served as a disciplinary marker of the identity of the scholars and scientists working in the field rather than remaining safely in their armchairs, in their studies scrutinizing objects, in their laboratories doing experiments, or in their museums studying objects collected by others. Historically, fieldwork and expeditions have also served to establish and consolidate new fields of study, making new territories available for future investigation and providing the means of enculturation for scientists and scholars. Within a multidisciplinary framework, this collection of essays arranged in chronological sequence provides a history of fieldwork and expeditions ranging from the voyages of exploration, educational journeys, and statistical surveys conducted from the fifteenth century and onwards through the institutionalization of fieldwork and the establishment of permanent field stations in the late nineteenth and early twentieth centuries, to activities that merge field-based research with contemporary policy issues and the increasing attention given to the public communication of science.

Fieldwork and expeditions have typically been closely associated with particular approaches to the object of study, often with mixed connotations of adventure, identity, mastery, and otherness. There are at least two series of metaphors describing the relation of scientists and scholars to fieldwork and expeditions: one agricultural and one military. Literally, "field" means open land as opposed to woodland, countryside as opposed to urban development, a piece of land set apart for pasture or tillage, a hunting ground, or a tract of ground covered with or containing some special natural product or formation. Cultivating the field in the agricultural sense, i.e. plowing, fertilizing, watering, and harvesting, is one source of the discursive imagination's attempt to make sense of the huge variety of practices in the field. The agricultural metaphor directs our attention toward the relationship between fieldworkers and the field itself.

Fieldworkers are often seen as acquiring their identity and knowledge from having extensive familiarity with the field. Knowing the field requires them to have physical intimacy with it, bordering on immersion. At the same time, they are active players in the transformation of the field from "uncultivated" to "cultivated" land. The process of cultivation calls for and possibly expands acquired competencies of fieldworkers, while also introducing changes to the field itself. Along the lines of the agricultural metaphor, and in order to connect the study of fieldworkers with the physical dimensions of the field, one might be tempted to speak about intensive vs. extensive fieldwork: where the cognitive yield per area or information unit is high, fieldwork can be highly intensive with fieldworkers returning again and again to exploit the resources of the field; whereas, in cases of low information density, fieldwork takes on a more extensive character. In this respect, the expedition can be compared to nomadic farming; the expedition, as the nomadic figure of the field sciences, keeps moving from place to place to profit from occasional fields of interest.

Understanding fieldwork and expeditions in the sense of military operations provides other meanings that apply more specifically to the interrelationship between fieldworkers. In military usage, fieldwork refers to keeping, taking, maintaining, holding, or conquering a field in battle or combat, while an expedition is a sending or setting forth of personnel and equipment with military intentions. These connotations also apply when the term "fieldwork" is used in connection with various forms of outdoor contest or sport, such as cricket and baseball. The martial and competitive implications of fieldwork address the field simply as the ground on which a battle is fought or a competition takes place. Fieldworkers may be seen as combating or competing against each other or, as representatives of scientific disciplines, as fighting for disciplinary boundaries. The question of how to define particular types of field science in opposition to other ones is, if not always immediately immanent, typically at stake when doing fieldwork.

On expeditions, which more often than fieldwork are defined as interdisciplinary, collaborative and mobile undertakings, the competition for time and resources used to study the field can turn out to be crucial. Another twist to the metaphorical juxtaposition of scientific, military, and sporting fieldwork is the view of field scientists as endeavoring to control the field in opposition to local actors and the natural conditions of the field. In order to obtain knowledge about the field, fieldworkers must control the field in some way: any kind of force that opposes their control has to be managed. Again, a difference in the way fieldwork and expeditions function in time and space results from the military interpretation: fieldwork is about securing knowledge from one and the same field on a semi-permanent basis, while an expedition enacts a transitory quest of knowledge in a number of dispersed fields.

Our discussion of agricultural and military metaphors for fieldwork and expeditions indicates that the concept of fieldwork marks epistemological, temporal, and spatial boundaries. The attraction of fieldwork arises from the multiple meanings of the field as an area of action, operation, play, and investigation; as a range of opportunities, or of objects; as a sphere for labour, study, or contemplation; and as a department or subject of activity or speculation. By defining and implementing fieldwork and expeditions, scholars and scientists chart the boundaries of their field in disciplinary, institutional, and territorial terms. In doing so, they specify the knowledge and methods necessary for the study of particular fields.

Historically, the botanical, zoological, geological, geographic, and social sciences were the first to enter the field in search of knowledge and discovery. A closer analysis of the historical changes in the field practices of disciplines as diverse as glaciology and the study of religions, palaeontology and linguistics, meteorology and ethnography, deep-sea biology and folklore studies will enhance our understanding of field practices in fieldwork and expeditions as social and cultural processes. While expeditions and fieldwork have been an important part of scholarly practice within many intellectual and scientific disciplines, the systematic discussion of the methodology, theory, and history of fieldwork has received little attention outside the disciplines of anthropology and geography. The essays in this collection are devoted to a broader study of the forms, practices, and applications of fieldwork as it has been conducted by scientists and scholars within a much wider range of disciplines in the humanities and the social and natural sciences.

Traditionally, fieldwork and expeditions have served as collective designations for a host of heterogeneous practices related to the collection and the production of data. Measuring, counting, mapping, excavating, interviewing, and experimenting are all activities that have been conducted "in the field" in order to transform information and artifacts into objects for scientific processing. Over the previous two decades, historians of science have increasingly subjected the many forms of scientific practice to studies of the social, cultural, and material conditions under which they occur. However, the majority of such studies have been directed towards the traditional cultural centers of science, such as laboratories, hospitals, museums, universities, and libraries.

The fact that until recently the study of the history of scientific field practices has received comparably sparse attention may well reflect the relatively low academic status usually ascribed to fieldwork. Since the nineteenth century, standards of scientific stringency, neutrality, and accountability have been defined according to ideals manifested in the limited, generalized, and carefully composed framework of the laboratory. Fieldwork, on the other hand, must by definition be conducted in intimate and often unpredictable interaction with specific geographical localities with specific characteristics that influence, shape, and to some degree even constitute results. Yet while this may have earned the field sciences a reputation for being further removed from the scientific ideal set up in the laboratory, fieldwork remains a crucial tool for making the world operational, comprehensible, and communicable for scholars and scientists.

On the basis of such considerations, a number of historians of science and the humanities were invited to join us in exploring the potential of a history of expeditions and fieldwork in as many different disciplines as possible and the framework of a network for the history and sociology of fieldwork and scientific expedition was established. The network activities included: a Field Studies Workshop at the Sandbjerg Estate, Denmark, 23-25 November 2006; a one day seminar in May 2007 at Aarhus University under the heading "Ideologies and Controversies in 20th Century Scientific Exploration"; a Session at the 4S Annual Meeting, Montreal, Canada, 11-13 October 2007; and an international research seminar on local knowledge and global consequences in August 2007 held at the Barents Institute in Kirkenæs, Norway. The network concluded its operation with a final conference called "Ways of knowing the field: International conference on the history of fieldwork, cartography, and scientific exploration," held at the Carlsberg Academy in Copenhagen, 13-15 August 2008. Most of the papers collected in this volume were presented at this conference. In what follows, we will briefly summarize the contributions to this volume under the themes that we have come to find important to studies of fieldwork and expeditions: disciplinary and interdisciplinary collaborations, public understanding of fieldwork, cartography and fieldwork, technologies of fieldwork, national identity, and encounters with locals and amateurs.

Disciplines and Interdisciplinarity
Scholars and scientists in the field often have found themselves immersed in settings characterized by the interaction and co-operation between many different disciplines that otherwise would not normally intermingle. The history of fieldwork and expeditions can therefore contribute not only to the history of individual scientific disciplines, but also to our understanding of the way in which co-operative relations between disciplines have developed over time. Studies in the history and sociology of fieldwork seem particularly promising approaches to understanding the emergence and consolidation of disciplinary boundaries and interdisciplinarity in the field sciences. In particular, such studies could serve to illustrate the various ways in which fieldworkers have tried to accommodate their experiences in the field into disciplinary or non-disciplinary frameworks.

Kasper Risbjerg Eskildsen's article shows how scholars of the pre-disciplinary European Republic of Letters travelled extensively in order to acquire knowledge from libraries and manuscript collections at famous universities and academies and also, as in the case of the later Göttingen professor Christoph August Heumann, information about underground freethinkers and sectarians. The travelling scholars established personal friendships and relationships of mutual trust with their colleagues. Their journals were filled with accounts of their academic experiences as well as detailed summaries of the conversations they had. Exploring the Republic of Letters implied crossing the boundaries between many different kinds of specializations, which were only just beginning to develop into different disciplines. At the same time, early modern scholars expanded the geography of the Republic by including knowledge from sources and places that had previously remained unexplored.

Anke Fischer-Kattner's contribution sketches the voyages of the French brothers Antoine and Arnauld d'Abbadie in the area of Africa then known to Europeans as Abyssinia, but which the brothers referred to as Ethiopia, the name preferred by the inhabitants of the region. The two brothers both made an effort to fit their complex and what Anke Fischer-Kattner refers to as "entangled experience of travel" into, on the one hand, the existing vocabulary of travelogues with a scientific twist that since the late Enlightenment had allowed for thick descriptions of personal and inter-personal uncertainties, and, on the other hand, the new framework of a more specialized, disciplinary, and impersonal rhetoric. Both brothers sensed that their travels and interactions with native people, local customs, and unexpected events called for hybrid stories and for possible changes to their own European way of thinking. Perhaps Arnauld more so than Antoine: Antoine always saw himself as a European man of knowledge, while Arnauld prolonged his stay in Ethiopia, allowing himself to become engaged in local politics and warfare. Upon their return, Antoine, the distanced scientist-voyager, wrote specialized books on the geography and linguistics of Ethiopia, while Arnauld, the immersed scientist-participant, published a lively and colourful account of his travels to meet the growing demand for entertaining books of adventure. The two brothers used separate genres, the specialized scientific book and the broader adventurous account respectively, thus effectively "disentangling" their entangled experiences of Ethiopia.

Paraphrasing Fischer-Kattner, we might understand the contribution of Mikkel Bunkenborg and Morten Axel Pedersen as attempting a "re-entanglement" of anthropological fieldwork with the somewhat old-fashioned notion of expeditions. Reflecting on their own field experiences in Mongolia, they come to the conclusion that what they are doing is inadequately captured by the dominant conception of "multi-sited fieldwork." Multi-sited anthropology seeks to follow persons, objects, meanings, and identities from one place to another rather than relying on the pre-understanding that such entities are tied to specific sites; yet the cosmopolitan fieldworker remains confident that his or her methodological grasp on multi-sited realities is firm and epistemologically sound. In contrast, Mikkel Bunkenborg and Morten Axel Pedersen found that their own identities as anthropological fieldworkers resembled their accidentally mobile and deliberately arbitrary object of study too much for them to feel entirely at home in the concept of multi-sited fieldwork. Instead, they identified much more with the expeditions of the past, arguing that (some) fieldworkers need to leave behind all presumptions about well-defined and stationary field sites, whether they are single- or multi-sited, and to embrace the collaborative and mobile nature of the "expedition 2.0." The reader thus might be tempted to see Bunkenborg and Pedersen as contemporary, inverted versions of the d'Abbadie brothers: as they try to grasp the implications of their entangled experiences of travel, they find themselves challenging contemporary anthropological and disentangled understandings of the field in terms of a location or a series of locations. They insist on the entanglement of the field and its fieldworkers.

Whereas Bunkenborg and Pedersen confront contemporary meanings of anthropological fieldwork, Michael Harbsmeier takes on the historiography of fieldwork and the making of anthropology as a discipline. In what is normally seen as the pre-disciplinary period of anthropological fieldwork, i.e. before the late nineteenth century, Harbsmeier traces detailed instructions of ways of knowing the field that can be compared to modern ones in terms of appropriate research questions, methods, interactions with local informants etc. In particular, he challenges the notion that, at the time, instructions were only produced by armchair anthropologists with little or no experience in fieldwork. It seems that field instructions of the eighteenth century were far from figments of the imagination; rather, they were based on deep understandings of the cutting edge of knowledge and of the numerous practicalities involved in doing fieldwork. In short, such instructions are not only evidence of anthropological fieldwork before it became institutionalized in disciplinary history: they also show us that, at the time, fieldwork and writing were closely related practices.

Fieldwork in the Public Eye
Expeditions and the research generated by them are commonly subject to considerable public interest and attention. The history of fieldwork and expeditions can therefore be expected to further our understanding of various modes of interaction between science, policy-making, and the public. Expeditions can be very costly affairs and

often involve raising private or public funds, not only in preparation for the actual field operation, but also afterwards for the processing and publication of results. Kristian H. Nielsen studies the making of public narratives about the Danish Galathea 3 Expedition from 2006 to 2007, showing how scientific ambitions, political agendas, and media interests were co-constructed. The Galathea 3 Expedition combined intensive live coverage of fieldwork with public appreciation of science: covering the expedition online for the public was seen as a way of promoting both the role of the researcher, while promoting first-grade research in Denmark. Public attention played an important part in making the expedition possible, and it also increased the complexity of portraying and understanding what the expedition was all about. Nielsen argues that, by spanning many different spheres of influence, the expedition was embedded in many social contexts not only in the sense that it acquired a high degree of interpretative flexibility and thus joined together multiple social groupings; it also became vulnerable to conflicting accusations of being too popular, or not popular enough. Close affiliations between the fields of science and government can sometimes become precarious.

Similarly, Jenny Beckman shows in her study of the Swedish Taxonomy Initiative (STI) that the popular rhetoric of STI, which emphasises the national heritage of Carl Linnaeus and thus establishes historical affiliations between Sweden and taxonomy, was important in the initiative's receiving national funding. Like the Galathea 3 Expedition, the STI became possible in consequence of contingent links between the organisers and the government. Circumventing traditional routes of research funding, both the expedition and the taxonomy initiative were viewed with scepticism by some scientists, while also enabling new scientific projects and collaborations. Whereas the STI was presented to the public as a national venture into biodiversity and inventory-making, the Galathea 3 Expedition was seen as a way of promoting Danish research by providing extra funding to researchers based in Denmark and by improving the public image of scientific research. Revealing to the general public and to young people in particular the fascination and attraction of scientific field studies played an important role in legitimizing the expedition. Because of its national starting point, the STI ran into problems defining national boundaries for the natural world, while the Galathea 3 Expedition encountered public criticism because it was unable to demarcate clearly the boundary between scientific research and public communication of science. The fields of research defined by the researchers of the STI and the Expedition were closely linked to the popular origins of the two projects.

Mapping Fields

Maps, images, figures and other types of visual representations often play a prominent role in the field sciences, be it for registration, organisation, and preservation of observations in the field, or for communication of results in books, journals, newspapers, movies, or on television. Maps define the spatial divide between the known and the unknown. They are tools of domestication, employed to represent spaces and territories so as to make them familiar and potentially inhabitable for us. Jeppe Strandsbjerg suggests that Bruno Latour's metaphor of "drawing things together" might be apt for understanding what it is cartographers do when they produce maps that mark the spatial divide between known (home) territory and unknown fields. Cartography, in this sense, is the practice of going out into the field in order to make immutable mobiles (signs, inscriptions, minute transformations of the field, object collections etc.) that can be returned to the home base, the centers of calculation, where the maps are being produced. The process establishes a network of circulating references and practices that produce knowledge about territories that were previously foreign while also enhancing the power of the calculative center.

Christopher J. Ries' paper about scientific credibility, territorial claims, and geological surveys in Greenland gives ample evidence for the power of maps to enable, sustain, and/or obstruct access to the field and the credibility of those speaking in the name of the field. In the 1930s Lauge Koch, the protagonist of the paper, produced and circulated many maps that were helpful to Denmark in the territorial case against Norway in the international court of The Hague, but also sparked controversy in the Danish community of geologists. Impatient with his Danish colleagues, but eager to introduce new means of drawing together information about the topography and geology of Greenland, Koch held an influential, yet somewhat uncertain position in the geo-scientific discipline. He was instrumental in transforming geological practices in the field, but also caused scientific "fault lines" that would take many decades for Danish geology to overcome.

The narratives of Strandsbjerg and Ries situate maps and mapmaking in a larger socio-epistemological framework. Similarly, in his essay on the historiography of cartography, Matthew Edney argues that maps have all too often been understood simply as generic devices for spatial representation. Just as historians of science until recently have failed to appreciate the heterogeneity of scientific field practices and thus the diversity of ways of knowing the field, so historians of cartography have based their studies on ideal conceptions of mapping and the map. Edney identifies three phases or approaches in the historiography of cartography, each of which conceptualized the relation between map and the field in different ways:

- Traditional map history emphasised map content and maps as evidence of historical worldviews. Map history was construed as a whiggish account of progress. Needless to say, this approach took no interest in actual cartographic practices or any kind of fieldwork.
- Internal map history enforced a shift of interest from map content to map form or design. The narratives produced were those of different cartographic professions or specialities. Although the internal approach produced a break with the notion of unified progress within map history, it shared the commitment of traditional map history to maps and mapmaking as representation. Fieldwork still remained outside the horizon of internal map historians.
- Socio-cultural map history primarily sought to appreciate the multiplicities of social and cultural contexts of maps, their production, and their function. This approach allowed for new studies of all kinds of mapping practices, although many proponents saw mapmaking as entirely embedded within existing power structures.

Edney envisages a fourth approach: processual map history. Taking seriously the basic contextual insights of socio-cultural map history, this fourth approach would instigate more detailed studies of the production, distribution, and circulation of maps. Basically, he suggests that historians of cartography see mapmaking as the coming together of a number of different processes: discourses or semiotic networks of spatial representations, technologies of cartography, materiality and design of maps, map usages, cartographic fieldwork etc. The relation of map to field or territory is contingent upon the discourses of space and spatial representation and the cartographic techniques by which mapmakers and others make sense of spaces and maps.

Technologies of Travel
Just as mapping depends on discourses as well as technology, so fieldwork in general requires the merging together of discursive and material practices. It is an obvious yet sometimes overlooked fact that fieldwork can be distinguished from work in the laboratory on the level of instrumentation and technology: fieldworkers simply require technologies of travel in order to enter into the field and to carry out their research. Fieldwork under a variety of different climatic, geographical, and social circumstances has stimulated the development of a wide spectrum of technologies ranging from clothing, nutrition, and means of transport to medicine, instrumentation, and various platforms for fieldwork. In some cases such

technologies may be ascribed to the ingenuity of a single person, in others they arise from international co-operation between several actors, or they may result from the adaptation of technologies used by the local population in a specific locality. Despite such differences, the technologies of travel that enable the physical mobility of fieldworkers all are carefully chosen, if not constructed, in a way that means they not only mobilise researchers and their equipment, but also facilitate the epistemological requirements of fieldwork. Investigations of the material culture of the field sciences will provide new knowledge about the ways in which the field in terms of mobility and episteme has been shaped and shared among different cultures, and contribute to our understanding of the role of logistics and instrumentation in the knowledge production of field scientists.

In her essay about two Danish expeditions to Central Asia in the late nineteenth century led by Ole Olufsen, Esther Fihl finds that new technologies of travel encouraged field researchers to adopt new ways of studying the field. Ethnographic expeditions have a long history as an important mode of exploring non-Western fields in particular. However, during the nineteenth century, and partly due to the invention or adoption of new technologies of travel, expeditions gradually changed their pace, so to speak. On an old-fashioned expedition, the dominant way of entering remote fields was by horseback, using tents for stopovers. These technologies, if one may use the term technology about horses and tents, provided for a slow expedition pace, but also necessitated being more or less constantly on the move. This was Olufsen's preferred way of exploring the field. It made possible certain types of observation, while also making it impossible to observe a single location in a specific field over a longer period of time. Olufsen's companion Paulsen wanted to do more of this other kind of field observation. To this end, he found that the rolling field station, a train wagon made available to the researchers on their second expedition, was perfect. It provided for easy and convenient transportation between different fields, making longer stays in one place possible and perhaps even enjoyable. In effect, it was a "rolling field station." As Fihl also stresses, technologies of travel available to the fieldworker include more than just means of transportation. In order to assess the epistemological consequences of the technologies brought to the field by researchers and others fully, we also have to consider research questions, conceptual frameworks, instructions, national narratives, and much more. These immaterial technologies of travel are important as they guide the fieldworker in understanding specific fields in certain ways.

Fihl gives particular attention to technological constraints and possibilities when it comes to knowing the field. So does Jenny Beckman in her article about the STI, which takes place in a completely different historical context, more than a hundred years after the two Olufsen expeditions. Whereas Fihl deals with technologies of travel, Beckman studies technologies of knowing by identifying the taxonomy inventory as crucial to the purpose of the STI, or, more specifically, the All Taxa Biodiversity Inventory (ATBI) of which only a few national ones exist in the world. Despite the fact that the fauna of Sweden is no diversity hot spot, the STI specifically aimed at constructing a Swedish ATBI and funded projects accordingly. The way in which different biological fields are found worthy of exploration depends to a large extent on the perceived shortages in the ATBI.

In the case of the geological mapping of Greenland, studied by Christopher J. Ries, technologies of travel and modes of knowing the field seem to be closely related. Lauge Koch's Three Year Expedition 1931-34 heralded a transformation of Greenland exploration, which had traditionally been conducted using ships and dog sleds. On this expedition, Koch introduced new technologies of travel such as permanent wintering stations, airplanes, radios, and more. He also organised international teams of highly specialized scientists, all of which brought instruments particular to their field of study. All of a sudden, the expedition could no longer be seen as the heroic achievement of an individual who embodied a special knowledge of Arctic conditions. In effect, Koch modernized the scientific expedition in Greenland, while also making possible new, more specialized ways of surveying the country's geology and geography.

Whereas Ries emphasizes the intimate connection between technology and knowing, Casper Andersen deals with the technological dimension of fieldwork on the institutional and infrastructural levels. Andersen examines the role of civil engineering in the construction of empire, noting that engineers often followed immediately after imperial explorers in order to secure infrastructure and thus help create political stability in the emerging empire. More specifically, he looks at how, in their writings about engineering and empire, civil engineers fed on cultural discourses of exploration, which made it possible for them to construct the public image of the explorer-engineer. In their autobiographies and articles, civil engineers sensationalized the foreign environments in which they were working, stressing that heroism combined with a strong morality and endurance were necessary for travel and exploration. The image of the explorer-engineer was the recurring motive in the narratives produced by civil engineers and others who sought public recognition and state support of their travels and engineering (or literary) projects.

Jeremy Vetter takes a systematic approach to the study of field practices in the U.S. Central West during the railroad era of the late nineteenth and early twentieth centuries, dividing emerging modes of knowledge production in the field into four categories: surveys, networks, stations, and quarries, each of which depended on institutional settings and technical equipment. Surveys entailed a high degree of mobilisation of fieldworkers and high standards for surveying practice. Field teams organised by government agencies and state universities undertook geological and biological surveys, following specifically selected routes in order to answer highly specific questions about the field. Like surveys, networks of lay observers were used to obtain and collect information about phenomena on a vast scale. Because of the cost of travelling into the field, surveys were often complemented by questionnaires and circulars, which were distributed along with proper instructions to individuals or local authorities. Unlike surveys and networks, stations and quarries were used to produce knowledge about particular environmental locations and thus were less dependent upon means of transportation. As railroad companies extended their tracks into the West and as telegraph-based communication was promoted, the mobility of fieldworkers and their knowledge increased, making possible new interactions between modes of knowledge production in the field.

Nationalizing the Field
Andersen's study of the engineers of the British Empire serves as a reminder that fieldwork and expeditions, especially the larger and more expensive expeditions, have been marked by the challenge of reconciling national ambition with the need for international anchorage and recognition. This challenge exists within many forms of modern science, but in relation to scientific expeditions it is often intensified and condensed in a mixed agenda of scientific investigation, political diplomacy, and intelligence work. A systematic analysis of the interaction between national and international players and interests in the preparation, execution, and subsequent processing of expedition work can shed light on differences, similarities, and developments in the practical realisation of this interaction.

As shown in Daniel Clinkman's study of eighteenth century co-operations between the Royal Society and the Royal Navy and the Board of Ordnance, scientific and military interests go well together for a number of reasons, such as joint national ambitions. From its inception, the Royal Society had been of assistance to the Navy in training officers in mathematics, thus translating scientific practices from the confines of a learned society to the (military) field. Many of the Society's most active members wore uniform. They and others were pivotal in making science a significant part of the pursuit of empire, providing the military with tools for organising, surveying, navigating, calculating etc. Tracing four stages in the co-operation between the Royal Society and the Navy, Clinkman shows that scientific knowledge influenced military operations, but also that the Navy provided for scientific activities with no immediate link to naval interest, and naval officers contributed scientific articles to the *Transactions of the Royal Society*. In view of its status as an independent and internationally minded society, the Royal Society balanced on the edge of patriotism vs. cosmopolitanism. The joint British scientific-naval field research of the eighteenth century, Clinkman argues, laid the foundation for the Society's role as advisor to the government in the nineteenth century.

The Galathea 3 Expedition of the twenty-first century, studied by Kristian H. Nielsen, also had to strike a balance between international and national aspirations. On the one hand, the expedition was designed to promote first-class research in Danish research institutions, which demanded both collaboration and comparison with international research communities. On the other hand, the expedition aimed to promote national interests and, moreover, fed on national narratives by creating links with the two previous Galathea expeditions in 1845-47 and 1950-52 respectively. The national rhetoric was bombastic at times, leading some commentators to portray the expedition as a thing of the past. Still, this national ambition along with the generous support of the government was essential to the expedition and a way in which nationalist aspirations could be reconciled with internationally recognised research agendas.

Neha Gupta revisits colonial archaeology in Imperial India in the first half of the twentieth century. Excavated before the creation of the independent nation states of India and Pakistan, Harappa and Mohenjodaro, the premier sites of the Indus Valley civilization, served as field sites for competing excavators and institutions, while also fuelling popular interest. The search for the pre-history of the region became a focal point for colonial interests and for emerging national movements in what later became India and Pakistan. Sir John H. Marshall, Director-General of the Archaeological Survey of India, headed the first excavation campaign in 1921-22 at Harappa, prompted by the findings of Harappan seals. In December 1924, he announced to the British public that an indigenous Golden Age had been discovered with possible links to the Sumerian civilization. Four years later, Marshall changed the name from the Indo-Sumerian civilization to the Indus civilization. He argued that the Sumerian and the Indus cultures had been distinct, although certain similarities suggested intimate commercial and other types of relationship. The image of an independent Indus culture confronted the received view that civilization had been imported by Aryan immigrants; yet it also stirred the imaginations of the British public, who at the time were following Leonard Woolley's excavations at Ur. Taking place in the wake of a crumbling empire torn apart by fierce territorial competition and economic depression, the excavations at Harappa and Mohenjodaro served to extend the westward expansion of Indus civilization towards the Kingdom of Iraq, while also opening up new venues for archaeological field studies in India and Pakistan.

As has already been mentioned above, maps are not only means of representing cartographic knowledge, but often also serve as instruments of national sovereignty. The most obvious examples are maps used to employ military or imperial power, and both Jeppe Strandsbjerg's and Christopher J. Ries' papers provide good examples in this respect. Strandsbjerg argues, for example, that mapping Denmark served as a precondition for the construction and successful operation of the absolutist Kingdom of Denmark. Before the first attempts to map Danish territory, knowledge and power of the kingdom were based on networks of allegiance and jurisdiction, which often found their expression in a textual form as well as personal experience. The first cartographic representations of Denmark, although incomplete and faulty, served to make the country knowable in a centralized and visual way. Whereas previously, Denmark had been tied together by local agricultural practices and dispersed social ties, the new cartographic representations made it possible to think about – and rule – Denmark as a homogenous and abstract national space. Noting that the first general map of Denmark completed in 1650 by Johannes Mejer coincides with the introduction of absolutism, Strandsbjerg claims that such maps effectively destabilised local authorities and served the interest of centralized government.

Encounters in the Field

Fieldwork abroad often involves interaction and co-operation between, on the one hand, scientists and scholars and, on the other, local inhabitants in the role of hosts, carriers, assistants, and informants. All these forms of interaction provide evidence for scientific and scholarly practices as inter-cultural and social processes, which are frequently considered in colonial and post-colonial contexts. Expeditions are often so well-documented by the sources (e.g. narratives, diaries, correspondence etc.) that the scientific processes can be followed at a level of intimacy resembling a sort of historical participant observation. A systematic analysis of individual expeditions may thus yield a better understanding of social and cultural aspects of the various scientific practices applied in the field.

Around 1920, anthropology witnessed a revolutionary change in its field methods from extensive to intensive field studies, primarily caused by the writings of Bronislaw Malinowski. At least, this is the standard view, found in textbooks and other kinds of presentist notions of the history of anthropology. In his article, Serge Reubi studies two Swiss anthropologists, Felix Speiser and Paul Wirz, both contemporaries of Malinowski. Although Speiser and Wirz both recognised the importance of doing intensive fieldwork in the new Malinowskian vein, Wirz was the only one of the two to practise this kind of fieldwork. Speiser apparently lacked a flair for getting to know local people very well; accordingly, he seemed content with collecting objects for his home institution, the Basel Ethnographical Museum. This institutional framework of his fieldwork was important. From the museum's point of view, extensive fieldwork seemed like a better option than its intensive counterpart. In the case of Paul Wirz, for as long as he had the financial means, he preferred intensive fieldwork, staying and living for long periods of time in a small village in New Guinea. Financially ruined by the post-World War I crisis, he took on freelance work for several European museums and adopted the extensive mode of anthropological fieldwork.

Reubi points out that, since Malinowski, the difference between extensive and intensive fieldwork has been used to demarcate different paradigms of anthropology: ethnology, based on extensive fieldwork, which can only produce superficial and diachronic knowledge about cultures, and social anthropology, which is based on the functional, synchronic relations between society and people. This is a long-term change in anthropology that took place in the period from the late nineteenth century to the mid-twentieth century and which even today defines disciplinary borders in terms of separate methods and cognitive ambitions. However, it is an anachronism to apply such sharp divisions to fieldworkers such as Felix Speiser and Paul Wirz, who both conducted fieldwork at a relatively low level of academic specialization. To them, the field was defined not so much in terms of different pre-conceived methodological approaches, but rather in consequence of a simple ontological division of the world: anthropology as the study of primitive populations. Accordingly, anthropologists may choose to study small-scale communities or large-scale areas; the decision has little epistemological consequence. Anthropology is what they are doing.

As a critique of presentist representations of anthropology in terms of revolutionary change, Reubi's study stays at the level of actual practices in the field. Along the same lines, Palle O. Christiansen tells the story of the Danish folklorist Evald Tang Kristensen, who until his death in 1929 had spent 50 years collecting oral tradition, i.e., songs and tales, mainly from the Jutland peasantry. Tang Kristensen conducted intensive fieldwork as a means of gathering, accumulating, and publishing folklore. He worked in the tradition of local recorders and formed part of the nationwide network of recorders organised by the Danish philologist Svend Grundtvig. Through his correspondence with Grundtvig, Tang Kristensen learned to focus on ballads and legends as the primary object of study. In order to collect such textual evidence of folklore, Tang Kristensen conducted fieldwork in ever-widening circles among singers and story-tellers in Jutland. In 1888 he received a life annuity from the state for permanent folkloristic studies, and from then on, he began extending the length of his fieldtrips, most of which were usually conducted in the winter, when his informants were at home and had time to talk.

For the most part, Tang Kristensen travelled by foot, staying overnight with the local village teacher or parish pastor, or at the home of his informants. He became accustomed to the living conditions and mental attitude of the rural population. Keeping a detailed diary, noting names and provenances, and writing down the songs and tales of the local people in their presence, Tang Kristensen's collection of folklore was of a remarkable quantity and quality. In the 1880s, he reflected on his own fieldwork several times, publishing methodological rules of thumbs for folklorists based on his own experiences. He certainly identified with the role of a collector of folklore, describing what he did as tapping the source or extracting the mine of ballads and tales. However, his intimate knowledge of people's lives and living conditions led him to take an interest in culture more broadly conceived. Although he never broke with the received view of Romanticism that local or national culture was condensed in ballads and tales, he also expressed culture as regional differences in human life and social relations. However, his intensive field studies and the possible consequences for the object of study had no immediate bearing on the study of folklore. Folkloristic armchair ethnography, relying on a network of local correspondents and on material collections in urban centers, continued to be the most widespread way of studying folklore until the mid-twentieth century. When the new scientific fieldwork of American and British anthropologists began to make an impact on the study of folklore in Denmark and its neighbouring countries, pioneers such as Tang Kristensen seemed to have been long forgotten.

Reubi and Christiansen both study the actual practices of fieldwork, and both critically assess the anthropological myth of the invention of fieldwork in the early twentieth century. Casper Andersen, in his study of imperial engineers, argues that it is also important to scrutinize the self-representations of various kinds of fieldworkers in order to understand their sense of self and the ways in which they want others to see them and their fieldwork. Andersen refers to these engineers as explorer-engineers, as they, like other fieldworkers, emphasise the hardships and sufferings of the field. Unlike expeditions, large-scale engineering projects often involved hundreds or even thousands of people and typically took place in just one location or, in the case of railway construction, moved ahead at a steady pace. Still, the rhetoric was very much based on individual heroism, as was the case with most expedition narratives at the time. Heroic individuality was the appropriate cultural match for the difficulties and problems encountered in the field, but it was also needed to meet and overcome what was described as the inferiority and primitiveness of the indigenous people, who were employed as the workforce in most engineering projects. Engineering narratives reflected the racial and cultural prejudices of the time by the standards of which it was all too easy to cast local people as ignorant and incompetent. The lack of ability of "the natives" formed an important part of telling stories about the heroism of explorer-engineers.

Who's the Amateur?

Fieldwork and expeditions yield a particularly clear and instructive image of the interaction between amateurs, who are often experienced in fieldwork, and professional scientists, who only practice fieldwork on occasion. In some disciplines, such as ornithology and archaeology, highly specialised amateurs play a very prominent role. Just as it is instructive to look at encounters between fieldworkers and local people and communities, investigations of the interaction between amateurs and professionals in the field may potentially serve to enhance our understanding of the development of specific characteristics by which the scientific generation of knowledge in various disciplines can be set apart from other ways of knowing.

Rengenier C. Rittersma studies the extended network of correspondents and assistants mobilised by the Bolognese count and naturalist Luigi Ferdinando Marsili in the early eighteenth century in order to acquire knowledge about truffles. Marsili wanted to prove his hypothesis that truffles, indeed all mushrooms, are generated by external factors relating, in particular, to their host environments: stone, wood, soils etc. Around 1714-15 Marsili designed and carried out a large-scale field survey in selected regions of North and Central Italy. It seems as if he specifically chose his regions in order to cover more or less the whole field of Italian truffle production, and also to allow him to correlate different habitats with the occurrence of different truffle types. For reasons still unknown, Marsili clearly wanted to get to the bottom of the problem of the origin and reproductive nature of truffles. He circulated detailed lists of queries to his correspondents, who were all located in the truffle-producing regions of Italy. The questionnaire concerned issues of habitat, climate, and morphological aspects of truffle production, and also invited the respondents to include "further information, serving the truffle research in order to reveal their nature." At a point in the history of science where the identity of the naturalist continued to be defined by his relationship to the writings of the ancients, Marsili sought new ways of gathering information in order to corroborate theories about the natural world. In so doing, he remained at the center of naturalist calculation, using technologies of writing to gather new knowledge about truffles from local, more or less knowledgeable, and more or less eloquent informants.

Palle O. Christiansen's essay on early folklore studies in Denmark also takes as its point of departure an elaborate network of more than 300 correspondents established in the nineteenth century by Danish philologist Svend Grundtvig. Unlike Rittersma, who looks at the network organiser, Christiansen follows one of the correspondents, Evald Tang Kristensen. By pointing out the many ways in which Tang Kristensen refined the fieldwork of folklore studies, Christiansen questions the clear separation of amateurs and experts. Except for his correspondence with Grundtvig, which occasionally revolved around methodological issues pertaining to linguistic and folkloristic fieldwork, Tang Kristensen was self-taught. He developed distinctive collecting and working methods. Being a persistent and enduring fieldworker, he himself probably collected more non-material folk culture than any other one person. Since both Rittersma's and Christiansen's studies takes place at a time when biological and folkloristic field studies were still highly sporadic and pre-disciplinary, it seems reasonable to speculate that, among other things, the specialization of field studies is a process that separates those with the proper field competencies and expertise from those without.

Jeremy Vetter gives an account of different modes of field knowledge production in the Central American West in which the difference between experts and lay people plays an important role. Whereas the data gathered by people with little knowledge about systematic research and scientific information processing were vital to surveys and networks, stations and quarries depended on the separation of scientists and amateurs. As in Denmark, where the collection practices of people like Tang Kristensen proved important to the subsequent establishment of folklore studies as an academic field in its own right, intensive ethnographic fieldwork in the Central West was used to separate the scientific study of American Indians from traditional approaches to interpreting legends and narratives. Ethnographic studies were carried out in what Vetter describes as the quarry mode of fieldwork, a term which also includes fossil quarries. The quarry field study is highly localised and specialised, using local people as a workforce or as informants. The same goes for the field station, the main difference between station and quarry being that stations were often established to perform experiments and to manage and maintain natural habitats, whereas the quarry was more like a temporary observatory with the ability to disturb or even destroy the environment in which it was placed.

Conclusion
Recent discussions about fieldwork among anthropologists seem to be moving in two directions: Studying the history of anthropology, some scholars have arrived at a more nuanced understanding of the origins and early history of fieldwork,

while others have raised doubts about the very possibility of doing fieldwork in the classical sense in a globalised world. The former have paid increasingly close attention to the circumstances under which Franz Boas, W. H. R. Rivers, and few years later Bronislaw Malinowski developed and indeed invented the practices and methods which were to become the very foundation of modern social and cultural anthropology. In effect, this amounts to narrowing down the history of fieldwork to apparently involving only a handful of British and American pioneers and the contexts in which they were working. Trying to expand the range of issues to be studied by their colleagues and students, the latter kind of anthropologist has on the contrary widened and broadened the range of practices to be called fieldwork. In its contemporary versions, such as multi-sited anthropology or anthropology at home, fieldwork becomes much less well-defined than it appears in the accounts of its origins with Franz Boas in Baffinland in the 1880s or the Torres Straits Expedition with Haddon and Rivers in 1898.

We are tempted to see both developments as being fruitful to our purposes. This edited book is, we believe, an attempt to widen our understandings of fieldwork in its historical and contemporary versions. Looking at fieldwork in disciplines and national traditions different from the ones in which historians of anthropological fieldwork normally feel at home, our collection of papers explores other ways of working in the field, other "spatial practices" to use one of James Clifford's expressions, than the ones canonized by the disciplines which have made fieldwork the very foundation of their identity and legitimacy. Moreover, most papers in this book not only provide insight into actual field practices, but also contextualize fieldwork in its historical, social, and philosophical contexts. Spanning several centuries and continents, the contributions make us aware that fieldwork has always already been (and still is) the subject of investigation, reflection, transformation, politicization, and controversy among fieldworkers.

With this volume, we would like to make a case for an extended and expansive historical research program with respect to studying scholars and scientists in the field. We specifically chose to reflect on the following themes: disciplinary and interdisciplinary collaborations, public understanding of fieldwork, cartography and fieldwork, technologies of fieldwork, national identity, and encounters with locals and amateurs. The dimensions of fieldwork are numerous, indeed too numerous to be explored in a volume like this. The huge diversity of fieldwork demonstrated by the book's many cases has made us painfully aware of the immense task involved in providing historical narratives of fieldwork throughout the ages and across the globe. The multiplicity and heterogeneity of fieldwork notwithstanding, we still see the concept of fieldwork as being fruitful to historical analysis. There are many different aspects of fieldwork that still need historical illumination. Currently, there is a contradiction between the historical details provided by careful historical analysis and the historiography of fieldwork as such. We would like to encourage historians in future explorations of fieldwork to move beyond thinking about it in the singular. We believe it is time to develop more nuanced conceptual frameworks that will allow us to become even more attentive to actual historical and geographical differences.

We offer no easy way out of the dilemma between historical particularity and conceptual comprehensiveness. At the start of this introduction, we contemplated two different meanings of the concepts of fieldwork and expeditions, one agricultural and one military. The agricultural metaphor engages the relations between fieldworkers and the field, whereas the military metaphor directs our attention to the relationships between different fieldworkers and between fieldworkers and their patrons. This distinction leaves historians of fieldwork with at least two paths to follow: each metaphor may be separately expanded upon, or they may be combined to discover overlapping meanings. Although we would applaud moves in both directions, we see the contributions of this volume as a joint historical effort to discover common ground between the agricultural and the military metaphors. Simply speaking, we hope that this book will add to our understanding of the multiple ways in which scholars and scientists have conducted fieldwork and expeditions. We look forward to seeing novel approaches to the study of fieldwork and expeditions, and we trust that fieldwork and expeditions will remain crucial and controversial topics for scholars and scientists alike.

Acknowledgements
This book was made possible by network grant no. 273-06-0106 of the Danish Councils of Independent Research and grant no. F-2009-FLS-4-35 of Aarhus University Research Foundation. The editors wish to acknowledge the contributions of all members of the Field Studies network and, in particular, the other members of the co-ordinating committee: Esther Fihl, Keld Nielsen, Henry Nielsen, and Søren Christensen (see: http://fieldstudies.dk/, retrieved July 7, 2011). Also, the kind cooperation of the Danish Research Training Programme in Philosophy, History of Ideas and History of Science, particularly Stig Andur Pedersen, is acknowledged.

Michael Harbsmeier
Fieldwork avant la lettre
Practicing Instructions in the Eighteenth Century

Historians of the discipline have always been in notorious disagreement about the origins and earliest beginnings of anthropology. Going back to either Herodotus, the Enlightenment, the evolutionism of the nineteenth century or perhaps even the earliest gossip at the bonfire about the strange and ridiculous customs of their neighbors, textbooks about the history of anthropology all seem unanimously to agree, however, about the fundamental break or revolution brought about by the invention of fieldwork around the turn from the 19th to the 20th century. Telling the stories about the fieldwork of Franz Boas on the Baffin Islands in the 1880s, Baldwin Spencer and Frank Gillen in Central Australia in 1896-7, the Torres Strait Expedition with W. H. R. Rivers, Alfred Cort Haddon, and Charles Seligman in 1898, Alfred Radcliffe-Brown among the Andaman Islanders in 1906 and then, most famous of all, Bronislaw Malinowski on the Trobriand Islands from 1915 to 1918, rarely goes without reference by contrast to the figure of the armchair anthropologist who, instead of living and working with and among the objects of his scholarly curiosity and attention, prefers to stay safely in his study at home. Except for Lewis Morgan (see White 1951), none of the major evolutionist anthropologists of the nineteenth century ever went to the field – Sir James Frazer is often quoted for having answered "Thank Heavens, no!" when asked whether he had ever met a savage in the flesh.

For historians of anthropology the history of fieldwork thus is rather short, going no further back than the beginnings of the Malinowskian revolution at the end of the nineteenth century.[1] One could also argue, however, that fieldwork before

---

1   This does not only apply to anthropological fieldwork. According to Henrika Kuklick, it is a more general process of "differentiation of labor among the various types of naturalists – from anthropologists to zoologists" that can be traced back to the middle of the 19th century, which is repsonsible for the emergence of modern fieldwork. "But the end of the century," she continues, "was the moment at which quantitative change became qualitative, the time when the disciplines we now recognize emerged and individuals' scientific interests were delimited by disciplinary boundaries. Moreover, as the natural history specialties differentiated, their practitioners determined that naturalists must break their long-established habit of relying on theories articulated by armchair scholars, that scientists could not do credible analysis unless they had themselves gathered the data on which their generalisation rested" (Kuklick 1997: 49).

this revolution was merely conducted by people other than the armchair scholars and scientists, by those merchants and adventurers, missionaries and explorers, sailors and captains, soldiers and colonial officials, amateurs and professionals who went abroad for all sorts of other purposes, but who brought home the material objects, the data, and the information on which the armchair scholars had to depend and survive. If fieldwork by definition only can be practiced by professional scholars and scientists who go to and stay in other places for exclusively scholarly and scientific purposes, historians of anthropology are right to claim that fieldwork has probably only become a regular practice from the last decades of the nineteenth century onwards. If, however, fieldwork can be defined as so many ways of practicing the instructions and answering the questionnaires issued by academies and other scientific and scholarly institutions and individuals, then we can add at least a couple of centuries to the history of fieldwork. This history started, one could say, around the time of the "crisis of the European conscience 1680-1715" described so eloquently by Paul Hazard (1935). Among the travellers presented in the first chapter of Hazard's book we can find some of the earliest real fieldworkers.

Writing a history of all these kinds of fieldwork is of course an impossible undertaking considering the overwhelming number of publications of travel accounts, collections in museums, and other evidence testifying to the influence and impact of the countless published and unpublished instructions and questionnaires. Starting from the history and development of the instructions and questionnaires so thoroughly scrutinized in a whole series of recent studies (Urry 1973, Copans & Jamin 1978, Puccini 1995, Blanckaert 1996a+b, Rubiès 1996, Kury 2001: 91-146, Bucher 2002, Chappey 2002, Bossi & Greppi 2005, Collini & Vannoni 2005, Vaccari 2007), I will in what follows take a closer look at how fieldwork has actually been practiced by readers and surprisingly also some of the authors of these instructions. Drawing attention to cases outside of the Anglophone world will hopefully also nuance what we already know about the early history of fieldwork in anthropology and other disciplines from the work of George W. Stocking (1971, 1973, 1983, 1987, 1991, 1995), Robert Kohler (2002), Henrika Kuklick (1991, 1997), and others.

In anticipation of one of my conclusions I would like to begin with the example of Joseph-François Lafitau (1681-1741) who, as a missionary of the Societas Jesu, had to work under instructions which only to a very limited extent aimed at scientific and scholarly purposes (see Pizzarusso 2005). During his nearly six years as a missionary in Sault-Saint-Louis (Caughnawaga) on the south shore of the St. Lawrence, Lafitau managed to collect enough information and observations about the Iroquois to write what often has been described as the first systematically comparative treatise of ethnology, published in Paris in 1724 under the title *Mœurs des Sauvages américains comparées aux mœurs des premiers temps.* Unlike his evolutionist successors of the nineteenth century, Lafitau did not issue instructions from his armchair, nor did he follow or practice the instructions of others to any recognizable extent when doing his fieldwork among these "French-praying Indians," about which he fails to tell us much either (Fenton 1969: 176). Precisely for this reason he can here stand as a reminder that fieldwork before the Malinowskian revolution was carried out perfectly well outside the confines of a co-operation and correspondence between a theoretician in his armchair at home and a practitioner out in the field. Very often, as I hope to be able to show, these two roles have actually been performed by one and the same person.

Robert Boyle's *General Heads for the Natural History of a Country, great or small, drawn out for the use of travelers and navigators* from the *Philosophical Transactions* 1666 (Collini & Vannoni 2005: 61-69), the unpublished instructions for the astronomical French expedition to Gorée and the Antilles 1681-1683 by the Académie Royale des Sciences in Paris studied by Nicholas Dew (Dew 2009: 6), and John Woodward's *Brief Instructions for Making Observations in all Parts of the World: as also for Collecting, Preserving, and Sending over Natural Things. Being an Attempt to settle an Universal Correspondence for the Advancement of Knowledge both Natural and Civil* (Collini & Vannoni 2005: 71-75) from 1696 are among the very earliest occurrences of the genre of instructions and both confirm that fieldwork at this early stage invariably aimed at various aspects of natural history, which was later to differentiate into botany and zoology, geography and mineralogy, geology and astronomy etc. They can also both be quoted as evidence for changes in the relationships between the theoretically interested academies and their members on the one hand and the practitioners of their instructions out in the field on the other. It is also worth noticing, as Daniel Carey has pointed out, that a number of the travellers targeted by these instructions actually became fellows of the society issuing these instructions, such as Sir Paul Rycaut, one of the travellers Paul Hazard referred to, who served as Consul in Smyrna (Carey 1997: 274).

One further aspect of fieldwork, however, is conspicuously absent not only from the text of the instructions, but also from much of the secondary literature about early scientific travels, and that is the role of the fieldworking travelers' hosts and informants. Making this aspect part of the very definition of fieldwork as a practice of executing the instructions issued by primarily scholarly and scientific agencies while staying somewhere away from home directs our attention towards what I would like to call the intrinsically "ethnographic" nature of fieldwork in any discipline: interacting with the inhabitants of the field both as informants and as objects of observation, both as hosts with whom to share knowledge (and necessities) and as individuals and populations to be described and written and reported about. In this very general sense, then, Malinowski was right to describe fieldwork as both participation and observation – as both sharing and extracting knowledge, as both interaction and dissociation.

But let us be more concrete. Leonhard Rauwolf (1535-1596), a botanist and physician from Augsburg posthumously famous for having given his name to the general class of medicines called antihypertensives as Rauwolfia alkaloids, took advantage of the trading network of his family when he went to the Near East from 1573 to 1576 (Dannenfeldt 1968: 31-2). His motivation, however, both for going and for writing a book about his voyage, was undoubtedly scholarly and scientific. "From my youth I had the strong desire to go to foreign lands," he explains at the beginning of his account,

> especially those of the Orient, as these were more famous and fruitful than others … ; not only to verify the life, manners, and customs of the inhabitants, but also much more to discover and to learn to know the beautiful plants and herbs described by Theophrastus, Dioscorides, Avicenna, Serapion etc. in the locations and places where they grow (quoted from Dannenfeldt 1968: 31).

Like many of his contemporaries, Rauwolf travelled in search of signs and evidence of a truth to be found not out in the field, but in books: in Holy Scripture, in the case of the pilgrims to the Holy Land among whom Rauwolf can also be counted, or in the texts invoked as authorities by Rauwolf himself. The closest we get to Rauwolf's practice as fieldwork in the sense of making use of informants can be found characteristically not in the more than 500 pages of Rauwolf's own text, but in the account of another traveller whom he met on his way, the merchant Hans Ulrich Krafft, who tells us how he brought Rauwolf in touch with an "intelligent Arab" in Aleppo, to whom Rauwolf presented his collection of dried plants. According to Krafft, surprise was on the side of the informant, whom Rauwolf impressed with his own knowledge of the classical names of the plants![2]

One example is not enough, of course, but Rauwolf seems to have followed much the same practices as many other humanist students of botany during the Renaissance (Reeds 1976, Ogilvie 2006: 141-150), practices which also faithfully

---

2   "Sein Reisegefahrte Krafft berichtet ergänzend, er habe Rauwolf in Alep- po mit einem verständigen Araber zusammengebracht, dem Rauwolf seine Sammlung getrockneter Pflanzen vorgelegt habe" (Krafft 1862: 209). „Überrascht habe der Einheimische zur Kenntnis genommen, dass Rauwolf … die korrekten „alten" (nämlich griechisch-römischen) Namen der Pflanzen kannte, die dem einheimischen Kräuterkenner demzufolge ebenfalls geläufig waren" (Walter 2009: 369).

reflect the norms and advice of the long series of apodemic treatises of the sixteenth to eighteenth century studied by Justin Stagl (1995) and Joan-Pau Rubiés (1996), which should be clearly distinguished from the instructions issued by academies and scholarly societies from the last decades of the seventeenth century onwards.

A striking example of how precise an effect such instructions actually had can be found in Captain Robert Knox's *Historical Relation of the Island Ceylon in the East Indies* from 1681 (see Rubies 1996: 139-41 and more recently and with much more detail Winterbottom 2009). Robert Knox (1641-1720) was sailing with his father to Persia in the service of the East India Company, but a storm forced them to go ashore on the island of Ceylon, modern Sri Lanka, in 1659. Here they were held captive for many years; his father having escaped in 1661, Robert Knox managed to return home via Batavia to London in 1680. While writing about his long stay on Ceylon, Knox seems to have worked closely with Robert Hooke (1635-1703), the well-known curator of experiments of the Royal Society. And it is from Hooke's preface to Knox's book that we are informed about the significance and importance of the instructions issued by this society for the coming into existence of the kind of scholarly and learned travel account written by Captain Robert Knox. Beginning with a complaint about the many discoveries which in spite of the invention of writing and the art of printing "have been lost, to the great Detriment of the Publick," Hooke continues:

> It were very desirable therefore that the Causes of these and other Defects being known, some Remedies might be found to prevent the like Losses for the future. The principal Causes I conceive may be these; The want of sufficient Instructions (to Seamen and Travellers) to shew them what is pertinent and considerable, to be observ'd in their Voyages and Abodes, and how to make their Observations and keep Registers or Accounts of them. Next, The want of some Publick Incouragement for such as shall perform such Instructions. Thirdly, The want of fit Persons both to Promote and Disperse such Instructions to Persons fitted to engage, and careful to Collect Returns; and Compose them into Histories (Knox 1681: preface).

The main result of Hooke's active participation in the preparation of the text of Knox's account of Ceylon has probably been the very structure of the account: rather than a narrative of the captain's voyage and captivity, it has the form of a systematic description of Ceylon and its inhabitants starting with *A General Description of the Island*, a chapter *Concerning the chief Cities and Towns of this Island*, a third *Of their Corn, with their manner of Husbandry*, a fourth *Of their Fruits and Trees*, a fifth *Of their Plants, Herbs, Flowers*, a sixth *Of their Beasts Tame and Wild. Insects*, and a seventh

Of their Birds, Fish, Serpents, and Commodities, this first part being mainly devoted to natural history. The second part then deals with *the present King of Cande*, his *Manner, Vices, Recreation, Religion*, and his *Tyrannical Reign*, his *Revenues and Treasure*, his *great Officers*, his *Strength and War*s and finally *A Relation of the Rebellion made against the King*. The third part is devoted to various aspects of the ethnography of *the Inhabitants of this Island*, their *different Honours, Ranks, and Qualities*, their *Religion, Gods, Temples, Priests*, their *Worship and Festivals*, their *Religious Doctrines, Opinions and Practices*, their *Houses, Diet, Housewifery, Salutation, Apparel*, their *Lodging, Bedding, Whoredome, Marriages, Children*, their *Employments and Recreations*, their *Lawes and Language*, their *Learning, Astronomy and Art Magick* and finally *their Sickness, Death and Burial*. It is thus only towards the end of the book, in the final part after the exhaustive descriptions of the island's natural history and what later would be called politics and ethnography, that we come to a narrative about how the author happened to get there, how he survived his captivity and how he finally escaped to return home again.

The co-operation between Knox and Hooke was certainly decisive for the shape of Knox's *Historical Relation*. The instructions, however, which Hooke deemed so deeply important, entered into the process *post festum*: not as instructions for how to act and behave out in field, but for how to organize and write up the experiences and observations thereafter. Knox's *Historical Relation* was perhaps one of the earliest scholarly ethnographies; but it was not yet based on fieldwork. In December 1689, Robert Hooke gave an address to the Royal Society in which he provided what was the first detailed description of cannabis in English, commending its possible curative properties and noting that Knox "has so often experimented it himself, that there is no Cause of Fear, tho' possibly there may be of Laughter" (Bennet 2003: 205-6). Knox's experiments with cannabis were not yet classifiable as anthropological fieldwork; the Ceylonese with whom he interacted only became his informants after he returned from the field.

A first example of fieldwork in the sense of extensive use of local informants in pursuit of botanical as well as many other kinds of knowledge can be found in the works of Engelbert Kaempfer (1651-1716), who a few generations later than Rauwolf and few years later than Robert Knox went to Persia and to Japan similarly motivated by scholarly interests in botany and other elements of natural history as well as many other fields. During his stay as secretary of a Swedish delegation in Isfahan from 1684 to 1685 Kaempfer had very close contact with Raphäel du Mans, the Capuchin monk and missionary of French origin who stayed in Isfahan from 1647 until his death in 1696, serving as an informant in many different ways not only for Kaempfer, but for a whole series of French and English travellers as well, including the abovementioned Sir Paul Rycault, British consul and fellow of the Royal Society.

Travelling through Japan as a member of a delegation of the *Vereenigde Oost-Indische Compagnie* 1690-92, Kaempfer managed to use a number of interpreters as his informants not only when botanizing, but more importantly also when pretending to botanize while actually pursuing the many different investigations which finally went into his monumental *Geschichte und Beschreibung von Japan* (first published in two volumes 1777-1779) and which showed Kaempfer as a virtuoso of what later would qualify as multidisciplinary fieldwork. Kaempfer does acknowledge his debt to his informants explicitly and we can only speculate whether he did so in an attempt to protect them against their superiors. Thanks to recent archival studies, however, we now know a great deal more about the interpreters Imamura Genemon Eisei, Namura Gonpachi, Narabayashi Chinzan and others, with whom Kaempfer shared and exchanged knowledge and information about natural history as well as many other disciplines (for details see Michel 2001: 76-88). Published in an English translation by J. G. Scheuchzer in 1727 in London, the very title of Kaempfer's work clearly indicates that we are concerned with a monograph fundamentally based on fieldwork in the sense of – in this case self-instructed – practices of sharing and exchanging substantial bodies of knowledge and information with native informants: Kaempfer could only give *an Account of the ancient and present State and Government of that Empire; of Its Temples, Palaces, Castles and other Buildings; of its Metals, Minerals, Trees, Plants, Animals, Birds and Fishes; of The Chronology and Succession of the Emperors, Ecclesiastical and Secular; of The Original Descent, Religions, Customs, and Manufactures of the Natives, and of their Trade and Commerce with the Dutch and Chinese* – we recognize these headings from Robert Knox! – on the basis of having practiced fieldwork.

Peter Kolb (1675-1726) and his *Caput Bonae Spei Hodiernum. Das ist: Vollständige Beschreibung des Afrikanischen Vorgebürges der Guten Hoffnung* ... , first published in Nürnberg in 1719 and subsequently in many later versions and translations, can be counted as yet another monograph in natural history and ethnography based on an extensive interaction with the traveller's hosts as informants and thus on what deserves to be called scholarly fieldwork. Kolb went to South Africa in 1705 for a job as astronomer at an observatory in Cape Town, but after two years he moved to Stellenbosch to work as secretary for the colony. Returning to his native Bavaria in 1713, he produced a monograph consisting of a series of letters amounting to almost 900 pages divided into three parts of almost exactly the same size (as noticed by Huigen 2009: 38): a first discussing natural history under the heading "Physicalia," a second devoted to ethnography and a third part describing the European colonies in the area. In his preface, Kolb explicitly mentions Robert Knox as one of his models.

Rather than receiving instructions after returning from his voyage, however, Kolb appears to have practiced such instructions during his stay in the field.

Kolb's book, "from an ethnographic point of view ... the most important work on the Cape to appear between 1500 and 1800" (Huigen 2009: 33), represents a major break in a long tradition of writing about the Hottentots as the most bestial and miserable of the many different nations described by the almost endless series of travellers who in one way or another passed by the Cape of Good Hope on their way to Batavia and other destinations in what was then called East India. Adding an account of how he got there and home again as a kind of appendix to his monograph, Kolb was the first to write about the "Hottentots" or Khoikhoi. Secondly he did so – very much in contrast to Robert Knox who had almost no predecessors – by systematically questioning and scrutinizing the "stereotypical" and "extremely depreciatory" descriptions (Huigen 2009: 32) by his many predecessors on the basis of his own experiences and observations. Kolb finally also included a comparative dimension into his ethnography by systematically looking for similarities and differences between his Khoikhoi on the one hand and what he thought he knew about the manners and customs of the Troglodytes and of the Jews on the other. A lot has been written about Kolb and his *Caput Bonae Spei Hodiernum* elsewhere (Pratt 1992, Harbsmeier 1994, Beinart 1998, Good 2006, Johnson 2007 and – authoritatively – Huigen 2009: 33-58); for our present purposes it must be enough (following Harbsmeier 1986) to take a closer look at his treatment of the language of the Khoikhoi as an example of how he performed his fieldwork.

Kolb explicitly discusses various opinions on the question of the Hottentot language at the beginning of the ethnographic second part of his book. "It will not be without some use," he states,

> also to relate something of the Hottentot language. The more so because opinions on this issue vary so much. Those who have written hitherto on this subject can be divided into three groups: some people think that the language cannot be learned at all; others assert that it can easily be learned and therefore also written and reduced to certain rules; the third group, finally, believes that one can learn the language, but that it is difficult to write and even more difficult to bring into any order (Kolb 1719: 29).

From his own experience Kolb knows that many Europeans do in fact master the Hottentot language, if they have lived among the Hottentots since childhood. But, Kolb continues, even if we could imagine an adult foreigner becoming as fluent as a native speaker, then again he soon faces the problem that he cannot translate it. The reason is that his teacher has no command of German. But even if one speaks it more or less perfectly, how is he to go about teaching it to anyone else, when he himself cannot tell him that he must form his tongue thus or thus when he wants to make this or that sound? Yet even when he teaches him this, still he is only over the first hurdle, when the next immediately follows: namely to take care once more to distinguish for oneself between all the smacks with the tongue and the mouth as exactly as possible, in order to communicate them to the other person. He who overcomes these difficulties and many others, and has no other resources than those he can get from the Hottentots, will certainly not claim any more that the Hottentot language is easy; on the contrary it is more probable that a woeful lament would be heard. For the Hottentots have no letters or anything of the sort, but one must learn everything from their mouth (Kolb 1719: 30-31).

Estimating that it will take at least half a century "before we will see a printed sheet in their language," Kolb has an almost Gargantuan vision of the future exploration of what goes on inside the mouths of the Hottentots:

If at last one knew how they move their tongues and bend them, and knew the place in the mouth where they make one or the other smack with the tongue, then perhaps it could happen sooner that one could gain a better knowledge of their language, and in time, if not actually write it down, at least more easily be able to teach it to others (Kolb 1719: 31).

One last illustration of Kolb's fieldwork is to be found in the last chapter of the ethnographic part of his book, which contains a detailed description of the mortuary ritual of the Khoikhoi. About one hour after coming back from the burial, the gathering suddenly stops talking and weeping:

Thereupon the very oldest man of the whole Kraal stands up, steps into the circle of men and women, takes off the little piece of hide that they wear over their member and which, as is well known, they call a kulkross, and pisses with his own water on all those sitting around. I was astonished, when I saw this the first time, over such an immodest deed and was minded to run off, so as not to see still greater irregularities. Yet my desire to know everything well prevented me from my intention. Therefore I stayed and watched the whole procedure, until they broke up (Kolb 1719: 300).

Kolb thus does not leave; on the contrary he tries to find out about the meaning of these strange customs:

> All these ceremonies are strange and remarkable and noteworthy, and so I was very anxious also to know the reasons for them. The wetting with urine, they say, means to say that the oldest man, who does this, thanks all the others for the last respects that they have paid the deceased. But when I objected, asking why he did not do it, which would have been more seemly, since he so shamefully exhibited his own member to all and sundry, great and small, young and old, I could however get no other answer from him than that it was an old custom that no Hottentot could have the heart to change, if he did not want to become weary of his life (Kolb 1719: 301).

Following the instructions of Robert Hooke by taking Robert Knox as a model, Kolb, like Engelbert Kaempfer, came to practice what I suggest may be called fieldwork: sharing and exchanging knowledge and information with the inhabitants of what through this very process becomes a field. Other examples from the first decades of the eighteenth century could easily be added. One of these, Hans Egede's *Det gamle Grønlands ny Perlustration eller Naturel-historie*, Copenhagen 1741, is quoted here to show that missionaries at that time were also deeply involved in combining studies of natural history with ethnography, which was so closely linked to these early forms of fieldwork. Hans Egede (1686-1758) went to Greenland in 1721 in search for the presumably still Catholic descendants of the Old Norse, only to return to Copenhagen after 15 years as a missionary among the Inuit. The book that resulted from his endeavors, the title of which in English would be something like *The New Illumination or Natural History of Old Greenland* (Egede 1741), does very much the same for the Inuit of Greenland as Robert Knox did for the Ceylonese, Engelbert Kaempfer for the Japanese, and Peter Kolb for the Khoikhoi. In the English translation from 1818, the book contains the following headings: *the situation and extent of Greenland, the First Settlement of Greenland, the nature of the soil, plants and minerals, the nature of the climate and the temperament of the air, the land animals, and land fowls or birds, the sea animals and sea fowls and fishes*, covering natural history, and *the ordinary occupations of the Greenlanders, the houses and house furniture, persons, complexion and temperament, customs, virtues and vices, manners and way of life, habits and way of dressing, diet, marriage and education of children, mourning and how they bury their dead, pastimes and diversions as also their poetry, and their language*, covering the ethnographic part (Egede 1818). And as Allan Sortkær has recently shown, there is an intimate link between these two parts because it is the cold climate of Greenland which for Egede explains the mentality of the inhabitants and their willingness to receive the evangelical message and salvation (Sortkær 2010).

None of the cases discussed so far can count as an example of how the instructions issued by an authority such as the Royal Society have directly shaped the behavior of travellers out in the field. If we turn to Russia and the series of great north- and eastwards expeditions under the responsibility of the Academy of Sciences of St. Petersburg founded in 1725, we can get a much more detailed image of how such instructions have been followed – in this case as in many others by travellers who themselves helped formulate their own instructions. Since many of these instructions have remained unpublished and even today are only accessible in extracts in Russian or German, they have so far been largely ignored even by scholars specializing in the history of travel writing, scientific expeditions, and ethnography. Thanks to the work of Gudrun Bucher (Bucher 2002, very shortly summarized in Bucher 2003) we now know a great deal more about what seems to have been the earliest case of multidisciplinary fieldwork on an almost industrial scale. Following Bucher, we can here content ourselves with the many instructions issued by the geographer and historian Gerhard Friedrich Müller (1705–1783), who after his education at the universities of Rinteln and Leipzig moved to St. Petersburg to become a member of the Academy and took part in the second expedition to Kamchatka under the command of Vitus Bering from 1733 to 1743. It was during this expedition that he produced a text which rightly can be considered as the most detailed and substantial set of instructions for conducting fieldwork of the eighteenth century. During this expedition, Müller authored (and co-authored) no less than six sets of instructions of various lengths, by far the most extensive of them in the city of Surgut in Siberia in the summer of 1740 when he, due to health problems from which he had suffered throughout the last three years, was finally permitted to return home, leaving his older colleague Johann Eberhard Fischer to continue his fieldwork. Despite the fact that Müller's instructions had to be approved by higher authorities, they were not meant for a general public, but much more specifically for his successor in the field.

Under the heading *Instruction for what is needed for the geographical and historical description of Siberia* ("Instruction Was zu geographischer und historischer Beschreibung von Sibirien erfordert wird"), Müller presents us with a list of no less than 1228 numbered questions consisting of rarely more than a single sentence over a couple of lines. According to Bucher's summary and translation, the first five sections of these instructions deal with keeping a detailed journal (20 paragraphs), making geographical descriptions (75 paragraphs), describing the present state of the towns, forts and regions (88 paragraphs), writing a detailed history of Siberia and doing research in the archives for this purpose (22 paragraphs) and describing and collecting antiquities (100 paragraphs) (Bucher 2003: 136-7).

More interesting, however, is the sixth and final section of Müller's instructions, which was published separately in German in full in St. Petersburg more than 150 years later (Müller 1900). Though they lack headings, these more than 900 ethnographic questions are clearly structured. After a few more general considerations about how to determine the name and identity of the different *Völker* (peoples) of Siberia and the similarities and relationships between them (1-10), there is a section about how to collect information about their languages (10 through to 16) by means of asking for their version of a Lord's Prayer (11) or by collecting "declination" and "conjugations" in their language (12), a section about the appearance and outer shape of their bodies (18-49), about how they wash themselves and embellish their bodies (50-71), the clothes of their men, women, and children and what these are made of (72-94), their dwellings, furniture, and utensils (95-117), followed by sections about various aspects of the inner character (Beschaffenheit) of a people covering such issues as literacy (121 to 125), cosmogony, cosmology, astronomy, calendars, and astrology (126 to 147), and medicine (148-168), summarized in a couple of paragraphs about the virtues and vices with a special call for observations about honesty, fraud, theft, murder, suicide, sense of shame, lasciviousness, whoredom, sodomy, adultery jealousy, bravery, timidity etc. (171). The list continues with questions about politics and law (172-185), weight and measures (186-187), courtesy and modes of greeting (188-195), conflicts, violence, and war (196-217), agriculture (219-227), animal husbandry particularly concerning reindeer (228-288) and dogs (285-290), ways and means of transportation (293-336), hunting (337-408), fishing and sealhunting (409-433), arts and crafts (435-72), food and narcotic drugs (453-549), pastimes and games (550-559), marriage, sexuality, childbirth, and upbringing of children (561-657), and finally old age, death, and mourning (658-701). The last 200 questions deal with religion, more generally with concepts of God, sin, prayer, life after death and the end of the world (702 to 712) and more specifically with what Müller calls heathen religion and shamanism (713-778), Islam (779-829), Buddhism or what Muller calls the Dalai-Lamic religion (830 to 905), and finally a mere nine headings (906 to 914) about the advancement of Christianity and the question of "whether it is advantageous to convert more people to the Christian religion" (914).

This list of topics is in itself enough to show that anybody trying to follow Müller's instructions would have to do a lot of things, which by many definitions would qualify as fieldwork requiring an extensive use of informants. But Müller also adds a series of reflections about how to combine observation by participation with the use of the statements and performances of carefully selected informants. "In order to obtain and describe the above information more thoroughly and carefully," Müller writes in paragraph 915:

> it is first of all necessary not to miss any opportunity ... to visit neighboring heathen and other non-Russian people in their houses in order to study their way of life, customs and habits and their religious ceremonies with your own eyes (Müller 1900: 82).

And in paragraph 916:

> One has to make the most distinguished and bright of these people obliging though kindness, friendly familiarity, treating them with spirits and tobacco and also minor presents. Only in this way can we gain a confidence in them, so that they will not shy away from showing and discovering whatever one wants to see and know (Müller 1900: 82).

For the study of their ceremonies at weddings, funerals, and "Schamanereyen," Müller continues, "one can call the most distinguished among them and order them to give notice when anything of that kind should occur among them." If one doesn't succeed in this way, "one also can visit them without such an opportunity, and make them do and show the ceremonies and customs by way of example" (918). The rest one will get to know "from old, experienced, stable and trustworthy people among them through familiar conversations" (919).

Müller's advice about how to choose proper informants probably also derives from his own experiences. Thus he particularly recommends those who have converted to Christianity when dealing with religious matters, even though, as he says, "many of them in their hearts still cling to Heathendom" and therefore tend to "hide even more than those, who have no reason to be ashamed of this" (920). Advising against the use of interpreters, Müller takes care to point out that the people generally are much more open-minded towards "us foreigners ... than in presence of an interpreter chosen from among the Cossacks, whom they fear very much because of their repressions" (921).

Muller's instructions of 1740 are undoubtedly the closest we ever will get to a description of how fieldwork was practiced in the eighteenth century. Originally existing only in the shape of handwritten advice for his immediate successor, Müller's instructions were certainly a close reflection of what he himself had been doing during his preceding seven years as a member of the second Kamchatka expedition. One further proof of this can be found in the very last paragraphs of his instructions, in which he not only calls for comparisons with "the customs and habits of the other Asiatic, African and American peoples" (922), but also urges the "comprehensive treatise" to come out of all this fieldwork not to describe each people and nation separately, but in order not to lose sight of all the similarities between them to deal with all peoples "in einem Zusammenhang" (923). At this point I think we can safely conclude that Müller's instructions deal much more with how to work in the field than how to work out the results of such fieldwork in one's study or armchair at home!

Historians of anthropology often tend to assume that if anything like fieldwork happened before the last decades of the nineteenth century, it would certainly not have been the scholar-scientist himself who went out to work in the field, but rather somebody who just followed and practiced the instructions given by some sedentary scholarly and scientific authority. Looking at some of the many ethnographically productive travellers of the second half of the eighteenth century, however, one can easily point out a whole series of examples of travellers who actually authored and issued the instructions which both they themselves as well as other travellers followed.

Peter Simon Pallas (1741-1811), for example, who in 1767 was invited by Catherine II of Russia to become a professor at the St. Petersburg Academy, led an expedition between 1768 and 1774 to the central Russian provinces and West Siberia, Altay, and Transbaikal, collecting natural history specimens. The regular reports which Pallas sent to St. Petersburg were collected together and published as *Reise durch verschiedene Provinzen des Russischen Reichs* (3 vols., 1771-1776), which covered a wide range of topics, including geology, mineralogy, botany, and zoology in addition to substantial reports on the native peoples and their religion. Nine years later, he published his instructions for another traveller: *Instructions pour M. Patrin, naturaliste, à qui est enjoint d'accompagner l'expedition destinée pour la Kovima et la mère glaciale* (extracts in Collini & Vannoni 2005: 139-143), and as if to follow up on his own instructions, between 1793 and 1794 he led a second expedition to southern Russia, visiting the Crimea and the Black Sea, of which he gave an account in his *Bemerkungen auf einer Reise in die Südlichen Statthalterschaften des Russischen Reichs* (Pallas 1799-1801 – for more see Collini 1995).

Another obvious example is the Venetian naturalist and traveller Alberto Fortis (1741-1803), who described his field methods in great detail in his "Preliminary Notes deemed necessary to serve as directions for travels aiming at illustrating the natural history and the geography of provinces adjacent to the Adriatic and particularly Istria, Morlacchia, Dalmazia, Albania and connected islands," published in Venice in 1771 (extracts in French translation in Collini & Vannoni 2005: 85-93). Fortis' European fame derives from his *Viaggio in Dalmazia*, which was published in two volumes three years later. Immediately translated into German, French, and English, it was his detailed ethnographic account of the Morlacchi, a pastoral people living in the mountains of Dalmatia close to the Adriatic coast, which according to Larry Wolff qualifies as "a pioneering effort in the emergence of modern anthropology":

> On the one hand, Fortis was fully versed in the philosophical writings of Rousseau, and familiar with the model of the noble savage, which shaped the account of the Morlacchi. On the other hand, unlike Rousseau, Fortis was committed to the labor of empirical observation, both as a natural historian and as a witness of customs, so that his philosophical reflections were applied to carefully observed phenomena. Montesquieu created Persians but never went to Persia. Rousseau conjured the noble figure of the savage Carib but never came close to the Caribbean. Fortis's account of the Morlacchi, however, was based on something like modern anthropological field research, and the Morlacchi of Dalmatia were accessible to his observations, because they were to be found just across the Adriatic Sea from Padua and Venice (Wolff 2005: 5).

One could probably find many more examples for the coincidence between the roles of scholar and of traveller, which may be said to lay at the foundation of fieldwork; here it must be enough to mention Johann Reinhold Forster (1729-1798), who in his youth like Müller went to Russia in hope of a career as scholar and explorer, and later became famous as a participant, together with his son Georg, in Cook's second voyage from 1772 to 1775. Forster's *A catalogue of the animals of North America ... to which are added short directions for collecting, preserving and transporting all kinds of natural history curiosities* was published in London 1771. In 1778 he published his own *Observations made during a Voyage round the World*, which in addition to a series of chapters about various aspects of natural history he develops, as Nicholas Thomas has argued, a complete comparative ethnology of the inhabitants of the various islands in the South Pacific (Thomas 1996). Fieldwork once again turns out to have been practiced by a scholar who at the same time nourished theoretical and comparative ambitions.

Three other series of instructions for travellers from the second half of the eighteenth century are much better known among historians of science and of anthropology than the examples quoted so far. Quite a lot has been written already about Carl Linnaeus (1707-1778) and the influential *Instructio Peregrinatoris* of 1759 ascribed to him,[3] about Johann David Michaelis (1717-1791) and his almost 500 pages of *Fragen an eine Gesellschaft gelehrter Männer, die auf Befehl Ihro Majestät des Königes von Dännemark nach Arabien reisen* from 1762, and even more about Joseph-Marie de Gérando (1772-1842) and his *Considérations sur les diverses méthodes à suivre dans l'observation des peuples sauvages* from the year 1800.

Common to all three of these books is that firstly their authors were not themselves among the travellers the instructions were meant for. Linnaeus was famous for his many botanical expeditions in Sweden undertaken both before and after the inaugural address he gave about "the necessity of travelling in one's own home country" when, shortly after returning from his tour to the Netherlands, England, and France, he was appointed as professor of medicine in Uppsala in 1741. His *Instructio Peregrinatoris* from 1759, however, was obviously meant for his many students who between them visited four continents while Linnaeus never left Sweden again. And neither Johann Michaelis, professor at the University of Göttingen, nor the ideologue Joseph-Marie De Gérando had any intention whatsoever of joining the expeditions for which they wrote their instructions.

Secondly, however, much more needs to be done in order to find out how these instructions have been followed by some of their readers. De Gérando's instructions apparently had no effect on the Australian expedition they were meant for. Much more can be learned about the history of fieldwork by taking a broader look at some of the travellers under the influence of this group of ideologues, for example Constantin François de Volney (1757-1820), who himself published a series of *Questions de statistique de statistique à l'usage des voyageurs* in 1795, a few years before his *Voyage en Egypte et en Syrie* (published 1787), and shortly before he went to the United States. His *Tableau du climat et du sol des Etats-Unis* from 1803 contains an appendix about the American Indians (reproduced in Moravia 1970: 397-439), which according to Moravia can be interpreted as an attempt to practice the instructions of De Gérando (Moravia 1970). In the same vein and as Siegfried Huigen has recently suggested, Lodewyk Alberti's (1768-1812) monograph from 1810 about *De Kaffers aan de Zuidkust van Afrika, Natuur- en Geschieedkundig beschreven* deserves a closer

---

3 Recorded by Stagl and Kaempfer as apodemic text authored by Erick Nordblad – "Sie steht somit gleichsam an der Wasserscheide zwischen der klassischen Apodemik und der modernen Methodik der Forschungsreise" (Kämpfer & Stagl 1983: 79).

look; Huigen proclaims it to be "the only practical application of the ethnographic questionnaire" of De Gérando (Huigen 2009: 192).

If we define fieldwork with Clifford Geertz as "going out to places, coming back with information about how people live there, and making that information available to the professional community in practical form," Linnaeus was certainly one of its practitioners among the Sami, as Zorgdrager has recently argued (Zorgdrager 2009 – himself quoting Geertz). Much more interestingly, however, one should also take a closer look at the field practices of some of the Linnaean apostles such as Anders Sparrmann (1748-1820), whose account of the Khoikhoi Mary Louise Pratt has compared so unfavorably with that of Peter Kolb as evidence of the imperial gaze brought about by what she calls the Linnaean watershed. Precisely such an investigation appears even more urgent in light of the fact brought forward by Huigen that Pratt's analysis in her influential Imperial Eyes seems to be based on serious misreading and misunderstandings of both these sources.

In the case of Michaelis and his one hundred questions for the *Gesellschaft gelehrter Männer* sent off to Arabia Felix and adjacent countries by the King of Denmark in 1761, the choice is more than obvious: it is in the books of Carsten Niebuhr (1733-1815), the only one of the learned young men to survive and return home seven years later, that we can learn more about what was arguably the most advanced case of systematic fieldwork of the eighteenth century. But this must wait for another occasion.

References

Beinart, W. 1998. "Men, science, travel and nature in the eighteenth and nineteenth-century Cape." *Journal of Southern African Studies* 24(4): 775-799.

Bennet, J. 2003. *London's Leonardo: The Life and Work of Robert Hooke.* Oxford: Oxford University Press.

Blanckaert, C. (ed.) 1996a. *Le terrain des sciences humains: instructions et enquêtes (XVIIIe – XXe siècle).* Paris: L'Harmattan.

Blanckaert, C. 1996b. "Histoires du terrain. Entre savoirs et savoir-faire." In: C. Blanckaert (ed.) *Le terrain des sciences humains: instructions et enquêtes (XVIIIe – XXe siècle).* Paris: L'Harmattan, 9-55.

Bossi, M. & C. Greppi (eds.) 2005. *Viaggi e scienza. Le istruzioni scientifiche per i viaggiatori neo secoli XVII-XIX.* Firenze: Gabinetto scientifico letterario G. P. Vieusseux Studie 13.

Bucher, G. 2002. *"Von Beschreibung der Sitten und Gebräuche der Völcker." Die Instruktionen Gerhard Friedrich Müllers und ihre Bedeutung für die Geschichte der Ethnologie und Geschichtswissenschaft.* Stuttgart: Steiner.

Bucher, G. 2003. "Gerhard Friedrich Müller's Instructions and the Beginning of Scientific Ethnography." In: P. U. Møller & N. O. Lind (eds.) *Under Vitus Bering's Command. New Perspectives on the Russian Kamchatka Expeditions.* Aarhus: Aarhus University Press 2003, 135-144.

Carey, D. 1997. "Compiling Nature's History: Travelers and Travel Narratives in the Early Royal Society." *Annals of Science* 54(3): 269-292.

Chappey, J.-L. 2002. *La société des observateurs de l'homme, 1799-1804: des anthropologues au temps de Bonaparte.* Paris: Société des études robespierristes.

Ciancio, L. 1995. "Alberto Fortis e la pratica del viaggio naturalistico. Stile di ricerca e modalità di prova." *Nuncius. Annali di Storia della Scienza* X(2): 617-644.

Collini, S. 1995. "Il viaggio possibile. Istruzioni e relazioni di viaggio nelle esperienze di Billings et Pallas." *La Ricerca Folklorica* 32(ottobre): 21-28.

Collini, S. & A. Vannoni (ed.) 2005. *Les instructions scientifiques pour les voyageurs (XVIIe – XIXe siècle).* Paris: L'Harmattan.

Copans, J. & J. Jamin (eds.) 1978. *Aux origines de l'anthropologie française. Les Mémoires de la Société des Observateurs de l'Homme en l'an VIII.* Paris: Le Sycomore.

Cowley, P., J. Fornasiero & M. Sankey 2004. "The Baudin Expedition in Review: Old Quarrels and New Approaches." *Australian Journal of French Studies* XLI(2): 4-14.

Dannenfeldt, K. H. 1968. *Leonhard Rauwolf. Sixteenth Century Physician, Botanist, and Traveler.* Cambridge, MA: Harvard University Press.

Egede, H. 1741. *Det gamle Grønlands ny Perslustration eller Naturel-Historie.* Copenhagen: Johann Christoph Groth.

Egede, H. 1818. *A Description of Greenland.* London: T. & J. Allmann.

Fauvelle-Aymar, F.-X. 2002. *L'Invention Du Hottentot. Histoire Du Regard Occidental Sur Les Khoisan, Xve-Xixe Siècle.* Paris: Publications De La Sorbonne.

Fenton, W. N. & E. L. Moore 1969. "N. J.-F. Lafitau (1681-1746), Precursor of Scientific Anthropology." *Southwestern Journal of Anthropology,* 25(2): 173-187.

Gaulmier, J. 1951. *L'idéologue Volney (1757-1820). Contribution à l'étude de l'Orientalisme en France.* Beirut: Imprimerie catholique.

Good, A. 2006. "The Construction of an Authoritative Text: Peter Kolb's Description of the Khoikhoi at the Cape of Good Hope in the Eighteenth Century." *Journal of Early Modern History* 10(1-2): 61-94.

Gupta, A. & J. Ferguson 1992. "Beyond "Culture": Space, Identity and the Politics of Difference." *Cultural Anthropology,* 7(1): 6-23.

Gupta, A. & J. Ferguson (eds.) 1997. *Anthropological Locations. Boundaries and Grounds of a Field Science.* Berkeley: University of California Press.

Harbsmeier, M. 1989. "Writing and the Other. Travelers' Literacy or towards an archaeology of orality." In: M. T. Larsen & K. Schousboe (eds.) *Literacy and Society.* Copenhagen: Akademisk forlag, 197-228.

Harbsmeier, M. 1994. *Wilde Völkerkunde. Andere Welten in deutschen Reisebertichten der frühen Neuzeit.* Frankfurt/New York: Campus.

Hazard, P. 1935. *La crise de la conscience européenne 1680-1715.* Paris: Boivin.

Huigen, S. 2009. *Knowledge and colonialism: eighteenth-century travelers in South Africa.* Leiden: Brill.

Johnson, D. 2007. "Representing the Cape "Hottentots" from the French Enlightenment to Post-Apartheid South Africa." *Eighteenth Century Studies,* 40(4): 525-552.

Kämpfer, C. & Stagel, J. *Apodemiken. Eine räsonnierte Bibliographie der reisetheoretischen Literatur des 16., 17. und 18. Jahrhunderts.* Paderborn: Ferdinand Schöningh.

Knox, R. 1681. *An Historical Relation Of The Island Ceylon In The East Indies Together With An Account Of The Detaining In Captivity The Author And Divers Other Englishmen Now Living There, And Of The Author's Miraculous Escape.* London: Richard Chiswell.

Kohler, R. E. 2002. *Landcapes and Labscapes: Exploring the Lab-Field Border in Biology.* Chicago: University of Chicago Press.

Kolb, P. 1719. *Caput Bonae Spei hodiernum. Das ist: vollständige Beschreibung des Afrikanischen Vorgebürges der Guten Hoffnung,* [...] Nürnberg: Peter Conrad Monath.

Krafft, H. U. 1861. *Reisen und Gefangenschaft des Hans Ulrich Krafft*, hrsg. von Konrad Dietrich Haßler. Stuttgart: Bibliothek des Literarischen Vereins, vol. LXI.

Kuklick, H. 1991. *The Savage Within: The Social History of British Anthropology, 1885-1945*. Cambridge: Cambridge University Press.

Kuklick, H. 1997. "After Ishmael: the fieldwork tradition and its future." In: A. Gupta, A. & J. Ferguson (eds.) *Anthropological Locations: Boundaries and Grounds of a Science*. Berkeley: University of California Press, 47-65.

Kury, L. 2001. *Histoire Naturelle et Voyages Scientifiques 1780-1830*. Paris: L'Harmattan.

Michel, W. 2001. "Einleitung zur Edition." In: E. Kaempfer. *Heutiges Japan*. Munich: Iudicium, 3-179.

Moravia, S. 1970. *La scienza dell'uomo nel settecento*. Bari: Editori Laterza.

Müller, G. F. 1900 [1740]. "Instruction was zu geographischer und historischer Beschreibung von Sibirien erfordert wird. Für den Herrn Adjunctum Joh. Eberh. Fischer." In: F. K. Russow (ed.) *Publications du Musée d'Anthropologie et d'Ethnographie de l'Académie Impériale des Sciences de St. Pétersbourg* = Beiträge zur Geschichte der ethnographischen und anthropologischen Sammlungen der Kaiserlichen Akademie der Wissenschaften zu St.Petersburg, 37-109.

Ogilvie, B. W. 2006. *The Science of Describing. Natural History in Renaissance Europe*. Chicago: Chicago University Press.

Pallas, S. 1771-76. *Reise durch verschiedene Provinzen des Russischen Reichs*. 3 vols. St. Petersburg. Reprint, Graz 1967: Akademische Verlagsanstalt.

Pizzorusso, G. 2005. "L'indagine geo-etnografica nelle istruzioni ai missionari della Congregazione 'de Propaganda Fide' (secoli XVII-XIX)." In: M. Bossi & C. Greppi (eds.) *Viaggi e scienza. Le istruzioni scientifiche per i viaggiatori neo secoli XVII-XIX*. Florence: Gabinetto scientifico letterario G. P. Vieusseux Studie 13, 287-308.

Pratt, M. L. 1992. *Imperial Eyes. Travel Writing and Transculturation*. London: Routledge.

Puccini, S. (ed.) 1995. "Alle origini della ricerca sul campo. Questionari, guide e istruzioni di viaggio dal XVIII al XX secolo." *La Ricerca Folklorica* no. 32.

Reeds, K. M. 1976. "Renaissance Humanism and Botany." *Annals of Science*, 33(6): 519–542.

Rubiés, J.-P. 1996. "Instructions for travelers: teaching the eye to see." *History and anthropology*, 9 (2-3): 139-190.

Sortkær, A. 2010. *Forestil dig verden. En beretning om udviklingen af dansk-norsk rejselitteratur i 1700-tallet*. Unpublished PhD dissertation, Aarhus University.

Stagl, J. 1995. *A History of Curiosity. The Theory of Travel 1550 – 1800*. Chur: Harwood Academic Publishers.

Stocking, G. W. 1971. "French Anthropology in 1800." In G. W. Stocking. *Race, Culture and Evolution. Essays in the History of Anthropology*. New York: The Free Press, 13-41.

Stocking, G. W. 1973. "From Chronology to Ethnology. James Cowles Prichard and British Anthropology 1800-1850." In: J. C. Prichard. *Researches into the Physical History of Man* (1813). Chicago & London: University of Chicago Press, ix-cx.

Stocking, G. W. 1983. "The ethnographer's magic: fieldwork in British anthropology from Tylor to Malinowski." In: G. W. Stocking (ed.) *Observers Observed. Essays on Ethnographic Fieldwork*. Wisconisn, University of Wisconsin Press, 70-120.

Stocking, G. W. 1987. *Victorian Anthropology*. New York: The Free Press.

Stocking, G. W. 1991. "Maclay, Kubary, Malinowski: Archetypes from the Dreamtime of Anthropology." In: G. W. Stocking (ed.) *Colonial Situations. Essays on the Contextualization of Ethnographic Knowledge (History of Anthropology 7)*. Madison, WI: University of Wisconsin Press, 212-275.

Stocking, G. W. 1995. *After Tylor. British Social Anthropology 1888 –1951*. London: Athlone.

Thomas, N. 1996. "'On the Varieties of Human Species': Forster's Comparative Ethnology." In: J. R. Forster. *Observations Made during a Voyage round the World*. Honolulu, University of Hawai'i Press, XXIII-XL.

Urry, J. 1973. "Notes and Queries on anthropology and the development of field methods in British anthropology (1870-1920)." *Proceedings of the Royal Anthropological Insitute of Great Britain and Ireland*, no. 1972, 45-72.

Urry, J. 1984. "A history of field methods." In: R. F. Ellen (ed.) *Ethnographic Research. A Guide to General Conduct*. London: Academic Press, 35-61.

Vaccari, E. 2007. "The organized traveler: scientific instructions for geological travels in Italy and Europe during the eighteenth and nineteenth centuries." In: P. N. W. Jackson (ed.) *Four Centuries of Geological Travel. The Search for Knowledge on Foot, Bicycle, Sledge and Camel*. London: Geological Society, Special Publications 287, 7-17.

Walter, T. 2009. "Eine Reise ins (Un-)Bekannte. Grenzräume des Wissens bei Leonhard Rauwolf." *Zeitschrift für Geschichte der Wissenschaften, Technik und Medizin*, 17(4): 359-385.

White, L. A. 1951. "Lewis H. Morgan's Western Field Trips." *American Anthropologist*, 53(1): 11-18.

Winterbottom, A. 2009. "Producing and using the *Historical Relation of Ceylon*: Robert Knox, the East India Company and the Royal Society." *British Journal for the History of Science*, 42(4): 515-538.

Wolff, L. 2005. "The Adriatic Origins of European Anthropology." Cromohs, 10: 1-5. Retrieved July 7, 2011, from http://www.cromohs.unifi.it/10_2005/wolff_adriatic.html.

Zorgdrager, N. 2009. "Linnaeus as Ethnographer of Sami Culture." *TijdSchrift voor Skandinavistiek*, 29(1/2): 45-76.

Jeppe Strandsbjerg
Surveying the Field of State Territory

Gaston Bachelard once noted how "[o]utside and inside form a dialectic of division, the obvious geometry of which blinds us as soon as we bring it into play in metaphorical domains. It has the sharpness of the dialectics of yes and no, which decides everything. Unless one is careful, it is made into a basis of images that govern all thought of positive and negative" (Bachelard 1994: 211). The spatial divide between inside (the domestic, the known) and outside (the foreign, the unknown) has remained persistent in social thought. The social sciences have been divided into domestic disciplines, such as sociology and political science, dealing with well-ordered societies where progress is possible, and International Relations and security studies dealing with an anarchic domain characterized by war and insecurity where no progress is possible. Here, the "outside", the international, is separated as a different sphere of study calling upon theories and concepts different from those used by "domestic" disciplines (for a Bachelard-inspired critique of International Relations, see Walker 1993). The process of dividing the political world into insides and outsides has been closely intertwined with knowledge production, both in terms of providing the knowledge categories with which to make sense of the world but also in a more specific manner. The idea of going into "the field" to study the unknown in order to bring this knowledge back home plays a central role in data collection within the social sciences. Indeed, the notion of the field often rests on an assumption about exploring that which is outside. As noted by Gupta and Ferguson, the romance of fieldwork "has rested on its exploration of the remote" (Gupta & Ferguson 1992: 6).

At the time when anthropology was institutionalized as a discipline the unknown, or the foreign, was situated in foreign cultures that were deemed primitive from a European vantage point. Yet the notion of going abroad to learn the unfamiliar easily ignores the fact that it was conceptually necessary to start at home by creating a familiar domestic space. That the territorially defined society should provide the domestic basis of national identity, national politics, and knowledge production of society is by no means obvious seen from a historic perspective. As I will argue in this chapter, the notion of field studies not only relies on a divide between the inside and the outside but field studies have, historically, been constitutive of this divide. As such, the argument follows an agenda set by Bruno Latour in the attempt to explain the historical constitution of "great divides" between inside and outside which has come to inform the division of labor in social science (Latour 1993, 1999, 2005).

The purpose of this chapter, then, is to present an interpretation of the fieldwork of early modern cartographers and surveyors as being constitutive of a modern understanding of state territory, while at the same time this understanding of territory came to define the foreign/the outside and the domestic/the inside. I argue that the process of territorializing the sovereign state; that is, making the territory the defining characteristic of the state, was conditioned by cartographic surveys of the field belonging to the state. Through such surveys the knowledge of the territory was transformed in a manner that made it possible to locate sovereignty in the territory rather than in the person of the king. This ascribed a timeless feature to the territory and it also turned the territory into a conceptually homogenous space which pushed the boundaries of what was known to the boundaries of the territory. In that respect there came to be an overlap between the inside and the outside of politics and knowledge. As such it was the field studies of the surveyors and cartographers in the service of the crown that made it possible to construct the territory as a domestic space serving as the conceptual basis for going abroad.

In order to describe how localized knowledge production and spatial assemblage create and govern spaces within which lives unfold, David Turnbull has introduced the term "knowledge spaces" (Turnbull 2000: 5-12). The key question that he asks is how local knowledge production moves beyond the site of its production (Turnbull 2000: 13). He sees cartography as a classic expression of modern science and investigates how the interrelationship between knowledge production and the development of political institutions, particularly the conjuncture between scientific maps and the state, produce a particular cartographic knowledge space (Turnbull 2000: 89-121). This notion alerts us to the fact that the joint history of cartographic developments and state formation was not only about fashioning territory in a new way; it was also about altering the boundaries between what was known and unknown.

Whereas the rationale of scientific cartography is commonly considered to be the accurate representation of space,[1] it has been firmly established that the map does much more than this. The performative power of the map in shaping the spatial reality which it supposedly represents has been combined with critical power studies underscoring the link between cartography and state power as well as imperial

---

1   Scientific cartography refers to what is conventionally labeled modern cartography. Scientific or geometric are preferable descriptions because they describe the rationale of this mode of cartographic representation rather than signifying that it was a result of a "European modernity".

practices (Wood 1992, Edney 1997, Harley & Laxton 2001, Pickles 2004). Several scholars have scrutinized the specific link between territory and cartography through analyses of how modes of cartographic practice and calculation technologies have played a central role for the development of modern political territory (King 1996, Häkli 1998, Biggs 1999, Strandsbjerg 2008, Crampton 2010, Elden 2010). Where most of these studies emphasize the ability of cartography to represent or perform space in a particular manner that allows the conceptual grasp of territory as a homogenous and bounded object of state power, there has been less focus on the way geometric cartography assembles a particular spatial reality which allows territories to be established in a particular way. This, in other words, is to advocate not only an analysis of how cartography becomes part of the exercise of state power but also how it produces space in a particular way that conditions boundary making both in its concrete manifestation between states and also in terms of thinking the borders between what is known and what is unknown.

The argument that follows is based on a broad analysis of mapping practices in the fifteenth to the seventeenth century, informed by a Latourian understanding of cartographic practice. This framework allows us to see early modern cartographic surveys as constitutive of the inside/outside division that has played such a significant role for social and political thought, dividing the world into a known home territory and an unknown field abroad. During the chosen period it is possible to trace a significant change in cartographic theory and practice that altered the possibilities of political rulers to co-ordinate and govern space. I investigate the interaction between state formation and cartography, focusing on how cartographic practice works to create a specific spatial reality with important consequences for social practices: in this particular case, the transformation of state territory. These arguments are developed through a case study of the mapping of Danish territory in 1450-1650. To begin, I present cartography as a networked field and laboratory practice.

Cartography as Field – and Laboratory – Practice
Associated with the field of Science and Technology Studies, authors such as Bruno Latour and David Turnbull have emphasized how all knowledge is necessarily a local enterprise even if it is considered to carry a universal "truth-value" (Latour 1999a, Turnbull 2000). The key here is that instead of accepting the self-proclaimed universality of scientific knowledge and practices, Latour and others seek to analyze and deconstruct how scientific facts are produced at specific sites of knowledge production. This is, in a sense, to denaturalize scientific practice as obvious and necessary, instead treating it as a historically contingent practice like any other social practice. Importantly, however, this does not mean that scientific knowledge is arbitrary,

accidental, or completely relative but rather that it is only through production that knowledge can achieve the status of the factual (Latour 1988, 1999b). To analyze this, Latour focuses on what he calls laboratory practice. The emphasis on lab science does not only constitute an argument for how science is always a local, and therefore particular, enterprise but significantly it also points to a particular relationship between the laboratory and the world about which knowledge is produced. The most important element of this relationship is, put in somewhat simplistic terms, that for the scientist to be able to say something about the world (outside the laboratory), the world has to be brought back into the lab in the shape of "data" which can be worked through in the laboratory. As such, the world has to be made moveable.

For an unknown place to be made familiar, knowledge of this place has to be combined with knowledge at another site (since all knowledge production is local). Hence a system of referents has to be employed in order to make sites *mobile* and to keep them *stable* and *combinable*, in order to combine various places. Latour calls the place where such referents are combined a *center of calculation* (Latour 1987: 223). The center of calculation is a site, a laboratory for example, where circulating referents are combined, calculated, and developed. And indeed, scientific cartography allows places to be rendered mobile, and kept constant and combinable. Hence, the mapmaker's workshop, or a royal court that controls such a process, will act as a center of calculation. They constitute a hub in a network of navigators, surveyors, cartographers, printers, etc., all of which are necessary for geometric cartography to develop. "Cartography is one network cumulating traces in a few centers which by themselves are [...] local" (Latour 1987: 229).

To understand the notion of circulating referents it is necessary to move away from the assumption that cartography simply represents an independent stable reality. It is the language and the ideas of knowledge production of the scientists that frame the way spatial data and representation are produced. But at the same time these language and knowledge categories are affected by their encounters with the sites and places the surveyors, in this case, seek to capture. In this sense, circulating referents move back and forth between the observed site and the laboratory, and are modified by each journey (Latour 1999: 24-79). These circulating referents, or spatial data put simply, are brought back to the laboratory.[2] If we consider space, not as a natural given, but rather as a phenomenon, then it is one that circulates "all along the reversible chain of transformations, at each step losing some properties to gain others that render them compatible with already established centers of calculation"

---

2   In its most simple expression, the laboratory is where the scientist works (Latour 1987: 64), but in this chapter I use the term interchangeably with the notion of the "centre of calculation."

(Latour 1999: 71-2). In this case, the "reversible chain of transformations" describes the process by which the surveyor enters the field, and through several stages of observations, effaces "locality, particularity, materiality, multiplicity, and continuity" (Latour 1999b: 70). This process of plotting observations on a piece of paper or a computer is called reduction. Simultaneously, at each stage of reduction the representation obtains greater "compability, standardisation, text, calculation, and relative universality" (Latour 1999b: 70). Hence a phenomenon circulates in between particular observations and a generally "universal" body of knowledge with which it is made compatible in its transformation into a fact. Space, then, is for Latour something that is fabricated within networks. Rather than being a natural frame, "space is something generated *inside* the observatory" (Latour 1987: 229). This implies that space achieves "reality" at the specific sites in which measurements and observations are assembled and represented.

In the historiography of European cartography it is debatable when cartographic practice turned modern – or scientific – but it is certain that it changed quite dramatically from the fifteenth century onwards; at this point it began to correspond with Latour's notion of scientific practice as a particular relationship between laboratory work and fieldwork. To an increasing extent cartography became governed by abstract geometrical principles of representation combined with a growing body of spatial data collected by travellers, navigators, and the like.[3] The origin of this transition has usually been considered the introduction and translation of Ptolemy's geography into Latin Europe during the first decade of the fifteenth century. Ptolemy's *Geography* served to codify and inspire an existing debate concerning space and the organization of the cosmos. Initially it was not the technical prescriptions for cartographic representation that gave the *Geography* its status as an indispensable reference, but rather the desire to understand the world of the ancient authors so admired by Renaissance scholars (Dalché 2007: 287-298).

While this desire seems to have been a motive behind Ptolemy's popularity at first, the discovery of lands unknown to those writers created an incentive to modify the world map towards the end of the century. Portugal and Spain both ventured further into the Atlantic and down the African coast and this stimulated the demand for new forms of mapmaking to aid navigation and planning of journeys. To achieve this, it was necessary to find a way to combine newly sighted lands with the existing knowledge of the world; the Ptolemaic framework was suitable for this. The use

---

3   None of these principles were particularly new; they were influenced by Arabic and Byzantine writings, but they developed into a context of expanding European navigation and empire formation which created a profound demand for accurate maps, and as such, provided the institutional framework for the development of a particular mode of mapping the world.

of the co-ordinate system ensured, in the words of David Woodward, that "new places could be fitted in as their coordinates became available without "stretching" or extending the map [and therefore], the Ptolemaic frame could theoretically accommodate discoveries worldwide" (Woodward 2007: 13). Prioritizing location by longitudes and latitudes, the Ptolemaic cartography provided a model of the world based on pure abstraction. By imagining the world as an empty sphere of isotropic space organized along a neat grid system, the mathematical principles of this model preceded the symbolic significance and prior knowledge of places.

This also made all lands combinable into a single coherent framework. By knowing the co-ordinates of each coastline, estuary, hilltop, town, and so forth, they could all be added to the world map, thus slowly completing the image of the world. What is significant for our discussion of cartography as field practice is the emphasis on gathering spatial data. In order to expand the representation of the world, it was necessary to collect information of the new places in a way that made them combinable with the framework. This meant that observations had to be made at the various sites, turning pilots and travellers into fieldworkers observing and recording the trail of their journey. Subsequently, these observations had to be brought back home where they would be added to the existing framework in sites that would effectively work as spatial laboratories, digesting the input from the fieldworkers.

To illustrate this relationship it is instructive to look at how both the Portuguese and the Spanish courts, in the context of early imperial expansion, set up offices that served to create a new world map (the *Casa de Mina* in Lisbon and the *Casa de Contratación* in Seville respectively). These trading houses controlled overseas trade and colonization, and a key task was to maintain and expand a master map of the world on which all new sightings were recorded and which served as the basis for maps issued to pilots serving in the fleets of the two countries. In Spain, pilots returning from the Americas had to deliver a report with their observations and measurements on every return from an Atlantic crossing. At the *Casa de Contratación*, a group of cartographers digested these spatial data, adding new locations to, and updating data on, a master world map called *Padrón Real* (Lamb 1974). Partly to remedy the problem of poor duplicate maps being sold in the Spanish ports, Amerigo Vespucci, who became the first "pilot major" in 1508, was instructed to create an official master map to minimize errors (Haring 1964: 306):

> We command that a Padron General be made and, so that it should be more accurate, we command our officials of the Casa de la Contratación that they assemble all our pilots, the most skilled captains at the time, and that the said Amerigo Vespucci, our pilot major being present, a padron of all the lands and

islands of the Indies hither to discovered and belonging to our kingdoms and seignories be drawn up and made ... when they find new lands or islands or shoals or new harbours or anything that should be recorded in the said padron real on their return to Castile they go to report to you the said pilot major of the Casa de la Contratación so that all shall be registered in the proper place in the padron real, in order that navigators be better advised and cautious (quoted by Turnbull 1996: 11).

In that respect the *Padron* worked as a mediator between the pilots and the spatial locations they were exploring. From 1527 the pilots were ordered to keep detailed records of their journeys, keeping track of "the path which they followed every day and on which rhumbs, and which lands or islands or bays they reached, and how far away they were, and how the coasts ran, and what ports or rivers or capes there were in these places, and in what distance and latitude they lay" (official instruction quoted by Sandman 2007: 1101). In effect these offices worked as laboratories, making the world knowable and navigable.

Cartographic practice, however, was not only a matter of field and laboratory work; it also relied on a network in which cartographic material circulated between laboratories, publishers, courts etc. A market for maps and books of maps emerged around the burgeoning publishing industry in Europe. Publishers such as Abraham Ortelius compiled maps from various sources, bringing together and disseminating the works of various surveyors and cartographers. Ortelius had made a living out of the map trade. He travelled and collected the latest maps which he then colored and resold at book fairs and was very well connected to a network of the best map producers in Europe.

Due to the way in which Ortelius compiled his work, his office became a hub for a large network of map-engravers in Europe. From the beginning, it was this network which made possible the publication of his famous atlas *Theatrum Orbis Terrarum* (first published 1570) and each subsequent edition presented maps of new areas and updated maps of areas already covered. Cartographers were very eager to send Ortelius their latest maps and gave advice on how to make improvements. In return, Ortelius made careful dedications to the origins of all the maps he printed. The *Theatrum* set itself apart by representing a turning point away from Ptolemy, in that Ortelius did not present his work as an edition or an update of Ptolemy. In a message to the reader, the "limits of ancient knowledge were graphically and textually underlined so that the reader could see 'how maimed and imperfect' were ancient world views which comprised 'scarce one quarter of the whole globe now discovered to us'" (Eisenstein 1979: 193).

In contrast with the Iberian mapping projects, cartographers like Ortelius worked as private enterprises operating largely on market terms, and their work producing maps was structured differently from those working in dedicated state institutions. Where the Iberian projects relied on a network of navigators to collect spatial data, the publishers relied on a network of other publishers and surveyors and sometimes their own surveying, as in the case of Gerard Mercator and Marcus Jordan, whom we shall encounter subsequently. These networks of cartographers further relied on a support network of royal and noble courts around Europe, tying the work of the cartographers back to the interests of powerful noblemen and sovereigns. The role of the courts, as noted by John Robert Christianson, was to provide the material means, and sometimes also the demand, for the natural philosophers (Christianson 2006). The key to these relationships was the more or less formal networks between surveyors, cartographers, and the courts under whose auspices they were often working, directly controlled by the state. It was at the different courts of kings and the nobility that much fieldwork was planned and co-ordinated with existing knowledge.

In these networks, the most important divisions were those between inside the laboratory and outside in the field, and between the court as the locus of decision-making and the various sites that were subject to the decisions of the rulers. The Iberian cartographies assembled the globe into a coherent single space which served as a natural stage for European empires; the private cartographers did the same, but relied on a different network of surveyors and patronage. They all made the world navigable, and hence subject to domination. It became possible to plan "action at distance," to use another Latourian term – I will return to this notion subsequently. All this served the purpose of growing European empires because they could now travel and co-ordinate action across the globe from the courts in Europe, and this made it possible for them to expand and govern remote provinces. But it not only transformed the relationship between the capital and imperial provinces; it also transformed the conditions for state territoriality. By assembling a single global space where clear boundaries could be drawn between inside and outside, they transformed the relationship between the court and the domestic territory and provided an incentive for states to become more involved in surveying their territory. As such, they relocated the boundaries between what was inside and outside politics, as well as inside and outside knowledge.

Surveying State Territory
During the sixteenth century several states in Europe undertook national surveys in order to map their own territories (for the best collection of accounts of this history,

see Woodward 2007). Generally, these surveys followed a similar principle in terms of collecting data and uniting them through calculation in order to unify space as a coherent territory subject to the single authority of the king. I shall demonstrate this using the Danish case, in which the mapping of the territory was a condition, I suggest, for the centralization of sovereignty in the abstract body of the absolutist state. And not only did the cartographic assemblage of the territory condition the centralization of authority, it also had profound effects for the sovereign's ability to construct and govern a territorially defined society. As such, the territory and the population under the formal jurisdiction of the king were united as a domestic territorial space that gradually came to give meaning to the notion of the territorially defined society as a familiar inside, while beyond the territorial confines of one's society was the foreign, the unfamiliar. In the following, I will present the process of the involvement of the Danish court in mapping the territory.

It is not documented when exactly the Danish rulers turned their attention towards cartography. It has been suggested that it was a request by the Danish king Erik af Pommern (Pomerania) which led the Danish cartographer Claudius Claussøn Swart to produce his Scandinavian addition to the Ptolemy atlas in the fifteenth century (Bjørnbo & Petersen 1904: 8). It was through this addition that Nordic locations and place names first found their way into the standard publications on geography. Swart explicitly wrote himself into a tradition in which he, as an eye-witness, possessed knowledge unknown by "the authorities" In one of the key texts, he states that "I, the Dane Claudius Claussøn Swart, [...] have by meticulous drawing as well as written records aimed to give to the world a true picture of said countries known to me by personal experience, yet unknown to Ptolemy, Hipparch and Marinus."[4] The significance of Swart being from Denmark is not that he added to a Danish cartography, but that this gave him firsthand experience which he could add to a general European body of knowledge with universal scientific aspirations.

As previously mentioned, Dutch cartographers came to dominate the European map industry during the sixteenth century and in this respect they decided the cartographic image of the Danish territory. It is due to these networks that we know of the first map to be published in Denmark. It was made by Marcus Jordan in 1552, when he was professor of mathematics at the University of Copenhagen. This map was not based on his own surveys, but was to a large degree based on previous sea charts and the famous *Carte van Oostland* (Nørlund 1943: 26). Taking its place in

4   My translation of "Jeg Danskeren Claudius Claussøn Swart [...] har ved omhyggelig Tegning saa vel som ved skriftlig optegnelse søgt at give Efterverdenen et tro billede af de mig ved Selvsyn nøje bekendte nedennævnte Lande, som var Ptolemæus, Hipparch og Marinus ukendte" (Nørlund 1943: 13). For more on Swart's different sources see Bjørnbo and Petersen (1904).

the European cartographic network, Jordan's map provided the basis for the map of Denmark in Ortelius' *Theatrum*. These personal networks are also clearly noticeable in the other great atlas of the time: Mercator's *Atlas sive Cosmographicæ*. Heinrich Rantzau, who occupied a leading position within the nobility as head of the Danish king's administration in the Duchies of Schleswig and Holstein was very interested in the new geographic disciplines. Establishing contact with Rantzau, Mercator promised to give Denmark a prominent place in his atlas and that "the glory of your name will be celebrated from help of this kind" (quoted by Crane 2002: 267). And Denmark is indeed described in very detailed and glamorous terms. In an English translation from 1636 we read of "[t]he great and populous Kingdome of *Dania*" that "[s]uch was the forme of the ancient Politicke State , & the Monarchie of the Danes; which never any nation was able to conquere, nor make them loose their ancient priveleges, and customes" (punctuation kept as original, Ortelius and Skelton 1968: 95). The text contains a detailed description of the Danish political structure and customs, as well as numerous references to the beauty and richness of the country. Due to the connection between Rantzau and Mercator, the atlas contains four different maps of Denmark, whereas Denmark's rival, Sweden, is represented with a single map. Not unexpectedly, the "extra maps" show the areas where Rantzau had most of his possessions. In this respect, Denmark now occupied a significant position in the publication of the emerging globalized and unified space of the world. Various scientists and noblemen, like Rantzau, were involved in putting the state of Denmark on the map. However, Danish territory was represented as part of a universal enterprise to map the world, and thus not in a fashion controllable by the monarch. This, in itself, might have provided an incentive for the sovereign to seek control with the authorship of his own territory. But no less significantly, a more controlled and detailed mapping of the territory would provide new possibilities for the state to develop its territory.

In 1553 King Christian III asked the abovementioned Marcus Jordan to map "all the kingdom's provinces, islands, towns, castles, monasteries, estates, coastlines, capes and anything else worth noticing" (from Bramsen 1975: 52, my translation). The result of Jordan's engagement by the King is not fully known. He only produced maps of some of the regions in Denmark and the map he delivered to the King has not survived. But, in 1588, a map by Jordan, based on his own surveying, appears in Braun and Hogenberg's *Civitates Orbis Terrarum*. It was printed in the fourth volume of this first city atlas, which mirrored the title of Ortelius' famous atlas. However, it was not the King but Count Rantzau who asked Jordan to provide the map for the *Civitates*.

Figure 1. Jordan's 1588 map of Denmark from Braun and Hogenberg's Civitates Orbis Terrarum (courtesy of the Royal Library, Copenhagen).

Around 1580, a new project was launched involving Tycho Brahe, that was supposed to provide a new and improved map of Denmark as part of writing a new history of the country (Kragh 2005: 285). This was supported by the King Frederik II, who ordered that Brahe should have access to all the maps held in the King's library. Surrounded by a group of people who became prominent in the mapping of all the Scandinavian and North Atlantic countries (Mead 2007: 1790), Brahe was the first to utilize the method of triangulation, which later became standard for scientific map making. The only direct result of this, however, was a very precise map of the island of Hven, where Brahe had built his workshop and castle Uraniborg (Nørlund 1943: 45-46). Despite this meager result in terms of published maps, Brahe's work greatly influenced cartography both in Europe and in Denmark. Subsequent Danish cartographers working to map the territory would draw on the work of Brahe, who also influenced the famous Blaeu cartographers that came to dominate Dutch cartography during the seventeenth century.

After the accession of Christian IV in 1588, tensions grew between Brahe and the king, and the disputes eventually made Brahe leave his island fief. While the king was reluctant to fund Brahe's research unconditionally, the state generally became more active in acquiring an accurate cartographic representation of its territory. In 1622, the state employed Dutch copper engravers to publish the maps which had been authored under the auspices of royal authority (Nørlund 1943: 45), and in 1623 Hans Lauremberg was employed as a professor at *Sorø Akademi* where he taught geometry, surveying, and the art of fortification. He was simultaneously appointed to map the Danish realm (Lauridsen 1888: 56-7). Lauremberg was granted access to all areas of the kingdom, and received a regular payment from the state. The king followed the process closely and in 1639 the head of the financial administration, Corfitz Ulfeldt, received an order to make sure that Lauremberg's maps were engraved and published (Nørlund 1943: 48-50).

This indicates both a concern by the sovereign to centralize and control map publishing and the writing of the territory in a geometric form. It also indicates that Lauremberg produced at least some maps even though they were never printed – at least not in Denmark. A likely reason for this was Christian's insistence on publishing them in Denmark under his control, whereas Lauremberg wanted the Dutch master publishers to do the job. And it seems that indirectly Lauremberg had it his way, since both of the major Dutch publishing houses – Blaeu and Janssonius – began to produce new and more accurate maps of Denmark around this time. Several of these maps were dedicated to Jørgen Seefeldt, who was a friend of Lauremberg, and who could have provided the Dutch publishers with these maps (Nørlund 1943: 51). This would also explain the appearance of maps of Denmark produced by the Swedish cartographer Erik Dahlberg in Samuel von Pufendorf's writings on the Swedish King Karl Gustav (Lauridsen 1888: 60-61) as Seefeldt's library was seized by the Swedish army in 1658.

While the actual destiny of Lauremberg's body of work remains unknown, it is likely that it provided much of the cartographic information that came to inform the first general map of Denmark produced by the cartographer, mathematician, and surveyor Johannes Mejer. Mejer developed his career as a map maker in the Duchies in the 1640s. He had settled in Husum in 1629, after the end of Danish involvement in the 30-years war, where he started his work with almanacs, short accounts of world history, and the cartographic enterprise. Mejer's main institutional support came from Prince Frederik, who was residing at Gottorp castle, and his first maps were published in historical accounts by Peter Sax. The former was significant for providing means for subsistence and the latter for providing an outlet for Mejer's attempts to develop a cartographic practice (Lauridsen 1888: 6-12).

Mejer was building on the cartographic technology of Tycho Brahe, Longomontanus, and Lauremberg (Lauridsen 1888: 6), and as such became part of the network of cartographic knowledge circulating between the courts. Initially his maps were based on sources other than his own surveying; that is, other ways of getting spatial data. As was common practice, Mejer based his maps partly on existing maps and data (calculations of location and so forth) and gradually added his own observations to this. His first (known) map based on his own surveys appeared in 1638, and in 1642 he got his first order from the king to map areas around Western Schleswig. Mejer rarely described the instruments he used but since he adopted the techniques of people like Brahe and Lauremberg it is likely that he worked in a similar manner.[5] In one instance he notes how "he measures the distance between locations with mathematical instruments from church tower to church tower" and on this basis recreates the maps (Lauridsen 1888: 29-30). On such field-trips he would usually travel by cart accompanied by one or two men with local knowledge (Lauridsen 1888: 34).

Following royal orders, Mejer was issued with a pass that requested all official personnel to provide horses, carts, people, and anything else he needed to complete his map (Lauridsen 1888: 24). These passes were significant; previously Lauremberg was only issued with passes that granted access to church towers and similar highpoints that would enable the survey. And later, Jens Sørensen received a pass stating that all subjects of the king were obliged to provide place names and necessary information to Sørensen so that he could complete his cartographic endeavor (Knudsen 1918: 46-9). While the royal passes granted access to all areas of the territory and obliged the locals to help the royal cartographer with access and local knowledge, it was the role of the cartographer to utilize local knowledge in order to bring the particular places back to the center of calculation in order to provide a uniform territory in terms of spatial knowledge. Hence, the king equipped the cartographer with prerogatives that would ensure access to local knowledge either through people guiding him or by ensuring access to church towers and the like that were required to carry out the survey. All this was done in order to enable the cartographer to enter the field, turn it into spatial data that could be brought back into the lab, and finally create a coherent spatial assemblage of the territory.

---

5   See Brown (1979) for a general description of the techniques and instruments that were used for surveys and the measurement of latitude and longitude in the period.

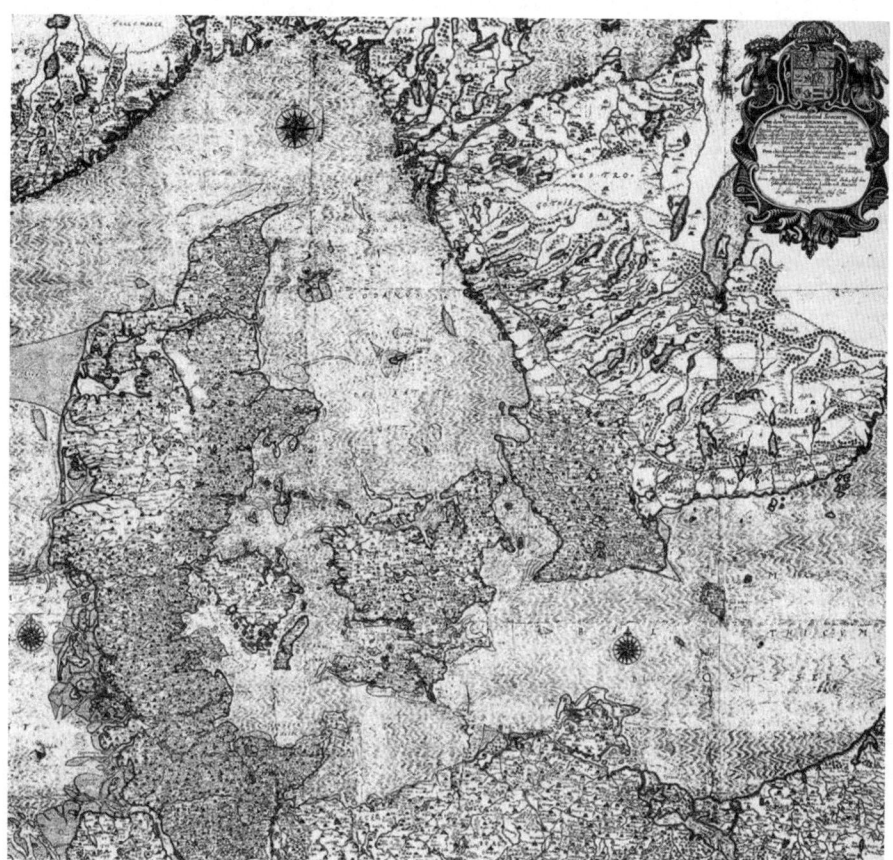

Figure 2. Johannes Mejer's 1650 general map of Denmark (courtesy of the Royal Library, Copenhagen).

In 1647, the yet unfinished task of providing a new general map of Denmark was handed over to Mejer and a small crew. After only a couple of years, Mejer could present a general map of Denmark to Frederik III, who had succeeded Christian IV as king. The map was printed in 1650 and, although never generally published (Nørlund 1943: 53), finally provided the king with a uniform map of the territory, and established a new standard for maps up until the eighteenth century. It would not have been possible, however, for Mejer to produce a general map based entirely on his own surveying in this fairly short time span – especially because the map represents a dramatic improvement in terms of its geometric accuracy compared to anything published earlier. The map (see figure 2) was probably based partly on Mejer's own surveying but also to a large extent on the works of Lauremberg (Lauridsen 1888:

56-61). As such the map was a result of both field laboratory practice and network cartography.

It should be noted, of course, that this was not a map of the entire territory of the Danish state. This would have required it to include Norway, Iceland, Greenland, the Faroese Islands, and various islands in the Baltic Sea such as Gotland, as well as the small trading colony Trankebar in India. While this fact has been somewhat ignored by the traditional historiography of the mapping of Denmark, I have maintained this focus in order to keep to a more stringent narrative about the process of establishing a cartographic territory. It should be mentioned, however, that mapping projects did take place in the other parts of the realm as well, sometimes by different cartographers, and sometimes by the same. Johannes Mejer, for example, had an ambition to complete a Scandinavian atlas; he undertook surveys in Norway and published new maps of Iceland and the Faroese Islands (Nørlund 1944b: 33, 1944c: 18). With these caveats in mind, Mejer's map provided the state with a novel guide to the territory. As I will go on to explain, Mejer's map completed the transition from a literary mode of knowing the territory to a uniform cartographic one which could abstract from local knowledge and combine – assemble – these into a coherent framework at the center. This contributed to a centralization of knowledge of the territory, and hence unified authorship and the state (Turnbull 2000: 116-117). The process of mapping the Danish territory has illustrated that the production of a scientific-cartographic representation of a territory was no straightforward process or one strictly controlled by the monarch, but rather one in which the King sought to control the networks of courts, surveyors, and cartographers in order to obtain a detailed map of the state territory.

Transforming the Territory
A historical atlas would usually represent the extent of the Danish realm in the thirteenth century by coloring the area on a contemporary atlas that was supposedly subject to the ruler. This, however, would not correspond with the socio-political reality at the time and few historical leaders would probably be able to show their realm on a contemporary atlas (Schäfer 2005, Smith 2005: 834). Before the territory was turned into a field for surveyors and cartographers, it was conceptualized and constituted in different ways. Thirteenth century boundaries were often ambiguous but this was not a problem in and of itself because sovereignty and control over the territory was based on a network of loyalties between the monarch on the one hand and the church, nobility, and towns on the other (Sahlins 1989). When a new king was elected it was pivotal for him to confirm previous oaths of allegiance as these were given to the actual person of the king, and not an abstract notion of the crown

or the state. As such, it was personal networks of loyalties that constituted the extent of the kingdom rather than the other way around, where today it is the territory that defines supposed ties of allegiance and jurisdiction.

Prior to the advent of geometric cartography, knowledge of the territory was based on descriptions of special characteristics of the landscape, and was generally contained in a textual form, as well as in personal experience. The earliest known record of the Danish territory is King Valdemar's *Cadastre* (*Kong Valdemars Jordebog*) of 1231, which contains a very specific description of the tax obligations of each shire and the market towns (*købstæder*) and, furthermore, lists the amount of cultivated land in each shire as well as a description of the king's demesne (Aakjær 1980, Ulsig & Sørensen 1981, Fenger 2000). There was thus a clear understanding of the territory even if this was not founded on a cartographic representation. It provided a comprehensive description of the territory available to all literate people, but the listing of islands, for example, would not give much of an idea about their location, shape, and size unless one knew them already. On the geometry-based map, however, the location, the name, and the shape were all linked and instantaneously present. In this way, the appearance of space itself, rather than the contents of that space in terms of population, production, and obligations, became accessible to the viewer. The new mode of knowing space made it possible to define spatial boundaries and, consequently, it became possible to create a rigid understanding of what was inside and what was outside the realm. In this sense, geometric cartography enabled the historical process of transforming boundaries from ambiguous frontier zones into strictly defined lines.

The literary way of knowing space coincided with a functional mode of measuring space in everyday practices. Larger distances were often measured in travel times (see for example Revel 1991: 148).[6] Distances at sea were usually measured either in *kending* which represented the distance that could be seen from the ship towards the horizon, or *uge-søs* (weeks-at-sea), representing the distance that could be sailed in a week. Both measures were tied strictly to the function and use of the ship (Nørlund 1944a). In a parallel fashion, agricultural land was measured in terms of labor time or the seeds that were necessary to grow crops; *tønde* (barrel) indicated the area which could be sown with the seeds of one barrel and *plovland* (carrucate) denoted the area which could be ploughed in one day (Kula 1986: 29-42). When more uniform measurements of land were needed, for example in order to share the land between

---

6   However, variations of the Roman mile were used in several Europeans contexts as well, calculated either as a fraction of the estimated circuit of Equator or as a multiplication of several foot, alen (the length of the lower arm), etc. (Nørlund 1944a: 60-1).

peasants, this would usually be measured out with ropes, sticks, and chains (Balslev & Jensen 1975: 16).

All these examples indicate that space, in general, was derived from a social functionality: labor time and capital input in the shape of seeds for agricultural space, travel time for longer distances, and personal ties of loyalty for the scope of the territory. In consequence of the introduction of geometric cartography, knowledge production of space served to establish space as an autonomous category. This means that, in principle, space could be perceived without reference to functional time or the immediate experience of the environment. Instead of referring to the social usage of space, it became a matter of the relationship between celestial features and the Earth; distances were calculated by degrees, triangles, and the use of geometry. And instead of locations being constitutive of space, they became locations *in* space. Space came to be seen as a framework within which to locate places and social relations. On an abstract level, then, the mode of knowing space changed from functionality, literacy, and tradition to one of mathematics, observation, and visual representation. This way of representing space enabled a uniform visualization of territory which, so to speak, rendered it coherent and tangible in its own right as the knowledge of the territory changed from one based on textual description to one based on visualization. Hence what surveyors did was to transform the territory into a different kind of space from what had existed before.

This abstraction of space from its social use involved a conception of space as homogenous and infinite, only disrupted by neat lines demarcating property, the territory, or the surface of the earth conceptualized as a geometric form. By implication, it became possible to think of territory as a homogenous space. Territories were no longer constituted through networks of regional centers in the guise of castles, towns, and so forth but were constituted at the boundaries by reference to a single sovereign authority in the capital. It is curious that in the Danish case, the completion of Mejer's general map of Denmark (1650) preceded the introduction of absolutism in Denmark (1660) by a single decade. Absolutism meant that sovereignty was re-located to the king alone, thus undermining the power of the council of nobles who would previously have to elect the king, and hold him accountable. Supposedly the sovereign authority of the new absolutist monarch derived from God and referred to a territory defined by the boundaries and not by the loyalty of subjects. On the contrary, loyalty now followed from location within the territory. Now a hereditary title, the institution of the crown was rendered abstract from the physical body of the king and thus became a permanent institution that existed in and of itself. In parallel, the territory was abstracted from the social network that constituted it previously and, through cartographic representation, it was given a timeless existence in and of itself. Danish territory was the same regardless of who governed it.

The undermining of the nobility's authority and the homogenization of the territory did not take place only on a conceptual level; after absolutism several initiatives were launched to centralize authority. Following a four year campaign to regain lost territories from Sweden, the state engaged in a comprehensive new survey of the countryside during the years 1681-1687 to improve its ability to raise taxes (Heering 1932: 14). The result was Christian V's famous Land Register, which provided a uniform registration of the entire territory (Henriksen 1971: 10). Traditionally, the crown had three sources of income apart from the royal demesne: toll, fief payment, and taxes. It was a very costly and arduous affair to tax the land compared to taxing easily controllable trade routes (Jespersen 2007: 73). Yet with the new surveying and mapping techniques it became possible to develop an administrative structure that improved the taxability of the land. By making the record of private property cartographic and systematic, geometric cartography changed the potential tax base of the country. The traditional functional measure of land, *tønde*, was retained, but with the new land register of 1683, it was standardized to cover 14.000 square *alen* (DSDE 1994, vol. 19: 425).

Such standardized reforms were made possible by the state's control of cartographic representations; the court environment in Copenhagen was competing to become the center of calculation, which would allow the state to expand control. Around the same time as Mejer was working on his general map of Denmark, he became involved in a project with Caspar Danckwerth to write a chorography and genealogy of the Duchies, illustrated with Mejer's maps. This became a politically controversial enterprise. Danckwerth's text spoke against the standpoint of the king with regard to jurisdiction and sovereignty disputes over municipalities, which Danckwerth described as being part of Schleswig. Frederik II was furious and subjected the book to censorship; Mejer began rewriting the history to save his relationship to the court (Lauridsen 1888: 67-81). This example illustrates how the new cartography became part of settlements of authority between the king and the nobility, often working in the king's favor to establish a uniform and transparent territory with more clearly settled demarcations of authority and jurisdiction than had previously been known.

In addition to these effects, cartography also enabled centralized authority to "act at distance" (Latour 1987: 219-32) and thus expand the space within direct reach of the capital. Whether it was through land surveyors issued with royal passes or early European expeditions to the rest of the world, the point was not primarily to find new land but to bring the new places back home in order to facilitate a return voyage. "By coding every sighting of any land in longitude and latitude [...] and by sending this code back, the shape of the sighted lands may be redrawn by those who have not sighted them" (Latour 1987: 224). The king would be able to capture and plan travels in the entire kingdom by looking at Mejer's and other maps decorating his walls. The abstract Ptolemaic representation of space allowed planning and travel independent of local intermediaries exactly because their spatial knowledge had now been recorded and taken back to the center of calculation.

The cartographic ability to act at a distance allowed sovereigns to instruct subjects to go into unknown, yet pre-defined spaces. In 1603, for example, the French King Henry granted royal letters to a Pierre de Guast appointing him to Lieutenant-General and to represent the king

> in the countries, territories, confines and coasts of La Cadie [Nova Scotia], commencing from the fortieth degree unto the forty sixth degree; and within the said limits or any part of them, as far and to such distance inland as may be possible, to establish and extend and make known our name, power and authority [...] (cited in Keller et al. 1967: 111).

This instruction contains a mix of abstract and practical demarcation. The starting point is defined by abstract co-ordinates and the scope of extending the king's authority into the foreign lands is made a matter of practical feasibility. Returning to the Danish case, several processes of centralizing authority, removing local intermediaries, and simplifying the territory followed the introduction of absolutism. A commission was established in 1685 to make a plan for how to improve and expand the connection between different parts of the country by *kongeveje*, or roads reserved for the king and his staff (Jørgensen 2001: 26-27). These were the first centrally planned roads in the country. And in the 1680s, the territory was unified through a single code of law (*Danske Lov*)[7]. By thus homogenizing the territorial space into a coherent unit, it was no longer defined only by the important sites, such as towns and castles, within it, but just as much by the spaces in between these sites. By rendering the previously unfamiliar spaces outside the towns and castles familiar, there were no longer significant divisions between inside and outside the town or the castle; the inside/outside divisions marked by the political boundaries to neighboring states were the ones that became increasingly significant. In effect, what the cartographers

---

[7] A commission to create a new legal code was established in 1661 and the law was signed by the king in 1683 (Scocozza 2003: 315-19). The law unified the territory of Denmark proper, excluding the Duchies, Norway, and the Atlantic possessions.

and surveyors did was to establish the spatial reality needed to demarcate a particular domestic realm distinguished from an external environment. In this respect, the fieldwork of the surveyors created a home territory.

Without the transition to a scientific cartography, the particular conceptualization of the state in terms of a uniform, clearly bounded territorial space would not have been possible. Now that the entire territory was available to the observer in an instant it was no longer necessary to travel for weeks to know the territory. Borrowing the words of the famous French cartographer, Nicolas de Nicolay, who was commissioned to map France by Catherine de Medici: "[i]n little time, in little space, and without great expense, to see with the eye and trace with the finger, in particular and in general, the whole extent, power, and status of the realm" (quoted and translated by Serchuk 2006: 146). This notion of mastering space through cartography was not unusual for the time. In the dedication from the Swedish cartographer Andreas Bureus we read: "This little map I thee give instead of the whole world, and then I shall give thee the whole of Scandinavia on a copperplate" (quoted in Mead 2007: 1794). Referring to the entire world, a customer wrote to Abraham Ortelius in 1579 and praised his endeavor: "You compress the immense structure of land and sea into a narrow space, and have made the earth portable, which a great many people assert to be immovable" (quoted in Brotton 1997: 175). As such, the "cartographic revolution" of the fifteenth and sixteenth centuries served to change the perception of terrestrial space into something that could be known and categorized as a unified totality. Within this totality it was possible to carve out particular places as distinct, differentiated polities; yet they were all part of a larger whole which was kept together by principles of geometry.

Conclusion

The division between inside and outside plays a significant role in knowledge production as well as in politics. The notion of going into the field in order to obtain data about the unknown relies on such a division, where the inside is where knowledge of the outside is assembled and computed based on information that is brought back home from the field. Drawing on a Latourian take on knowledge production, I have presented an interpretation of the practices of early modern cartographers that demonstrates how it was the early field studies of the territory that allowed it to be turned into a coherent domestic space. The argument contributes to our understanding of how the territory of the state was domesticated as a familiar knowledge space in a process in which science and politics worked together to establish the inside/outside distinction as one the great divides informing our understanding of society. Where the field, or the foreign, is often conceived as being situated in remote places, seen from the vantage point of one's own society, it was the change in cartographic practice in Europe during the fifteenth and sixteenth centuries that made the fieldwork of surveyors central for mapmaking in the pursuit of ever more efficient maps. Turning locations into spatial data that made them mobile, stable, and combinable, they were brought back to centers of calculation where a new cartographic reality of space was produced.

Where the new cartographic practices were encouraged by concerns with navigation and exploration on the one hand, and a growing private map publishing industry on the other, states became increasingly preoccupied with mapping their own territory. In the classic study by Richard Helgerson, maps are as much a representation of power as one of space because such maps indicate sovereignty and ownership of space (Helgerson 1992: 107). The central display of royal arms speaks of a relation of power to the land (Helgerson 1992: 111); it signifies the ownership of the King over the entire territory and renders, in theory at least, the nobility and the estates invisible. This was the case with Jordan's map from 1585 which confirmed the king's authority over a Danish territory, while also proclaiming the grandeur of the Rantzau family. This is to say that early mapping of the state in Europe made possible effective visual and conceptual possession of "the physical kingdom" (Helgerson 1992: 107).

Yet early cartography not only tied the state and the territory closer together in cognitive and practical terms, it also changed the boundaries between what was familiar and what was unfamiliar. In order to transform knowledge of the territory according to the new cartographic template it was necessary to send surveyors into the field of the territory in order to bring it back to the capital. Assembling the territory at the royal court allowed what Latour has called "action at distance," that is, it allowed the king's administration to contemplate the entire realm without leaving the capital. In other words, it became possible to govern the entire territory from the capital, thus eradicating the unfamiliarity of the territory; what was within the reach of government and knowledge was pushed to the boundaries of the territory. Emphasizing the process of mapping the territory as a big field studies exercise has served to problematize the widespread notion of territory as a historically unproblematic entity. It has demonstrated how the development of the of a so-called modern territory, conceptualized as homogenous and neatly demarcated at the borders, relied on the field- and laboratory-work of early surveyors and cartographers. As such it was not only the foreign that was unfamiliar; it was the surveying of the field at home that provided the basis for conceptualizing the national territory as a domestic inside, serving as the basis for going abroad.

Acknowledgments

I would like to thank the participants at the conference *Ways of Knowing the Field* at the Carlsberg Academy, August 2008, as well as the editors of this volume for some very constructive comments and suggestions on how to improve the manuscript.

References

Aakjær, S. 1980. *Kong Valdemars Jordebog udg. af Samfundet til Udgivelse af Gammel Nordisk Litteratur.* Copenhagen: Akademisk forlag.

Bachelard, G. 1994. *The Poetics of Space.* Boston, Mass.: Beacon Press.

Balslev, S. & H. E. Jensen 1975. *Landmåling og landmålere Danmarks økonomiske opmåling.* Copenhagen: Den danske Landinspektørforening.

Biggs, M. 1999. "Putting the State on the Map: Cartography, Territory, and European State Formation." *Comparative Studies in Society and History*, 412: 374-405.

Bjørnbo, A. A. & C. S. Petersen 1904. *Fyenboen Claudius Claussøn Swart "Claudius Clavus," Nordens ældste Kartograf.* Copenhagen: Det Kongelige Danske Videnskabernes Selskab, Høst & Søn.

Bramsen, B. 1975. *Gamle danmarkskort en historisk oversigt med bibliografiske noter for perioden 1570-1770.* Copenhagen: Rosenkilde og Bagger.

Brotton, J. 1997. *Trading Territories: Mapping the Early Modern World.* London, Reaktion.

Brown, L. A. 1979. *The Story of Maps.* New York, Dover Publications Inc.

Christianson, J. R. 2006. "Hoffet som formidler af naturvidenskaberne under Frederik II." *Renæssanceforum*, 2: 1-16.

Cosgrove, D. E. 2003. *Apollo's Eye: a Cartographic Genealogy of the Earth in the Western Imagination.* Baltimore: Johns Hopkins University Press.

Crampton, J. W. 2010. "Cartographic Calculations of Territory." *Progress in Human Geography*, 35 (1): 92-103.

Crane, N. 2002. *Mercator: the Man who Mapped the Planet.* London, Weidenfeld & Nicolson.

Dalché, P. G. 2007. "The Reception of Ptolemy's *Geography*." In: D. Woodward (ed.) *Cartography in the European Renaissance. The History of Cartography. Volume 3.* Chicago: University of Chicago Press, 285-364.

DSDE 1994. *Den store danske encyklopædi.* Copenhagen: Danmarks Nationalleksikon.

Edgerton, S. Y. 1975. *The Renaissance Rediscovery of Linear Perspective.* New York: Basic Books Inc.

Edney, M. H. 1997. *Mapping an Empire: the Geographical Construction of British India, 1765-1843.* Chicago: University of Chicago Press.

Eisenstein, E. L. 1979. *The Printing Press as an Agent of Change: Communications and Cultural Transformations in early-modern Europe.* Cambridge: Cambridge University Press.

Elden, S. 2010. "Land, Terrain, Territory." *Progress in Human Geography*, 34 (6): 799-817.

Fenger, O. 2000. Kongelev og krongods. *Historisk Tidsskrift*, 100 (2): 257-84.

Gupta, A. & J. Ferguson. 1992. "Beyond 'Culture': Space, Identity, and the Politics of Difference." *Cultural Anthropology*, 71: 6-23.

Haring, C. H. 1964. *Trade and Navigation between Spain and the Indies in the Time of the Hapsburgs*. Gloucester, Mass.: Peter Smith.

Harley, J. B. & P. Laxton 2001. *The New Nature of Maps: Essays in the History of Cartography*. Baltimore, MD: Johns Hopkins University Press.

Heering, H. T. 1932. "Knud Thott og Forhistorien til Kristian V's Matrikul." *Tidsskrift for Opmaalings- og Matrikulsvæsen*, 131: 1-19.

Helgerson, R. 1992. *Forms of Nationhood: the Elizabethan Writing of England*. Chicago: University of Chicago Press.

Henriksen, P. G. 1971. *Hærkort i Danmark og nabolande gennem tiderne*. Copenhagen: Geodætisk Institut.

Häkli, J. 1998. "Manufacturing Provinces: theorizing the encounters between governmental and popular 'geographs' in Finland." In S. Dalby & G. O'Tuathail (eds.) *Rethinking Geopolitics*. London: Routledge, 131-169.

Jespersen, M. L. 2007. "Administration og Statsdannelse i Danmark 1400-1600." *Den Jyske Historiker*, 116: 63-87.

Jørgensen, S. E. 2001. *Fra chaussé til motorvej det overordnede danske vejnets udvikling fra 1761*. Copenhagen: Dansk Vejhistorisk Selskab

Keller, A. S., O. Lissitzyn & F. J. Mann 1967. *Creation of Rights of Sovereignty through Symbolic Acts, 1400-1800*. New York: AMS Press, Inc.

King, G. 1996. *Mapping Reality: an Exploration of Cultural Cartographies*. Basingstoke: Macmillan.

Knudsen, J. 1918. *Søkortdirektør Jens Sørensen "den danske Hydrografis Fader" 1646-1723: et Bidrag til det danske Søkortsvæsens Historie*. Copenhagen: Det Kongelige Danske Søkort-Arkiv.

Kragh, H. 2005. *Dansk naturvidenskabs historie*. Aarhus: Aarhus Universitetsforlag.

Kula, W. 1986. *Measures and Men*. Princeton; Guildford: Princeton University Press.

Lamb, U. S. 1974. "The Spanish Cosmographic Juntas of the Sixteenth Century." *Terrae Incognitae*, VI: 51-64.

Latour, B. 1987. *Science in Action: how to follow Scientists and Engineers through Society*. Milton Keynes: Open University Press.

Latour, B. 1988. *The Pasteurization of France*. Cambridge, Mass.: Harvard University Press.

Latour, B. 1993. *We have never been Modern*. New York: Harvester Wheatsheaf.

Latour, B. 1999a. "Give me a laboratory and I will raise the world …" In: M. Biagioli (ed.) *The Science Studies Reader*. London: Routledge, 258-276.

Latour, B. 1999b. *Pandora's Hope: Essays on the Reality of Science Studies*. Cambridge, Mass.: Harvard University Press.

Latour, B. 2005. *Reassembling the Social: an introduction to actor-network-theory*. Oxford: Oxford University Press.

Lauridsen, P. 1888. *Kartografen Johannes Mejer: et Bidrag til ældre dansk Kaarthistorie*. Copenhagen: Bianco Lunos Kgl. Hof-Bogtrykkeri.

Mead, W. R. 2007. "Scandinavian Renaissance Cartography." In. D. Woodward (ed.) *Cartography in the European Renaissance. The History of Cartography. Volume 3.* Chicago: University of Chicago Press, 1781-1805.

Nørlund, N. E. 1943. *Danmarks Kortlægning*. Copenhagen: Geodætisk Institut.

Nørlund, N. E. 1944a. *De gamle danske Længdeenheder*. Copenhagen: Ejnar Munksgaard.

Nørlund, N. E. 1944b. *Islands Kortlægning: en historisk fremstilling*. Copenhagen: Ejnar Munksgaard.

Nørlund, N. E. 1944c. *Færøernes Kortlægning: en historisk fremstilling*. Copenhagen: Ejnar Munksgaard.

Ortelius, A. & R. A. Skelton 1968. *Abraham Ortelius: The Theatre of the Whole World. London 1606. With an introduction by R.A. Skelton*. Amsterdam: Theatrum Orbis Terrarum.

Pickles, J. 2004. *A History of Spaces: Cartographic Reason, Mapping, and the geo-coded World*. London: Routledge.

Revel, J. 1991. "Knowledge of the Territory." *Science in Context*, 4: 133-61.

Sahlins, P. 1989. *Boundaries: the Making of France and Spain in the Pyrenees*. Berkeley, University of California Press.

Sandman, A. 2007. "Spanish Nautical Cartography in the Renaissance." In D. Woodward (ed.) *Cartography in the European Renaissance. The History of Cartography. Volume 3*. Chicago: University of Chicago Press, 1095-1142.

Schäfer, W. 2005. "Ptolemy's Revenge: A Critique of historical Cartography." *Coordinates*, Series A3: 1-16.

Scocozza, B. 2003. *Ved afgrundens rand*. Copenhagen: Gyldendal & Politiken.

Serchuk, C. 2006. Picturing France in the Fifteenth Century: The Map in BNF MS Fr. 4991." *Imago Mundi*, 582: 133-149.

Skelton, R. A., H. Hexham & J. Johnson 1968. *Mercator-Hondius-Janssonius: Atlas or A Geographicke Description of the World*. Amsterdam: Theatrum Orbis Terrarum.

Smith, M. L. 2005. "Networks, Territories, and the Cartography of Ancient States." *Annals of the Association of American Geographers*, 954: 832-849.

Strandsbjerg, J. 2008. "The Cartographic Production of Territorial Space: Mapping and State Formation in Early Modern Denmark." *Geopolitics*, 132: 335-358.

Turnbull, D. 1996. "Cartography and Science in Early Modern Europe: Mapping the Construction of Knowledge Spaces." *Imago Mundi*, 48: 5-24.

Turnbull, D. 2000. *Masons, Tricksters and Cartographers: Comparative Studies in the sociology of scientific and indigenous knowledge*. Newark: Harwood Academic.

Ulsig, E. & A. K. Sørensen 1981. "Studier i Kong Valdemars Jordebog – Plovtalsliste og Møntskat." *Historisk Tidsskrift*, 81: 1-25.

Walker, R. B. J. 1993. *Inside/outside: International Relations as Political Theory*. Cambridge: Cambridge University Press.

Wood, D. 1992. *The Power of Maps*. New York: The Guilford Press.

Woodward, D. (ed.) 2007. *Cartography in the European Renaissance. The History of Cartography, Volume 3*. Chicago: University of Chicago Press.

Rengenier Rittersma

Subterranean Fieldwork
Marsili's Survey on the Biogeography and Ecobiology of Truffles in Eighteenth Century North and Central Italy

Since Antiquity, botanical interest in truffles has been a rich topic for speculation. There has been an abundance of fanciful theories bent on explaining the mysterious nature of this subterranean fungus. At the start of the eighteenth century, the perennial discussion about the origins of mushrooms experienced a revival, due to the first microscopic observations of what seemed to be seeds. As a result, some scholars formulated the hypothesis that fungi are generated by an internal reproductive element. These first proponents of the seminiferous theory were, however, opposed by scholars who still clung to the idea of spontaneous generation. Among the most prominent of them was the Bolognese count and naturalist Luigi Ferdinando Marsili (also Marsigli, 1658-1730), who in 1714 published his *Dissertatio de generatione fungorum*.

In this treatise, he argued that all mushrooms are generated by external factors and that fungi can actually be classified according to their different matrices, such as stone, wood, soil, etc. Truffles were, then, to be considered as products of the soil. In order to prove his hypothesis, around 1714-1715 Marsili designed a large-scale field survey of the subterranean mushrooms which lasted at least five years and for which he mobilized a huge network of correspondents and research assistants in different regions of North and Central Italy. This enquiry precipitated a huge manuscript, entitled *Schedae pro tuberorum historia*, possessed by the university library in Bologna.

The aim of this chapter is, firstly, to explain the reasons why Marsili eventually opted to conduct field research in order to shed new light on the vexed question of the origins and nature of fungi. Secondly, this article simultaneously seeks to show that Marsili's field research on the *Tuberi escolenti* was actually a quest, which involved the implementation of unprecedented research techniques. In this sense, Marsili's venture reflected the gradual rise of the experimental paradigm and in some respects even anticipated modern mycological research on the biogeographical diffusion and ecobiological preferences of truffles.

Count Luigi Ferdinando Marsili was born into a Bolognese patrician family. He studied under the Bolognese anatomist Marcello Malpighi, but never took a degree, since he embarked upon a military career. From 1682 onwards he served in the Imperial army and spent most of his time in the lower Danube river basin. His military career ended abruptly in the court-martial because of the loss of the Imperial fortress of Breisach (1703). The rest of his life was devoted to learning and collecting, in which he was equally successful. Marsili is, amongst others, recognized as a pioneer in military engineering and geological stratigraphy, as the founder of oceanography, and as one of the spirits behind the foundation of the Institute of Sciences and Arts of Bologna (1711), which also included the *Accademia delle Scienze* and the *Accademia di Belle Arti* (the *Clementina*). His scholarly prominence also became apparent in his election as fellow of the French and English Royal Academy of Sciences (for biographical information see Stoye 1994, Cavazza 2007, Gullino & Preti 2008).

This necessarily succinct overview of Marsili's scientific activity may be sufficient to conclude that the Bolognese count was fascinated by a virtually infinite range of natural phenomena. But in view of the immense efforts which he undertook during his lifetime, it can be said that he must have been completely obsessed by mushrooms and even more by their subterranean variant, the truffle. On the other hand, it is remarkable that his published and unpublished research on truffles barely revealed any interest in them as a gastronomical and economic resource. Marsili found truffles, fortunately, intriguing above all as a research object.

In this sense, he also exceeded all previous historical accounts on truffles in a threefold way. His survey was, firstly, by far the most extensive naturalist study on the truffle. Secondly, it was one of the first to deliberately combine laboratory and field research. And thirdly, Marsili's project was also unparalleled in that it made considerable use of the expertise of illiterate local connoisseurs from the truffle-producing regions. These three facets make the *Schedae pro tuberorum historia* a mine of information for historians of science and the environment, as well as for historical geographers and historical anthropologists, even more so since virtually no research has been done on this specific manuscript.[1]

The paper upon which your eyes rest is a first attempt to compensate for that undeserved lack of attention. The chapter is subdivided into two parts. The first section starts with a discussion of the pre-modern perception of fungi in general

---

1   Baldacci (1930) only touched upon it, while Cordella's more recent research (2003) rather concerned a presentation of some telling documents related to the area of Norcia-Spoleto. Nonetheless, I would like to thank Romano Cordella for his useful information and Gioia Filocamo and Fabrizio Ammetto for drawing my attention to this manuscript.

and truffles in particular. Its aim is to contextualize Marsili's ideas on these parasitic fungi, and, more specifically, to elucidate what actually urged Marsili to initiate his exploits. In the second section the focus will shift to the survey itself, providing an analytical overview of its backgrounds, its scope, and the research methods applied. It will be demonstrated that his enquiry considered an admirably wide variety of ecobiological determinants in relation to the appearance of truffles.

Mysterious Delicacy: Ideas about the Origins of truffles in pre-Michelian Mycology

A truffle – or at least the seven to eight European edible species that belong to the so-called Ascomycetes class – is a mushroom that, unlike other mushrooms, spends its whole lifecycle (including fructification) below the earth surface. They live, like fungi, in a symbiosis with other plants, mostly with trees. They reproduce themselves by means of their spores, which are dispersed after maturation (the process of primary mycelium), when they deteriorate or when they are consumed by, for example, game, insects, mice etc. These spores "conquer", so to speak, the root system of a hosting plant (the process of mycorrhisation) and induce the production of new truffles (the secondary mycelium). Fungi science is called mycology, whereas hydnology is the official term for biological research on truffles.

At least since Antiquity mankind has enjoyed truffles without actually knowing much about their nature and, more specifically, without having a very clear understanding of their reproductive cycle. How then were truffles perceived in pre-modern botany? The oldest hitherto known scholarly writings on truffles date from the ancient Greek and Roman botanists Theophrastus (3rd century BC), Dioskorides, and Pliny the Elder (both 1st century BC). They all wondered about the nature of truffles because in some aspects, such as their perfume and habitat, they seemed to belong to the vegetal world. However, truffles obviously lacked all the properties that characterize plants, such as seeds, flowers, roots, and fibers. For that reason, Pliny (*Historia Naturalis*, Lib. XIX.11.33) concluded that this *miraculum* was nothing more than *callum terrae*, i.e. a callous concretion of the earth. In medieval and early modern botany, this idea persisted and eventually resulted in the conception that truffles should be considered as products of soil deterioration.

This perception of truffles corresponded with the more general view that mushrooms originate from putrefaction (Ainsworth 1976, Dörfelt & Heklau 1998). Although nobody could actually *prove* the similar reproductive principle in fungi and truffles, all scholars intuitively assumed that above-ground and below-ground mushrooms were akin. For that reason the discussion of the pre-modern perception of the truffle automatically implies the concomitant question of how mushrooms were perceived. Until the end of the seventeenth century the view was held that, as Robert Hooke put it in his *Micrographica* (1665): "Mould and Mushrooms require no seminal property, but the former may be produc'd at anytime from any kind of *putrifying* Animal or Vegetable Substance, as Flesh, &c. kept moist and warm ... [The latter] seem to depend upon a convenient constitution of the matter out of which they are made, and a concurrence of either natural or artificial heat" (Ainsworth 1976: 15). In other words, fungi were considered to be products of spontaneous generation.

According to this theory, living entities, vegetal, mineral or animal, may, under appropriate conditions, arise suddenly from matter independently of any intrinsic pre-existent element. The term spontaneous generation (or alternatively equivocal generation) is a container term which actually represents a wide variety of different scenarios of how life can generate itself from non-living matter, but all its proponents shared the belief in the exogenous origins of animate and inanimate beings, including for example, eels, fleas, lice and worms, as well as mushrooms and corals. Adherents of this theory basically presupposed a certain amount of belief in the possibility of miracles, either in the guise of supra-natural intervention or in terms of an infra-natural coincidence of propitious factors. Spontaneous generation was actually a very helpful concept in clarifying the genesis of organisms whose reproductive cycles had not yet been made visible, such as insects and cryptogamic plants. This usefulness may also partly explain the durability of this idea, which was conceived and diffused by Aristotle and his pupil Theophrastus and consequently adopted by Roman, Christian, and scholastic naturalists. The backing of such prominent exponents obviously also contributed to its enduring success. But over the course of the seventeenth century, new ideas started to undermine its monopoly on explaining the ultimate *Arcana naturae* (Farley 1977: 1-8, McLaughlin 1994: 1-2, Harris 2002: 1-6. For an overview of pre- and early modern belief in wonders, see Habermas 1991, Mullin 1996, Campbell 1999, Daston & Park 2001, Goodich 2007, Signori 2007).

By the middle of the seventeenth century, at least two developments gave reason to question the adequateness of the explicatory model of spontaneous generation. Firstly, there was the rise of Cartesianism in the sciences: according to Cartesian explanations, nature is determined by immutable universal and uniform laws, which systematically govern the (re-)generation of organisms. Cartesian scholars consequently argued that plants and animals systematically originate from seeds (germism) and from eggs (ovism) or semen (spermism), respectively. These principles of reproduction presuppose that nature is predictable and categorically exclude chance from scientific explanation, thus leading to the outright rejection of the occurrence of spontaneous generation.

In much the same way, a second influential scientific trend also contributed to the decline of the theory of spontaneous generation. Like Cartesianism, the concept of Physico-Theology postulates that since nature is regular and uniform, vegetal and animal procreation must derive from a preexistent reproductive element. Unlike Cartesians, natural theologians did not explain the regularity and uniformity of nature as a mechanistic model but rather as a demonstration of divine foresight (Farley 1977: 10-18). But whether one adopted a purely physical or a metaphysical explicatory model, it was, after all, the very assumption that regeneration is triggered off by an *intrinsic* reproductive element that turned out to be crucial. In mycology this notion very soon provoked new debates on the origins of mushrooms, which would continue until the nineteenth century (Ainsworth 1976: 15-18).

It was the Italian botanist and preceptor of Luigi Ferdinando Marsili, Marcello Malpighi (1628-1694), who in his *Anatome plantarum* (1679) first hypothesized that "either fungi have their own seed, by which their species is propagated, or they sprout from the growth of fragments of themselves, as happens in other plants" (Ainsworth 1976: 15). Strictly speaking, before Malpighi wrote this the Neapolitan naturalist Giambattista Della Porta had already assumed the occurrence of seminiferous fungi, reporting in his book *Phytognomonica* (1591) that he had observed spore-dust in certain above-ground and below-ground (!) mushrooms. Significantly, the chapter at issue was headed "Contrary to the Ancients all Plants are Provided with Seed," in which he countered Porphyrius' assumption that fungi are *deorum filios*, as they do not arise from seed. However, Della Porta's observations were virtually ignored since he adhered in other respects to the speculative and denigrated Paracelsian signature doctrine (Buller 1915: 3-4, Dörfelt & Heklau 1998: 56-57, Ainsworth 1976: 13-14).

Be that as it may, at the start of the eighteenth century there were already some French authors, such as Joseph Pitton de Tournefort and Jean Marchant *fils*, who argued that mushrooms have their own reproductive bodies, but their ideas were based on speculation rather than on exact observation. However, these suppositions incited botanists and gardeners to try growing of mushrooms. Between 1707 and 1731 the first reports on successful experiments to grow fungi in Paris and London gardens appeared. But all these advancements were examples of trial-and-error processes, since nobody had actually proved the existence of a reproductive body in fungi. It was not until the introduction of the microscope that botanists began to have a better understanding of the reproductive cycle of fungi. But as long as the microscope was a novelty, "the apperception of microscopic plants by the early observers was naturally influenced by a phanerogamic bias [i.e. by those plants who have visible reproduction organs] and Malpighi, like his predecessor Hooke, failed to interpret the spores correctly" (Buller 1915: 6-8).

Dubito, Ergo Cogito: Puzzled by the Minute World
Apart from Giambattista della Porta's random observation of spores, the first to detect spores and to intimate their function was the French scholar Jean Marchant *fils*, when he examined the so-called *Xylaria polymorpha* fungus in 1711. However, it is neither Tournefort nor Marchant junior, but the Florentine botanist Pier Antonio Micheli who is considered the founder of mycology. Not only was he the first to grasp the rudiments of the reproductive cycle of fungi, he also collected a considerable amount of evidence to corroborate the hypothesis that reproductive bodies are present in all fungi, and last but not least, he successfully applied this theory by culturing fungi from spores. Although Micheli's major discoveries took place between 1710 and 1718, it was, due to different circumstances, only after the publication of his *Nova plantarum genera* (1729) that his groundbreaking research results were disseminated. And even then, his views were anything but taken for granted (Buller 1915: 10-25, Lazzari 1973: 96-132, Ainsworth 1976: 13-14). On the contrary, it was precisely in that intermediate period that the theory of spontaneous generation had a significant upsurge, since there was still a considerable group of scholars who persistently clung to the exogenous theory. Around 1714-1715 one of the most influential of them, count Luigi Ferdinando Marsili, started a new offensive against the theory of seminiferous fungi.

In 1714 Marsili issued his *Dissertatio de generatione fungorum*, on which he had presumably worked for more than 20 years. The treatise was dedicated to Giovanni Maria Lancisi, physician of Pope Innocent XI and Clement XI. It was published with a short exposition on the origins and nature of mushrooms, written by the same Lancisi, in which he briefly confirmed Marsili's opinion that fungi are devoid of *semi* and that they are products of heterogenesis, and especially of processes of organic degeneration. However, Marsili's own investigations were much more profound, since they were based on protracted observations and experiments in the field and in the laboratory. Lancisi was, by way of his high position, of course considered an authority, but in medical affairs rather than in natural history and botany.

But there was more at stake than a mere knowledge lag. Marsili's treatise is characterized by an overwhelming zeal; in his endeavor to reveal this *Arcanum naturae*, he tries to discuss all aspects of fungi. In barely 40 pages of his *Dissertatio de generatione fungorum* we find, consequently, a synthesis of historical thinking about mushrooms, a provisory classification, a synopsis of his field observations, and even a detailed description of an observation and an experiment which should once and for all crush the seminiferous theory.

During his military expeditions Marsili observed that a certain kind of mushroom grew upon horse dung, which induced him to conclude that it was generated by the excrement, since he excluded the possibility that the semen would remain intact during the whole digestive process. Marsili also described an experiment carried out by his friend Francesco Bartolucci (more about him here below): lichen, a kind of moss, was hermetically closed in a jar and after only 7 months some miniscule mushrooms appeared on it. His colleague waited another couple of months, but nothing happened; the mushrooms did not reproduce themselves. This was the ultimate proof that mushrooms lack seeds, because otherwise it would have produced more specimens of the same fungus. Furthermore, it was excluded that such an original seed, if present, would remain undamaged during the long fermentation process, nor was it likely that a seed from outside would have entered the hermetically closed jar. Ergo: mushrooms arise spontaneously as a result from degeneration (Marsili 1714: 6-7 resp. 37-39). However, just below this patina of positiveness there was an enduring undercurrent that, perhaps, there might be something true about the seed-theory.

So, paradoxically, in 1714 Marsili published a treatise containing a strong statement against the concept of seminiferous fungi, since this publication constrained him to take a clear position, but he nevertheless continued his quest for the fanciful spores. And in order to prove his own provisory theory, that mushrooms sprout from different putrefying substrata, and that truffles are essentially products of geological putrefaction, he conceived his truffle survey. Marsili presupposed an identical reproductive system in aboveground and belowground fungi: "Tubera similia sunt in omni fungo [Truffles equal fungi in all respects]" (Marsili 1714: 27, compare Baldacci 1930: 304). Or did he set off on this truffle expedition in search for spores in the subterranean mushrooms? Marsili, who was too accurate a scholar to accept something as truth without hard evidence, might have been in two minds about these possibilities.

There are several indications that Marsili was not truly convinced by his own theory on the origins of fungi. Why otherwise did he continue with a new research project on the genesis and nature of truffles immediately after the publication of his *Dissertatio*? Even in his mushroom treatise Marsili described that he did observe *granula* in the *fungilli calyciformes*. He rejected, however, the assumption that these little kernels were *semina* as a preconceived notion, concluding that they could never be *semina* because they were disproportionally big for such a tiny fungus (Marsili 1714: 19, 17-18; compare Baldacci 1930: 290). Moreover, the brief description of truffles in the *Dissertatio* also intimated Marsili's suspicion that there could be an alternative reproductive system. Examining a soil sample from the place where truffles grow, the Bolognese scholar discovered a kind of fleece ("lanugo") which at a more advanced stage observed through the *occhio armato* (as Marsili put it) turned out to be a sort of arachnoid ("telae araneae instar"). What Marsili describes here is nothing less than the first microscopic observation of the fibred network of the mycelium, about 170 years before its function was tentatively described by Giuseppe Gibelli and Albert Bernhard Frank (Marsili 1714: 26-27, compare Dörfelt & Heklau 1998: 185-187).

On the other hand, this negative attitude towards the theory of seminiferous fungi was partly also biased by social-cultural factors. Bolognese botanists especially, such as Lelio Trionfetti, Marsili's teacher and life-long correspondence partner, were known for denigrating Malpighi's micro-anatomical approach in general and his view of seminiferous mushrooms in particular. This might have rendered Marsili even more undecided, since he was emotionally closely related to both preceptors (Baldacci 1930: 290-291). But methodical preferences and aversions also played a role.

To give just one example: throughout the whole of the eighteenth century, microscopic observations, which were, of course, fundamental for those who clung to the idea of autogenetic fungi, were regarded with some skepticism. A related problem was the rather restricted and inductive way of conducting scientific research. Those who adhered to seminiferous theory "rarely extrapolated beyond the confines of their own experiments." Significantly, until Louis Pasteur nobody argued that the theory of spontaneous generation was complete nonsense, simply because scholars were reluctant to express universal claims without available evidence (Harris 2002: 32-34, 42). In other words, uncertainties and reservations were not typical for Marsili, but rather an example of the skeptical and sometimes even Pyrrhonist attitude of many contemporary scholars.

We must dwell on this aspect, because otherwise we may be unable to make sense of the dilemma of a scholar who had intuitively grasped the reproductive cycle of fungi without actually having a cogent explanation for this phenomenon. If we do not evaluate the profound meaning of his scientific doubts, we will not comprehend Marsili's zeal and titanic efforts to unveil the origins of above-ground and below-ground fungi. These were the very parameters in which Marsili and many of his colleagues were conducting their research. To overcome their doubts, "scholars sought to devise experimental conditions that eliminated or at least minimized extraneous sources of error" (Harris 2002: 33). Just to mention one example of this methodical rigorism: in order to be sure that his experimental vessels were hermetically closed, the Italian biologist Lazzaro Spallanzani exposed them to extreme conditions (Harris 2002: 44-47). Micheli was also perfectly aware of the crucial role played by material factors in scientific experiments and therefore endlessly repeated his tests (Buller 1915: 6-23). In much the same way, Marsili tried to find a way out of the impasse of the habitual laboratory research on the origins of fungi, by conceiving a field inquiry on the adjacent topic of the genesis of truffles.

Marsili's inquiry on the "sì obscura vegetazione de' tartufi": backgrounds, scope, & methods

Backgrounds

For as long as Marsili had studied fungi – and he examined them from his youth – truffles had also aroused his interest. There were several reasons that made the choice of this subterranean variant for a case study quite obvious. Firstly, at that time there was not much existing research on truffles and this topic therefore offered a possibility to exceed his preceptors, Malpighi and Trionfetti, and other scholars as well. A second positive factor was the availability of a rudimentary network of partners in the truffle-producing regions of North and Central Italy, which would facilitate the envisaged project. It seems that one of his associates in the countryside, Francesco Bartolucci, had even done some of his own research on truffles (Baldacci 1930: 303-310). Last but not least, Marsili, after his dishonorable discharge from the Imperial Army in 1703 and his return to Bologna five years later, also found himself in appropriate circumstances to co-ordinate such an inquiry, since he planned to settle for a while in his hometown.

Marsili's inquiry on the truffle has been conserved in the university library in Bologna in a convolute manuscript, entitled *Schedae pro tuberorum historia*, which consists of 17 folders, containing more than 400 pages of handwritten information, as well as three publications related to the topic.[2] The different folders include a wide variety of data, such as letters, drawings, excerpts, notes, sketches, fragments of chapters, provisory tables of contents etc. Originally, the manuscript must have also contained numerous soil and truffle samples, as Marsili asked his correspondents to provide him not only with written information but with material evidence as well.[3] In some cases, the same document has been passed through many hands, just as the whole manuscript is characterized by discontinuities in style and expression. In

---

2 Frati 1928: 102-103. This account is based on a rough calculation of mine, since Frati's catalogue does not always specify the number of pages in an archival folder. The books contained in Marsili's manuscript include a copy of Alfonso Ceccarelli's treatise on truffles entitled Opusculum de tuberibus (Padua 1564), and two copies of his own *Dissertatio de generatione fungorum* (including Lancisi's comments).

3 See question 18, 20, 26, and 30 in L. F. Marsili, *Interrogatorio a'i maestri cavatori delle tartufare de I Monti di Spoleto, o di Norcia*. Manuscript no. 86, Biblioteca Universitaria dell' Università degli Studi di Bologna (subsequently LFM86), fascimile 10.

short, the *Schedae* are a classic example of an *unvollendetes Projekt*, full of contradictory and fragmentary information, a circumstance which substantially complicates interpreting – let alone publishing – this manuscript.[4]

There is nevertheless an indication that Marsili originally planned to dedicate at least a part of the manuscript as an *opusculum* to the British Royal Society. In a letter from February 19, 1724 to Isaac Newton, Marsili expressed his short-term intention to publish this little tract– presumably jointly with a reprint of Alfonso Ceccarelli's truffle treatise – in a second edition of his own *Dissertatio de generatione fungorum*. But in a next letter from May 1724, the Bolognese count informed Newton of the upcoming publication of his *magnum opus* on the Danube without mentioning his original plans, but rather announcing a new project on the "organic structure of the earth": "I shall soon dedicate to this distinguished Society another little tract on the generation of tubers in the earth, which will appear with the second edition of *The Generation of Fungi* already known to the Society for some time" (Turnbull 1977: 283-285, compare Vai 2006: 95-129).

The highly unsystematic and illogical character of the *Schedae pro tuberorum historia* may, however, at the same time function as the score for its interpretation, as the headings of the different folders, for instance *Schedae* and *fragmenta*, suggest. What goes for Marsili's investigations on fungi in general applies *a fortiori* to his research on truffles: it is very much designed as a trial, as a thought experiment in which he expected to pin down the essential features of subterranean fungi, and indirectly, of supraterranean mushrooms as well. But in spite of the fragmentary and erratic research results, the approach of Marsili's case study on truffles was remarkably holistic with due attention to a wide variety of determinants, such as soil, surrounding flora, insects etc. It is this awareness of the numerous ecobiological factors that makes his truffle inquiry an inspiring model, even for modern mycologists.[5] In accordance with the general theme of this volume, a selection has

---

4   The publication of a scientific edition of Marsili's unpublished manuscripts seems to be envisaged, either in the framework of the "Progetto Marsili" or in the framework of a new research programme "TELLUREXPLOR" that was launched in Autumn 2010. See Biagi Maino 2005: 9-13, 129-159, Rittersma & Schmid 2009.

5   After decades of rather reductionist approaches in truffle scholarship and cultivation experiments, modern truffle researchers are, in view of the very complex, fragile, and unpredictable character of truffles, increasingly becoming aware of the necessity of a more holistic procedure. This at least could be concluded from the *Terzo Congresso Internazionale di Spoleto sul Tartufo*, Spoleto, Italy, November 25-28 2008. At a local level there are more recently even historicist tendencies, as the re-edition of the influential *Nouveau manuel de trufficulture. Réédition du manuel du Docteur Pradel (1914) réactualisé et amendé par Henri Dessolas, Gérard Chevalier et Jean-Claude Pargney* (Périgueux 2007) shows.

been made of different elements, which illustrate Marsili's admirably comprehensive approach.

Scope

Marsili was so obsessed by the question of the nature and origins of truffles that he wanted to get to the bottom of the problem. How else can we explain why he was willing to invest such enormous amounts of time and resources in this research project? This determination consequently resulted in a highly ambitious and sophisticated modus operandi, as is particularly clear from the wide geographical, thematic, and temporal range of his investigations.

In fact, the selection of the regions in which Marsili wanted to make inquiries may be read as a first indication of his intuitive awareness of the correlation between different habitats and the occurrence of different truffle types. Marsili chose the following four regions, three of which are situated along the mountain chain of the Apennines: the area of Bologna-Modena, Urbino-Acqualagna, Norcia-Spoleto, and the district Cori-Velletri in the Alban Hills, south of Rome. Marsili does not give an explicit explanation of why he chose these areas, but in two different texts which could best be described as a kind of preface, he says a bit more about his preference for these regions: "It is by the way true, that with regard to the nature and origins of truffles hitherto nobody could have known more than what I will ….. and what will be proved by the testimony of the eyewitness inspections of mine and of ancient truffle hunters *in the most fertile parts of our Italy.*" In another introductory document, Marsili refers to "(…) the most experienced hunters [literally 'diggers'] of these Tubers […unreadable word] in the *Papal State.*"[6] However, this selection also shows that Marsili started his inquiry well-prepared, since the regions at issue covered more or less the whole spectrum of Italian truffle production.

Nowadays the Apennine mountain chain between Emilia and Umbria produces all types of edible truffles, which can be subdivided into two groups. The first category concerns whitish truffles, such as *Tuber magnatum Pico* (white precious truffle) and *Tuber borchii Vitt.* (also known as *Tuber albidum Pico*). The second group contains dark-colored types, as for instance, *Tuber melanosporum Vitt.* (black precious truffle), *Tuber aestivum Vitt.*, *Tuber uncinatum Chatin*, *Tuber brumale Vitt.*, *Tuber*

---

6   Translations by the author (emphasis added). Original text in Italian: "Vero è, per altro, che fino ad ora non si è da veruno potuto mai giungere a trovare e conoscere in questo proposito più di quello che io ti porrò sotto gli occhi e che verrà autenticato dalla testimonianza delle oculari mie ispezioni e da quelle di antichi cavatori di tartufi *ne'più fertili siti della nostra Italia* […]." And: " i più esperti cavatori di questi Tuberi in queste Terre Pontificie" LFM86, fasc. 3, *Fragmenta adnotationum et observationum pro tuberorum historia*, cart. 11-12 resp. cart. 93-retro.

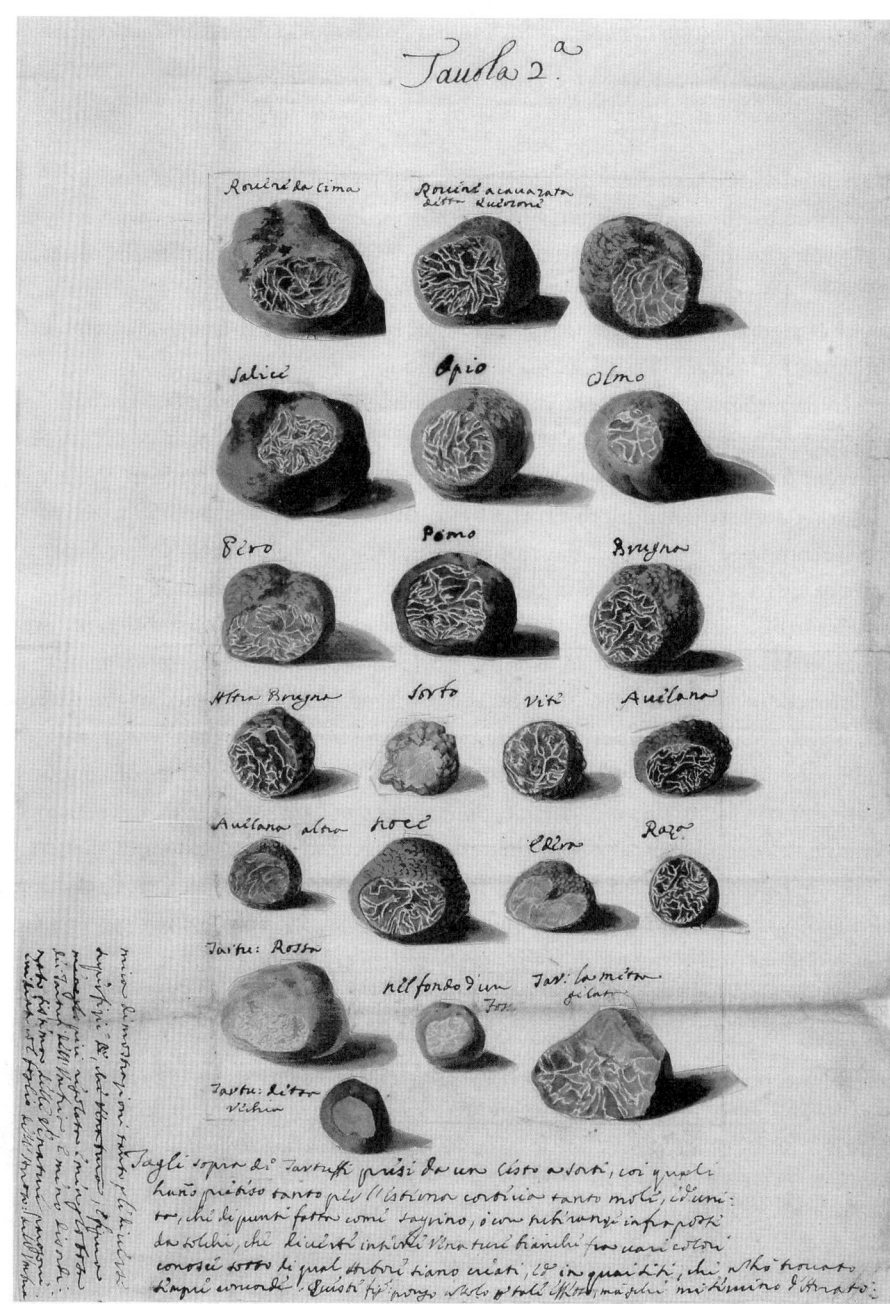

Figure 1. Drawing of cross-cut sections of truffles, showing that truffles change morphologically in accordance with their tree hosts (LFM86, fac. 12, STORIA NATURALE DE TUBERI ESCOLENTI *(…), cart. courtesy of University of Bologna Library).*

macrosporum Vitt., and *Tuber mesentericum* Vitt. It can be deduced from Marsili's notes and drawings that he examined most of these truffle types, although he was not always aware of their differences, since very often their specific features could only be observed with the aid of a sound microscope.[7]

In addition, with regard to the most prestigious truffle types, namely the *Tuber magnatum Pico* (white precious truffle) and the black exponent, *Tuber melanosporum* Vitt., Marsili chose two truffle producing territories which already had a transregional reputation at that time. The domains of the former duchy of Urbino were famous for their white truffles, whereas the black diamonds from the area Norcia-Spoleto were in high demand in Florence, Rome, and Venice (with regard to the area Gubbio-Urbino resp. Norcia-Spoleto: Campano 1495: *Epigrammatum libri viii*, Lib. quartus, xvii, Ancajani 1762: 87-88, Soderini [1526-1596] 1851: 330-331, Allegrucci 1992, Castellucci 1995: 24-25). However, Marsili's selection was not only based on criteria like representativeness and reputation. Significantly, his inquiry also bore in mind somewhat exceptional aspects.

The choice of the district Cori-Velletri in the Alban Hills, for example, was anything but obvious. Marsili's specific reason for involving Cori-Velletri in his research was the frequent occurrence of the so-called *pietra fungaia* – a kind of conglomerate of stony, telluric and fungal elements – in this region. It would go far beyond the framework of this chapter to discuss this aspect in detail, but Marsili expected that this particular fungus could corroborate his theory that some mushrooms – including truffles – should be considered as products of soil deterioration. The comparative case of the *pietra fungaia* is nevertheless noticeable, since it clearly demonstrates that Marsili tried to set up an all-inclusive inquiry.

The scope of Marsili's truffle investigations, which resulted in the *Schedae pro tuberorum historia*, was also quite wide from a temporal point of view. From the letters of one of his key correspondents, Francesco Bartolucci, we can deduce that he must have begun it in the year his *Dissertatio de generatione fungorum* (1714) was published.[8] The collective research project lasted at least six years, and, according to

---

7 It was no coincidence that the final classification of truffles took only place in the course of the 19th century by the Lombard botanist Carlo Vittadini (Lazzari: 207-220).

8 See Bartolucci's letter to Marsili from September 19 1714 in: LFM86, fasc. 5, *Ricerche ed Osservazioni fatte sopra li tartuffi* (...), cart. 15-retro. This was relatively quick after the publication of his Dissertatio, since the preface of Lancisi's comments on Marsili's treatise, which were jointly published, was dated 9 January 1714 (Lancisi 1714: xviii).

some scholars, Marsili continued with it until his death in 1730.[9] Moreover, Marsili's truffle research was not only a long-term enterprise but a very time intensive undertaking as well; he succeeded, by means of his huge network, in monitoring the developments in the sample regions throughout the different seasons of the year. This was made possible by the ingenious idea of designing a questionnaire and distributing it among his correspondents, which brings us to this chapter's final section on how Marsili gathered and organized his research data.

Methods

The list of queries (see appendix 1) that Marsili circulated to the survey takers located in the abovementioned truffle producing regions of North and Central Italy can be subdivided into five different sections. These groups correspond with specific ecobiological factors that determine the occurrence of truffles which, together with some other contextual aspects, were of interest to Marsili.

By far the largest part of the questionnaire concerned the issue of the *habitat* (1) of truffles. In this section Marsili tried to collect information about many different ecological circumstances which – he believed – in one way or another affect the occurrence of truffles, including soil conditions and hosting trees but also the presence of above-ground fungi and other flora as well as insects and other animals in the surroundings of the truffle site.[10] Likewise, Marsili paid due attention to the influence of the *climate* (2) on truffles, by posing several questions on the role of meteorological conditions during different periods of the year as well as on the correlation with the position of the moon (Camporesi 1989).[11]

Marsili also addressed rather specialized issues in his questionnaire, such as the *morphological* (3) and *organographical* (4) aspects of truffles. With his detailed queries on the external and internal features of truffles and particularly with his requests for

---

9   Compare Marsili's following remark "Nel corrente anno 1720, avendo notizia per il mio erudito compadre Monti che in Urbania (…) si fosse trovato a' 3 maggio nelle vicinanze del monte Nerone uno di questi detti smisurati tartufi (…)." LFM86, fasc. 3, *Fragmenta adnotationum* (…), cart. 97-retro. According to Baldacci, referring to a letter of July 23 1730, Marsili continued his research on above-ground and below-ground fungi until his death in 1730 (Baldacci 1930: 311).

10 See, for instance, questions 1-3, 6-8, 21, and 23-27 (soil conditions) resp. question 29 (hosting trees) resp. question 5, and 8-12 (fungi) resp. question 3-5, 29 (flora), and question 20 (fauna). In his questionnaire Marsili only asked about the presence of worms in truffles, but in his enquiry there are some indications that he might also have been aware of the ecological role of deer and insects with regard to the appearance of truffles. For drawings of insects, see LFM86, fasc. 12, *Storia naturale de' Tuberi escolenti per tentare lo scoprimento della loro generazione*, cart. 22 seq. According to Baldacci, Marsili even consulted foreign scholars about the role of insects in truffle sites (Baldacci 1930: 306).

11 See question 12, 15,16, 28, 30, 32 (weather and season) resp. question 15 and 22 (position of the moon).

samples of truffles in the different stages of their development, Marsili can even be considered as one of their first taxonomists (see also Figure 1).[12]

Significantly, Marsili's interest in truffles was not restricted to their biological aspects. Just as his hexamerous *Danubius Pannonico mysicus* (1726) discussed extensively various natural *and* cultural phenomena in the Danube river basin, his truffle enquiry in much the same way deliberately regarded the cultural dynamics of the subterranean mushroom. As a result, the questionnaire also addressed some practical issues with respect to truffle hunting.[13]

Marsili's *interrogatorio* may leave an almost exhaustive impression, especially since the final question on his list (number 34) concerned an open question or rather a kind invitation to his survey takers to gather "(…) further information, serving the truffle research in order to unveil their nature (Appendix 1)."[14]

This ever humble or even vacillating attitude was typical of Marsili, who never excluded the possibility that he could be ignorant of an essential aspect and who especially in this manuscript over and over again gave evidence of his own doubts and lack of knowledge. Conversely, Marsili was well aware that for the successful realization of his truffle survey he relied essentially on the life-long experience of the local connoisseurs. What is striking throughout the manuscript are his recurrent expressions of deep respect for the – as he called them – *periti cavatori* or *Maestri cavatori*.[15] It was, in fact, precisely the innovative combination of research *in labo* and *in situ* that made – and still makes – this enquiry an exceptionally interesting information source. Compared to the previous scholarly contributions on truffles, Marsili's project was actually the first to pay due attention to the insiders' view, and it seems that he was well aware of the ground-breaking and entertaining dimension of his research method: "(…) with my methodical questions I have been able, as you will see, to extract answers even from these rude persons [literally: from their rudeness], which you will

---

12 See question 13, 14, 17-19, 30 (morphology) resp. question 12, 21, 30 (organography).

13 For cultural aspects, see question 31-33.

14 "(…) ulteriori notizie per utile dello studio dei Tartufi e per rinvenire la loro natura (…)." This invitation seems to have yielded some unexpected research results. The manuscript contains, for instance, a drawing of cross-cut sections of truffles which show that truffles change morphologically in accordance with the tree that is actually hosting them (LFM86, fasc. 12, *Storia naturale de' Tuberi escolenti* (…), cart. 6-9).

15 See, for instance, the following fragment from what was conceived as a kind of preface to his treatise: "(…) vi rappresento su questi miei Foglii, che non è altro che l'osservazione di fatto, e l'informazioni avute con l'accuratezza vedrete da I più esperti cavatori di questi Tuberi in queste Terre Pontificie, già che non era a me possibile di vedere tutto nel corso di quattro anni che questi esperti hanno potuto osservare per più di trenta e chi di quarant' anni (…)" (LFM86, fasc. 12, *Storia naturale de' Tuberi escolenti* (…), cart. 3-retro).

all read in the form in which they were actually given to me. The reading will be entertaining, because in essential issues you will find uniformity and in a lot of things contrarieties and other diversities which (however) in any case will shed abundant light and make this reading less boring for those who are curious of the study of nature"[16]

Conversely, the truffle treatise written by the Umbrian scholar Alfonso Ceccarelli, *Opusculum de tuberibus* (Padova 1564), was predominantly based on other scholarly publications on truffles and only implicitly referred to the experiences of truffle hunters: "In fact, I learnt from experienced men that in the place one day truffles were found (...)."[17] Marsili, however, obviously often made use of local illiterate experts, as he also engaged fishermen in his oceanographic research. In this respect, Marsili's research perfectly mirrored the increasing importance of unlearned connoisseurs in accumulating empirical information (Sartori 2003, Eamon 2006: 217-221).

Conclusions

The eventual result of Marsili's truffle survey was nothing more than a question mark, or even worse, a series of *new* mysteries. In the end, Marsili acknowledged in what was designed to be the preface of his treatise that the reader may be surprised not to find " (...) any positive reflection," since the many observations did not embolden him to "formulate systematic thoughts that were watertight like laws."[18] But the following sentence revealed that he was still hopeful to resolve once and for all the vexed question of the origins of fungi: "If life bestows me to finish my observations on a large amount of molds, generated by plants, fruits and animals, and once I have completed my chemical experiments, then would I speak with more certainty about fungi, tubers and molds."[19] So the Bolognese mycologist Gilberto Govi rightly

---

16 " (....) con le mie domande ridotte al metodo, che vedrete, ho potuto anche dalla loro rozzezza riccavare risposte che tali quali mi furono date le leggerete con vostro piacere, perche nell'essenziale, trovarete uniformità ed in molte cose contrarietà ed altre diversità che ad ogni modo non lasciano di dar molti lumi e di render la lettura ai curiosi dello studio della Natura men noiosa" (LFM86, fasc. 12, *Storia naturale de' Tuberi escolenti* (...), cart. 3-retro-4-verso).

17 "Nam ego intellexi a viris spectatae fidei, quod in illo loco ubi sunt in uno die excavata tubera (...)" (Ceccarelli 1564, cap. XIX, see also cap. IX).

18 "(...) alcuna positive riflessione, e questo per che non mi trovo ancora con le tante e poi tante mie osservazioni assai forte da potermi rendere in quel comune ardire di pronunciare sistemi appunto con Autorità da Legislatore" (LFM86, fasc. 12, *Storia naturale de' Tuberi escolenti* (...), cart. 4-retro).

19 "Quando e vita e ozio Dio (?) m'accordasse per terminare le mie intraprese osservazioni circa una serie di gran numero delle Muffe nate da Piante, da Frutti, e da Animali, e che avessi compiti li miei esperimenti chimici e sopra de Funghi de Tuberi e delle Muffe con più sicurezza parlare" (LFM86, fasc. 12, *Storia naturale de' Tuberi escolenti* (...), cart. 4-retro).

argued that we should not consider Marsili's *Dissertatio de generatione fungorum* as the final result of his mycological studies, since he also left an unpublished manuscript on fungi (Govi 1984).

In the subsequent section, entitled *Della Denominazione de Tuberi*, which was already written out in full and presumably concerned a final draft made ready for publication, Marsili even rejected his own presupposition that truffles are tubers which originate from the soil as well as the alternative option that they represent a subterranean variety of fungi, whether seminiferous or not.[20]

In spite of the rather disenchanting outcomes of his research, Marsili did draw a kind of macro conclusion from the results of his truffle survey which was actually rather revealing for his philosophy of science. Since nature could not be considered to be regular, uniform, and predictable, the Bolognese count pleaded for a science which was based on experimental evidence rather than on theories: "I finish bothering you with this preface, saying that I am progressively getting aware that nature does not have one model (literally one way) for all things, but rather different ways, depending on what it generates and from where it takes its nutrition and the substance of which it is actually composed. Of this new method of doing research is the great Francis Bacon the father (...)"[21] Here Marsili advocated, in a nutshell, an empirical and inductive method, specifically its Baconian concretion (Cavazza 1985, 1990: 119-149, 2002: 3-10).

---

20 See, for instance, the following remarks: "Ogni Tubero o sia Ligneo o carneo hà tal nome per che sono la sostanza che lo compongono. Questi, che sono chiamati Tuberi o Callo della Terra per se stessi non partecipano minima cosa di terra che per accidente che la loro superficie sia macchiata di questa ma senza La minima conessione nella sostanza del Tubero, per che con l'acqua lavandosi uno d'essi coperti di Terra se ne va lasciando la struttura di esso intatta. Fungo sotterraneo e pure un nome che non corrisponde punto alla struttura de Funghi come in un ochiata si comprende" (LFM86, fasc. 12, *Storia naturale de' Tuberi escolenti* (...), cart. 5-retro, see also cart. 6-verso). The term "Callo della Terra" derives from Pliny; see section A in this chapter. So Marsili meanwhile also abandoned the idea that truffles could be seminiferous, a view which he seemed to hold in his letter to Isaac Newton from February 29 1724: "It happens that I have compiled a variety of observations concerning these same tubers of the earth, neglected hitherto, and [to the effect] that nature propagates all things by seeds, though these are of a different nature, form and substance from the seeds of ordinary plants. [Circa eadem terrae tubera varias observationes conserere mihi contigit, ignotas hactenus, et, quod natura propagat omnia p[er] semina, sed ea diversae naturae formae ac substantiae a seminibus regularium plantarum]" (Turnbull 1977: 265).

21 "Termino d'infastidirvi (...) con questa prefazione col dirvi che sempre più mi stabilisco a conoscere che la natura non ha una sol strada per tutte le cose ma che le ha diverse secondo il bisogno di quel sogetto che produce ed in quel luogo dove deve pigliar alimento e quella sostanza di cui deve essere composto. Di questo nuovo Metodo di studiare che ha per Padre il Gran Baccone" (LFM86, fasc. 12, *Storia naturale de' Tuberi escolenti* (...), cart. 5-verso).

The same fragment is quite significant for another reason, since it demonstrates that Marsili was indeed obsessed by his quest for spores throughout his whole life, or at least until 1727.[22] At that time he also acknowledged that it was up to other scholars to unveil the *sì obscura vegetazione de' tartufi*.[23] Luigi Ferdinando Marsili certainly would have been pleased to learn that it took quite a long time before biologists unanimously agreed upon the fact that truffles reproduce themselves by spores; he would have been even more amused that the holistic approach of his enquiry in some respects anticipated current hydnological research and that modern scholars, after all, are still puzzled by various aspects of truffles.[24]

Acknowledgements

I would like to thank the following people and institutions for having been of great help during the preparation of this chapter. First of all, Gioia Filocamo and Fabrizio Ammetto for drawing my attention to this manuscript as well as Romano Cordella for kindly and generously sharing with me all his information about this historical source. I am also grateful to the staff of the *Biblioteca Universitaria Bologna* and, in particular, *Dottoressa* Rita De Tata as well as her colleague Maria Cristina Bacchi for their help. This chapter benefited significantly from the stimulating observations of Allen J. Grieco, Marta Cavazza, Alessandra Zambonelli, and Martin Schmid, to whom I extend my sincerest thanks. Last but not least, I should also like to thank Richard Kirwan for his linguistic corrections. The remaining errors and inadequacies are, of course, all my own.

---

22 Marsili wrote that "(…) da Settanta anni in qua che nella Corte di Toscana benemerita di tutto L'Orbe Litterario cominciò e da essa si trasportò per tutto il rimanente dell'Europa che su quelle Orme ha attaccata la Natura e tutti sotto li Precetti del gran Baccone" (LFM86, fasc. 12, *Storia naturale de' Tuberi escolenti* (…), cart. 5-verso). Marsili presumably referred to the foundation of the *Accademia del Cimento* (1657) by Grand Duke Ferdinando and Prince (later cardinal) Leopoldo dei Medici, which also championed an experimental approach to science. I would like to thank Marta Cavazza for her assistance in interpreting this fragment.

23 "(…) acciò che altri, replico, più capaci di me valendosi di quanto ho detto e dico in hora possino con più facilità arrivare alla perfetta meta di quanto è stato a me impossibile di pervenire (…)" (LFM86, fasc. 12, *Storia naturale de' Tuberi escolenti* (…), cart. 3-verso). For a similar statement, see also LFM86, fasc. 3, *Fragmenta adnotationum* (…), cart. 1-11.

24 To mention just two examples: the papers of Jean Claude Pargney and Alexander Urban at the *Terzo Congresso Internazionale di Spoleto sul Tartufo*, Spoleto, Italy, November 25-28 2008 (proceedings are forthcoming) showed that modern truffle research is increasingly interested in the role of environmental determinants, such as earthworms and small mammals.

References

Ainsworth, G. C. 1976. *Introduction to the History of Mycology*. Cambridge: Cambridge University Press.

Allegrucci, F. 1992. *Un diamante in cucina: l'iconografia del tartufo nell'antichità*. Siena: Editioni Alsaba.

Ancajani, A. 1762. *Commercio attivo, e passivo della città di Spoleto e suo territorio secondo il calcolo formato nell'anno corrente 1761*. Unpublished manuscript, 2nd edition. Spoleto.

Baldacci, A. 1930. "I fondamenti botanici nell'opera di L. F. Marsili." In: C. Marsiliano (ed.) *Memorie intorno a Luigi Ferdinando Marsili*. Bologna: Zanichelli, 277-356.

Biagi Maino, D. 2005. *L'immagine del Settecento da L. F. Marsili a Benedetto XIV*. Torino: Archivi di Arte Antica, Umberto Allemandi Editone.

Buller, A. H. R. 1915. "Micheli and the Discovery of Reproduction in fungi." *Transactions of the Royal Society of Canada*, 9(IV): 1-25.

Campano, G. 1495. *Epigrammatum libri viii*. In: G. Campano. *Omnia Campani opera*. Venice: B. Vercellens.

Campbell, M. B. 1999. *Wonder and Science: Imagining Worlds in Early Modern Europe*. Ithaca: Cornell University Press.

Camporesi, P. 1989. *La terra e la luna. Alimentazione folclore società*. Milan: Mondadori.

Ceccarelli, A. 1564. *Opusculum de tuberi bus*. Unpublished manuscript. Padova.

Castellucci, L. 1995. *Il tartufo, Trüffel*. Munich: Droemer Knaur.

Cavazza, M. 1985. "Impact du concept baconien d'histoire naturelle dans les milieux savants de Bologne." *Les études philosophiques*, 3: 405-414.

Cavazza, M. 1990. *Settecento inquieto. Alle origine dell'istituto delle scienze di Bologna*. Bologna: Il Mulino.

Cavazza, M. 2002. "The Institute of Science of Bologna and the Royal Society in the 18th century." *Notes and Records of the Royal Society of London*, 56(1): 3-25.

Cavazza, M. 2007. "Marsili." In: N. Koertge (ed.) *New Dictionary of Scientific Biography*. Volume 7. Detroit: Scribner's, 36-38.

Cordella, R. 2003. "I tartufo nero di Spoleto e di Norcia negli inediti di L. F. Marsili (sec. XVIII)." *Spoletium*, 44: 74-86.

Daston, L. & K. Park. 2001. *Wonders and the Order of Nature, 1150-1750*. New York: Zone Books.

Dessolas, H., G. Chevalier & J.-C. Pargney. 2007 (eds.) *Nouveau manuel de trufficulture. Réédition du manuel du Docteur Pradel (1914) réactualisé et amendé (...)*. Périgueux: Bailliére.

Dörfelt, H. & H. Heklau. 1998. *Die Geschichte der Mykologie. Eine Übersicht von den Anfängen bis zur Gegenwart*. Schwäbisch Gmünd: Einhorn.

Eamon, W. 2006. "Markets, Piazzas and Villages." In: K. Park, K. & L. Daston (eds.) *The Cambridge History of Science. Volume 3: Early Modern Science*. Cambridge: Cambridge Universoty Press, 206-224.

Farley, J. 1977. *The Spontaneous Generation Controversy from Descartes to Oparin*. Baltimore: The Johns Hopkins University Press.

Frati, L. 1928. *Catalogo dei manoscritti di Luigi Ferdinando Marsili conservati nella Biblioteca Universitaria di Bologna*. Florence: Biblioteca Universitaria di Bologna.

Goodich, M. 2007. *Miracles and Wonders: The Development of the Concept of Miracle, 1150-1350*. Aldershot: Ashgate.

Govi, G. 1984. *Luigi Ferdinando Marsili. Micologo Bolognese (1658-1730)*. Bologna: ClueB.

Gullino, G. & C. Preti. 2008. "Marsili Luigi Ferdinando, v." In: *Dizionario Biografico degli Italiani*. Vol. 70. Rome: Istituto dell'Enciclopedia italiana, 771-781.

Habermas, R. 1991. *Wallfahrt und Aufruhr: Zur Geschichte des Wunderglaubens in der frühen Neuzeit*. Frankfurt am Main: Campus.

Harris, H. 2002. *Things come to life. Spontaneous Generation Revisited*. Oxford: Oxford University Press.

Lancisi, J. M. 1714. "Dissertatio epistolaris de ortu, vegetatione, ac textura fungorum (…)." In: L. F. Marsili. *Dissertatio de generatione fungo rum*. Rome: F. Gonzagae.

Lazzari, G. 1973. *Storia della micologia italiana*. Trento: Arti Grafiche Saturnia

Marsili, L. F. 1914. *Dissertatio de generatione fungo rum*. Rome: F. Gonzagae.

McLaughlin, P. 1994. *Spontaneous Generation vs. Equivocal Generation in Early Modern Science*. Berlin: MPI Wissenschaftsgeschichte.

Mullin, R. B. 2007. *Miracles and the Modern Religious Imagination*. New Haven: Yale University Press.

Pliny the Elder AD 77-79. *Historia Naturalis*. Second English translation by J. Bostock & H. T. Riley. Retrieved July 7, 2011, from http://www.perseus.tufts.edu/hopper/text?doc=Plin.+Nat.+toc&redirect=true

Rittersma, R. C. & M. Schmid. 2009. "Luigi Ferdinando Marsili (1658-1730) and the Contemporary Fascination for the Telluric Reign: Transdisciplinary Perspectives from History to Science (TELLUREXPLOR)." Unpublished research proposal (April).

Sartori, R. 2003. "L. F. Marsili: founding father of oceanography." In: G. B. Vai & W. Cavazza (eds.) *Four Centuries of the Word Geology. Ulisse Aldrovandi 1603 in Bologna*. Bologna: Minerva Edizioni, 169-179.

Signori, G. 2007. *Wunder: Eine historische Einführung*. Frankfurt am Main: Campus.

Soderini, G. 1851. *Della cultura degli arti e giardini*. Milan.
Stoye, J. 1994. *Marsigli's Europe: The Life and Times of Luigi Ferdinando Marsigli*. New Haven: Yale University Press.
Turnbull, H.W. 1977 (ed.) *The Correspondence of Isaac Newton*. 7 vols. Cambridge: Cambridge Universty Press.
Vai, G. B. "Isostasy in Luigi Ferdinando Marsili's manuscripts." In: G. B. Vai & G. Caldwell (eds.) *The origins of Geology in Italy*. Boulder, CO: The Geological Society of America. Special Papers, 95-129.

Appendix 1: Interrogatorio a'i maestri cavatori delle tartufare de I Monti di Spoleto, o di Norcia

1. Is the soil which generates truffles a long unbroken area or does it alternate with other pieces of soil?
2. Does it, with regard to its fertility, make any difference whether a *tartufaia* (truffle-site) is situated on an east-, west-, south-, or north-facing slope?
3. What are the external signs which indicate that a soil is producing truffles?
4. Is the soil which usually generates truffles fertile with grass or sterile and bare? And what about oats, do they grow if sown in this soil?
5. Are there any fungi which grow at the sites where truffles occur and do they have the aroma of truffles?
6. What are the color and the flavor of the truffle producing soil? Is it black, yellow, red, ash grey, odorous, insipid, or of any other quality?
7. Do the truffle-sites produce constantly or is the production sometimes interrupted for a year or a longer period?
8. How deep in the soil are truffles situated?
9. Did you ever find a truffle, which, like a fungus, grows up out of the soil?
10. Did you observe any particular piece of the tuber inserted in the soil, serving either as a fundament or like a root growing out of these truffles?
11. In which season does the development of the truffle start, and is it true that their first buds arise at the same time as the fig tree usually sprouts in spring?
12. How many months does the truffle grow until the winter comes, or does it after its millet-like origins grow in a few days, as we usually see happen with mushrooms, and maintain itself and subsequently obtain a mature state during the winter?
13. Is the internal color of the truffle white and the external color identical with that of the surrounding soil as soon as it starts to grow?

14. Is the truffle white and odorless during the entire summer?
15. In which lunar month and which lunar state do truffles take on their dark color and their aroma?
16. In which period of the year do they obtain a mature state?
17. How big and how heavy can a truffle tuber be?
18. How many sorts of truffle tubers exist, according to their form, structure, and other features which truffle hunters may observe? The coordinator would like to receive a sample of all these sorts, in order to draw them.
19. Did you ever find several truffle balls unified like gluten?
20. Do they, like fungi, have worms inside? The coordinator would like to receive two or three samples of these worm-eaten truffles.
21. Were there any truffles which were not extracted from the soil during the year in which they arose? The coordinator would like to know whether they are conserved in this way, whether they grow until the subsequent year or whether they deteriorate, and, in that case, if they make the soil more fertile for that truffle?
22. Is it true that the truffle harvest is very abundant when during the moon of August it rains a lot? And that, if it does not rain in that period, the harvest will be in vain?
23. More than anything else the coordinator would like to know how deep the matrix soil of truffles is situated?
24. Is it true that if you are digging deeper than this matrix, then you will not find any truffles in that place?
25. Beyond every belief the coordinator would like to know from the truffle hunters what this truffle matrix actually is. Is it a different kind of soil, or are there mold fibers, as can be observed around fungi, or any other similar thing?
26. The coordinator would like to receive some pieces of what truffle hunters consider the truffle matrix in a box by mail, in order to observe them properly with a microscope. He would then come to the site in order to determine better the matrix in its natural fresh state. The survey takers are kindly requested to act carefully, in order to find out, if possible, what this matrix is.
27. Do the truffle hunters believe or did they ever test out the belief that if you transport this earth matrix to another place with a different soil, that after covering this extract of earth the same truffles arise?
28. What effect do heavy snowfalls, rains, and frost have on truffles?
29. Do they (i.e. the survey takers) observe in the mountains what can be observed in Lombardy, namely that under oak trees there are more truffles?

30. The coordinator would like to have every month two or three balls of truffles in order to draw them and to describe their development until they are mature.
31. What is used in order to find truffle sites: dogs, pigs, or external signs of the soil?
32. In which period is the truffle harvesting abandoned?
33. Is it abandoned because the truffles are not good, or for what other reason do people give up harvesting?
34. The coordinator would like to have further information serving the truffle research in order to unveil their nature; just as I did when detecting mushrooms in the treatise I published [i.e. *Dissertatio de generatione fungorum*].

Source: LFM86, *Schedae pro tuberorum historia*, fasc. 10, *Interrogazioni e domande fatte al Co. L. F. Marsili a diversi cavatori di tartuffi de' monti di Spoleto, di Norcia, del territorio di Bologna (…)*, cart. 1-31.

Kasper Risbjerg Eskildsen

Exploring the Republic of Letters
German Travellers in the Dutch Underground, 1690-1720

Geographies of the Republic of Letters

To early moderns, the European Republic of Letters was a lived experience. The bonds that connected scholars across confessional and geographical boundaries were not just woven with words, but products of personal relationships and mutual trust (Goldgar 1995, Miller 2000, Shelford 2007). These connections not only enabled a feeling of community, but also established reliable channels for the exchange of information, arguments, and discoveries (Shapin 1995, Licoppe 1995, Chartier 2007). A normal starting point of an academic career was therefore a European journey. During their journey, young scholars established friendships for life, initiated correspondences, and positioned themselves in the Republic of Letters. Unlike students on the *peregrinatio academica*, academic travellers seldom remained at one university, but instead visited several important venues of learning and scholarship. The route normally included large libraries and manuscript collections, botanical gardens and anatomical theatres, cities with many or important scholars, and famous universities and academies (Becker 1980, Sauerland 1983, Dibon & Waquet 1984, Beyrer 1985, Kulczykowski 1989, Siebers 1991, Stagl 1995, de Ridder-Symoens 1996, Boedeker 2002, Roche 2003: 569-666). Thus, the primary destinations were not the ruins, osterias, and landscapes of Italy that attracted so many nobles on the *grand tour*. Early modern scholars, especially during the seventeenth and early eighteenth centuries, instead preferred Holland, with its many universities, academics, bookshops, printers, publishers, and journals.[1]

In the spring of 1705, the later Göttingen professor Christoph August Heumann travelled through Northern Germany and Holland together with his friend, the

---

1 For a bibliography of early European travellers to Holland, see Jacobsen Jensen 1919-39. For German academic travellers, see esp. Schneppen 1960, Bientjes 1967, with a bibliography of travel journals 248-93, and Laeven 2005. For French and English academic travellers, see also Murris 1925, van Strien 1993, 1998.

mathematician Bonifacius Heinrich Ehrenberg. In some ways, Heumann's journey resembled those of hundreds of young academics before him. He travelled to Holland shortly after defending his magister degree in Jena in 1703. He visited all major cities and universities, but only remained at each location for short periods of time. He met with important representatives of Dutch late humanism, such as Jacob Gronovius and Jacob Perizonius, but also encountered exiled Huguenots, such as Pierre Bayle and Jacques Basnage. Heumann's travel journal, published in part by Andreas Cassius in 1768, also reveals his attempts at establishing personal relationships and mutual trust (Cassius 1768: 32-137). Heumann briefly described which topics, books, and scholars had been discussed and noted how his conversation partners treated him. He catalogued friendly scholars as "polite," "very polite," or "humane," according to their willingness to engage in academic dialogue and to share information. He also noted if his encounters were impolite or uncommunicative. For example, Heumann scribbled this about the theology professor Salomon van Til in Leiden:

> It appeared as if he only wanted to show me, how it fitted a theologian with a long tobacco pipe in his mouth: Thus, he remarked that he soon again had to lecture in his private collegium; therefore he did not converse, and was also not humane (Cassius 1768: 69).

The majority of the notes in Heumann's journal, however, did not resemble those of other seventeenth century academic travellers. Heumann travelled to Holland not only to prepare his future academic career, but also to explore and experience firsthand the notorious Dutch underground. As a diligent fieldworker, he collected as much information as possible and filled his journal with detailed minutes of conversations with religious and intellectual dissenters. He often recorded exchanges word-by-word and noted questions as well as answers.

A few similar journals from the Early Enlightenment have survived. The authors of these journals were all recent candidates of protestant German universities and probably all connected to the Thomasius circle in Halle. They travelled together, in pairs, and visited either just Holland or Holland and England. The most voluminous of these journals is that of the later Jena professor Gottlieb Stolle, who used more than a thousand manuscript pages to detail his encounters in Holland, 1703-1704. Other examples are the journals of Johann Burkhard Mencke, 1698-99, Christoph August Lämmermann, 1709-11, Zacharias Conrad von Uffenbach, 1709-11, and Johann Gottlieb Deichsel, 1717-19. These travellers all wrote notes in the same meticulous style as Heumann, and often followed the same routes and visited the same scholars, but their interests differed.² Mencke almost exclusively visited academics and representatives of the moderate Enlightenment, probably preparing himself to inherit his father's position as editor of the *Acta Eruditorum*, the most significant academic journal in protestant Germany. Stolle wrote lengthy comments regarding the clothing and manners of academics and about the Spinozistic underground. Uffenbach was especially interested in manuscripts, books, and curiosity cabinets (see also Franke 1965, Laeven 2005, Mulsow 2007: 67-86). Heumann also had his own particular interests and the most extensive notes, which have survived in Cassius' selection, record conversations with lay preachers, mystics, and sectarians.

The interest of the Thomasius circle in the Dutch radical underground is not an indication that the Republic of Letters was splitting into two, as Dena Goodmann argues (1994: 22), or even into "an indefinite number of overlapping mini-Republics," as L. W. B. Brockliss claims (2002: 390). It is more likely that the minute accounts of interviews with dissenters in these travel journals indicate that the ideal of the Republic of Letters was changing from one of community into one of transparency. During the last decades of the seventeenth century, Thomasius, following Pufendorf's theory of functionalist personae, redefined the Republic of Letters as a realm of disembodied arguments (Eskildsen 2004, 2008). Here, every argument counted and should be heard, regardless of its origins. Only the free exchange of opinions could secure the vitality and integrity of scholarship. Thus, the voices of the underground deserved attention, even if the dissenters were of little help to the future academic careers of their German visitors. It is possible that the confrontation with the Dutch underground was even an integrated part of Thomasius' educational program aimed at transforming budding professors into tolerant and open-minded philosophers.

Travel Guides

Before leaving Jena, Heumann prepared for his journey. He studied the Dutch intellectual and religious landscape, carefully planned a route, and selected where to stop and whom to meet. His notes indicate that his most important source of information was Henrich Ludolff Benthem's *Holländischer Kirch- und Schulen-Staat* (1698). Benthem, born in Celle near Hanover, was a Lutheran theologian and later superintendent in Harburg (Kniggen 1727). His book was based upon years of

---

2   Not all German scholars of the period wrote in this style. Other academic travel journals only contained short descriptions of destinations, libraries, and curiosity cabinets, see, for example, [Erndl] 1711 and Curtius 1891.

study and two journeys to Holland, in approximately 1686 and 1693. It contained a comprehensive historical and institutional overview of the Dutch intellectual and religious life, but also offered practical advice to students and academics visiting Holland. Benthem even presented his book as a travel guide and had it printed in a handy *octavo* format, in two volumes, that allowed travellers to take a copy in their luggage; several of the German travellers did.[3]

The most obvious advantage of Benthem's book was the recommended travel route that ensured that one visited "the most significant places in these countries, in particular all universities, in the right order" (Benthem 1698: I:24). The route was accompanied with recommendations excursions to locations "where something remarkable can be seen, or great artists and scholars reside, or where people with unusual opinions in matters of religion can be encountered" (Benthem 1698: I:27). Some of Benthem's destinations were merely tourist sites of particular interest to scholars. A prominent example of such a site, and an obligatory stop on most academic journeys, was the statue of Erasmus in Rotterdam that immortalized the founder of the *respublica literaria*. Celebrating the shared European humanistic heritage while demarcating the academic tourist's view from that of common plebs, Benthem passionately described the site:

> A scholar rejoices, when he here on the great bridge sees the statue of Erasmus … He stands in his doctor robe on a marble column and has a book in his hand, as if he was reading and wanted to turn the page. Boys have thrown many small stones on it. One tells children and fools that Erasmus turns a page whenever he hears the bell ring and that when he turns the last page the world will end (Benthem 1698: I:120).

Heumann and Ehrenberger more or less followed Benthem's route, as did many other German academic travellers. Heumann also followed Benthem's recommendations about interesting scholars and places of knowledge. For example, Benthem explained that after the visit to the Erasmus statue, one could also find many well-stocked libraries in Rotterdam and that "that of the polite mister Henrici is not among the worst" (Benthem 1698: I:121). Heumann noted this remark in his journal and visited Henrici and his library (Cassius 1768: 47). On his way back to Jena, Heumann also briefly visited Benthem in Uelzen, but unfortunately Cassius' selection tells us nothing about their conversation.

---

3   On the preparations for the journey, see esp. Uffenbach 1753-1754: I:XLII, CXII.

Bentham's book, together with his *Engeländischer Kirch- und Schulen-Staat* (1694), probably also influenced the particular note-taking practices of German travellers. One important inspiration for seventeenth-century travel journals was the emergence of diaries, in which the author wrote short notes every evening concerning the events of the day, even if nothing particularly noteworthy had happened (Tschirnhauß auf Hackenau 1727: 124-127, Fostergill 1974, Hartmann 1991, Sherman 1996: 29-76, Bekker 2007). Benthem also recommended that his readers write down their impressions every night, after returning to their lodgings. To remember the content of academic conversations, however, even daily note-taking was not enough. Academic travellers, Benthem admonished, must therefore always keep a "small tablet at hand" and "already in the presence of the learned man write observations down" (Benthem 1694: Vorbericht). Benthem realized that the practice was unusual and therefore recommended that his readers ask permission first. But even the restrictions of academic decorum could be circumvented. Thus, Uffenbach developed "a particular skill" to write secretly with a pencil in a notebook hidden in his pocket and "to record memorable conversations and important relations" (von Uffenbach 1753-1754: I:LII) without his conversation partners noticing.

The German travellers did not always follow Benthem's advice. Benthem's book was more than a historical and institutional handbook and a travel guide for students and scholars in Holland; it was also a contribution to the reconciliation between Lutherans and Reformed in Brandenburg-Prussia. Benthem supported this effort in several of his writings. He even dedicated the *Holländischer Kirch- und Schulen-Staat* (1698) to Elector of Brandenburg Friedrich III and, in this dedication, expressed his hope and conviction that the protestant Churches could be reunited "through the ties of doctrine and love." His travel guide therefore also portrayed Reformed Holland as positively as possible and his route traversed an enlightened landscape in which Brandenburg scholars would feel at home.

Benthem openly admitted that his guided tour was selective, designed according to his own convictions. He wished "before we fully embark on the journey to show the bad and the good in the United Netherlands so that [the traveller] better can avoid the former and embrace the latter" (Benthem 1698: I:2). Students and young scholars should not only shun Dutch inns and whores, but also avoid the many sectarians, Jews, and Spinozists who had found refuge in the country. Benthem praised Dutch liberty and claimed "that so many scholars live in the United Netherlands because one not only provides well for them, but also allows everyone to speak, write, and live as suits him best" (Benthem 1698: II:169). He was nevertheless critical of the excessive religious freedom in the country. Tolerance and reconciliation between Lutherans and Reformed should not lead to tolerance of radical dissent, and Benthem lamented "that it is very difficult to combine the large freedom of religion in Holland with the rules of the school of Christ" (Benthem 1698: I:6). In his preface, he even expressed a pious hope that his book could serve the "defense against and extermination of [religious] enthusiasm" (Benthem 1698: I:Vorrede).

Heumann ignored Benthem's warnings. His travel journal only briefly mentioned the Jewish community in Holland, but he carefully recorded numerous conversations with Mennonites, Socians, radical Pietists, Labadists, and Quakers. Short remarks in Benthem's book, even if this was not the author's intention, also delivered clues to how and where one could find members of this religious underground. For example, there is a brief mention of the locations of the Quakers and Mennonites in Amsterdam (Benthem 1698: I:52-53). Heumann followed these hints and visited both congregations. Benthem also remarked that one could read and buy Quaker writings in Jacob Claus's bookshop in the Princenstraet in Amsterdam (Benthem 1698: I:621). Jacob Claus was not only a book dealer; he and his older brother Jan were key figures in the Dutch Quaker community. They had previously lived in England and accompanied William Penn and George Fox during their travels in Holland in 1677 and 1684 (Kannegieter 1971: 135-54, 206-208). Heumann sought out Jan Claus, had a two hour conversation with him, and wrote lengthy notes about the history and faith of the Quakers.

Benthem's book was not Heumann's only guide to the underground. In 1700, Gottfried Arnold published the second volume of his controversial *Unparteiische Kirchen- und Ketzerhistorie*, which detailed the history of heresy in the modern age and included comprehensive descriptions of religious sects in Holland. Arnold's book, especially his anti-clerical sentiments, resonated for Thomasius, who recommended it to his students as "the best and most useful book" in the genre (Thomasius 1705: I:227). Not surprisingly, the travel journals, especially those of Heumann and Stolle, frequently referred to Arnold. The travellers also utilized personal contacts and random encounters in Holland to track down interesting conversation partners. For example, Uffenbach, an ardent bibliophile, sought out local book dealers and asked them if "here among the preachers or others is a connoisseur of studies and books" (von Uffenbach 1753-1754: II:271). The future theologian Heumann instead employed his contacts in the Dutch Lutheran community. When he arrived in a new city, Heumann first visited the Lutheran minister, who often delivered important clues to local controversies and radical dissenters. The Lutheran minister in Rotterdam mentioned the quarrel between Pierre Jurieu and Pierre Bayle, and Heumann afterwards visited both. The Lutheran minister in The Haag praised the exemplary life of the radical Pietist Friedrich Breckling, but critiqued his problematic theology and Chiliastic ideas. Heumann's next visit was to Breckling. The Lutheran minister in Leiden warned Heumann against Thomas Crenius, a pseudonym for Brandenburg philologist Thomas Theodor Crusius, who had immigrated to the city and earned a paltry living as a textbook writer. The minister declared that Crenius "doesn't attend divine service, doesn't receive the Eucharist, and is a true rabble-rouser, who reject all our theologians and criticizes everyone" (Cassius 1768: 64). Heumann immediately went to Crenius.

History as Fieldwork

Many of Heumann's conversation partners, even among lay preachers and sectarians, knew the historical literature and had strong opinions about Benthem's and Arnold's descriptions of Dutch religious life. From the dissenters' point of view, the books were not impartial accounts of historical facts, but interventions in an ongoing battle between the Dutch Reformed majority and different religious minorities (see also Israel 1995: 637-676, 1019-1037, Marshall 2006: 138-193). Shortly after Heumann's arrival in Holland, Friedrich Breckling informed him that Benthem "on his Dutch journey had enjoyed himself, and often drunken a glass of wine with the Reformed, and, thus, was lured into judging according to the surface in his book and to pay no attention to the true condition of the Church, which he also did not realize" (Cassius 1768: 64).

In Holland, Heumann experienced how the production of books was part of the process of interpretation. The people he encountered often argued that an author's personal preferences and convictions had influenced the content of a book, but they also pointed to the significance of informers, readers, translators, and printers. One example is the interpretation of Quaker history. Benthem did not write much about the Quakers, but instead referred readers to Gerard Croese's *Historia Quakeriana* of 1695 (Benthem 1698: I:623, also Hull 1933: 115-144). During his long conversation with Jan Claus in Amsterdam, Heumann asked about this book. Claus reported that Croese had consulted him to find material for the book. Claus had agreed to help, and wrote to England to acquire copies of relevant sources, but Croese disappointed him. He had not allowed the Quakers to read the book before publication and had suppressed the truth, because, Claus claimed, he feared the reaction of the Synod. Claus, however, recommended the Latin edition over the German translation, which was even more hostile (Cassius 1768: 92-94).

The recent publication of the second volume of Arnold's *Kirchen- und Ketzerhistorie* had especially stirred emotions among dissenters in Holland. Heumann also asked Jan Claus about Arnold's account of Quaker history. Claus had not yet read it, but reported that the book had provoked controversy in the city and that one prominent member of the congregation, Casper Kohlhans, had written a letter to Arnold with grievances. Breckling, on the contrary, complained about German efforts to censure the book and had encouraged leading Pietists, Philipp Jacob Spener and Johann Winckler, to intervene on Arnold's behalf. Another interesting example is Heumann's conversation with a Labadist teacher, Thomas Varmod, in Franeker:

> I also asked if Arnold's history [of the Labadists] was accurate and recorded in accordance with the truth. He said that in the beginning Arnold had written many falsehoods, because he had trusted the reports of evil people. However, when his history of heretics was translated into Dutch and should be printed in Amsterdam, the printer, who was their friend, had informed them about this. Then, Arnold also himself had contacted them, and made an appendix, based upon exacter information, and corrected everything: so that now the true history of the Labadists could be found in [Arnold's book] (Cassius 1768: 105).

Heumann explored the Dutch underground, but not the Spinozist underground that has recently attracted so much attention (Israel 2001, 2005, van Bunge 2003, Wielema 2004). Spinoza and Spinozism were familiar topics in the Thomasius circle and also interested some of the travellers, especially Gottlieb Stolle (also, Mulsow 2002). But in Heumann's journal the notorious philosopher was only a minor concern. Neither was Heumann particularly interested in the controversies over Cartesianism, although Benthem described these over many pages in the second volume of his book (II: 55-116). Cassius' selection of Heumann's journal, for example, did not even mention Baltasar Bekker's *De Betoverde Weereld*. Heumann instead investigated religious ideas and beliefs that were less known in Germany and not easily available in print. For example, he had several conversations about the mystic Jacob Böhme, whose influence upon Dutch religious life Benthem had ignored. The minutes of some of these conversations even indicate that Heumann had prepared a questionnaire before leaving Jena. One remarkable example is the conversation with Albert Jansen, a shoemaker and Mennonite lay preacher in Emden who had troubles answering Heumann's questions and therefore repeated himself:

> I asked what he thought of Jacob Böhme's writings, to which he answered that he sometimes looked into them. However, he could not judge them since they were very unclear and hard to understand. He would also not condemn them, since he did not know if they were condemnable as he did not understand them, and maybe God had granted Böhme a larger light than himself, and he therefore could not understand them. He thus could not say that he agreed with Böhme, as he did not understand him (Cassius 1768: 129).

Other travellers were also interested in the religious underground, although no-one was as focused as Heumann. Stolle, for example, constantly interrogated theologians and lay preachers about Arnold's and Böhme's writings. The travellers also found inspiration and guidance in one another and sometimes repeated the same experiences in Holland. Collected information was not necessarily lost, but often transmitted from person to person. One example is Johann Gottlieb Deichsel's visit to Uffenbach in Frankfurt am Main in 1718: Uffenbach was not particularly interested in religious matters, but he nonetheless followed Benthem's hints and attended the divine services of the Mennonites and Quakers in Amsterdam. He also acquired manuscripts relating to the religious underground. Deichel visited Uffenbach on his way to Holland and England and studied his manuscript collection. He was particularly fascinated by Uffenbach's "beautiful collection of very many Quaker writings, mostly English, which he had bought collected from a Quaker in Rotterdam" (Deichsel 1786: 182-183). In Holland, Deichsel himself visited the divine services of Armenians, Quakers, Socians, and Mennonites, and interviewed members of these congregations.

Fieldwork as History?
After his return from Holland, Heumann became known as one of the most prominent historians of scholarship in protestant Germany. He founded the *Acta philosophorum*, the world's first journal for the history of philosophy, published in Halle between 1715 and 1723. His short 1718 *Conspectus reipublicae litterariae* served as the standard textbook for the history of scholarship at German universities for most of the eighteenth century. In this work, Heumann advocated a view of history that fitted Thomasius' redefinition of the Republic of Letters as a realm of disembodied arguments. *Historia literaria*, according to Heumann, should deliver a multitude of arguments that the eclectic philosopher could choose between to formulate his own philosophy. The institutional and personal authority of scholars, and their friendships and personal connections, were irrelevant for this purpose and belonged in an entirely different realm. In *Conspectus reipulicae litterariae*, Heumann therefore also clearly demarcated the history of scholarly ideas from the political history of scholars. In his Halle lectures on Heumann's book, Nikolaus Hieronimus Gundling explicated the ideal:

> We must ... show the origin of truth and opinion. This is the soul [*anima*] of the *Historia literaria*. Thus, one examines who had the opinion, by whom and how it has been contradicted, etc. The rest, when you tell something from the author's life, is only accidental. He who thinks he understands the *Historia literaria* when

he can account for the lives of many scholars deceives himself badly ... He has, in this way, only a corpse without soul (Gundling 1734-1736: I:17).

Many of Heumann's travel notes did not fit these ideals of *Historia literaria*. His commitment to eclectic philosophy may explain his interest in diverging opinions, but the undocumented anecdotes and personal impressions in the journal had little relevance to other eclectic philosophers. It was perhaps because of this that Heumann, like all the other travellers from the Thomasius circle, refrained from publishing his journal. It was only after his death in 1764 that Cassius (1768) discovered "a voluminous journal in manuscript" among the papers on his desk.

Heumann did not even recycle information that he had collected during his fieldwork in his later scholarly writings. In his historical works, he discussed scholars whom he had encountered in Holland with the same academic detachment as those he had never met. For example, in Rotterdam Heumann had a lengthy discussion with Pierre Bayle about the relationship between reason and revelation. A decade later he discussed the accusations of atheism against Bayle in *Acta Philosophorum* and, despite its immediate relevance, did not mention this conversation. Instead he emphasized that the "accusation is, so far as I can judge from his writings, false" (Heumann 1715-1723: I:604). In his first draft of a history of scholarship in *Acta philosophorum*, Jacob Böhme only surfaced in passing as one among other "dreamers and enthusiasts" (Heumann 1715-1723: I:582).

Neither did his many encounters in the religious underground influence Heumann's later theological writings (also, Mager 1987, Sparn 1988, Lehmann-Brauns 2004). Late in life, he claimed that as early as 1704, shortly before his journey to Holland, he had adopted the Reformed interpretation of the Eucharist (Heumann 1764: 78). In his theological writings, Heumann, like Benthem, defended the reconciliation of Lutherans and Reformed. He developed a historical-critical method of exegesis and a rationalist theology. The frontispiece of the first issue of his *Acta philosophorum* even depicted theology and philosophy sitting together under the motto *Concordia crescimus*. Heumann's rationalism was not only opposed to Lutheran orthodoxy, but also to the mystical and spiritual theology that filled the pages of his travel journal. The Dutch encounter that left the most lasting impression on his theology was probably that with Jean Le Clerc and his *Ars critica* of 1697. In his journal, however, Heumann only referred to a brief conversation with Le Clerc about his quarrels with the Wittenberg theologians.

There are nevertheless some important parallels between Heumann's travel journal and his later writings, especially his histories of philosophy and scholarship. The purpose of such histories, he explained in *Acta Philosophorum* and later in

Prolegomena historica, was not only philosophical, but also educational. Intellectual tolerance, impartiality, and open-mindedness were not just abstract principles, but had to be personally acquired and demanded practice. Confrontation with the historical plurality of contradictory arguments would develop such intellectual and moral habits (also, Scattola 2007). Prolonged engagement with the history of philosophy and scholarship habituated readers to "endure the brightness of paradoxical truths" (Heumann 1715-1723: I:26). Even the history of unsubstantiated arguments, faulty judgments, and wrong hypotheses served an educational purpose. The recollection of the mistakes of the past ensured that "we do not become stubborn in our opinions" and prepared the readers for "modesty in disputes" and tolerance of disagreements (Heumann 1715-1723: I:32-33).

The educational purpose of travelling was similar to that of historical studies. In *Engeländischer Kirch- und Schulen-Staat*, Benthem described academic travelling as a road to self-knowledge (Bentham 1694: Vorbericht). Thomasius and his followers shared this opinion. Even if their travel journals were unsuitable for publication, the experiences of their journeys and the process of writing of such journals served personal edification. If the most important virtues of the eclectic philosopher were open-mindedness and tolerance of dissent, no destination was more appropriate than Holland. Here, young promising academics encountered not only Dutch Reformed clergy, as Benthem wanted, but also a plurality of opinions that questioned their fundamental assumptions and beliefs. Heumann may have been asked to interview mystics and sectarians not only to gather new information and to verify recent historiography, but also to counter his personal inclinations towards rationalist theology. This may also help explain Heumann's many conversations with dissenters about the inner light, mystical and spiritual understanding, and the relationship between reason and revelation.

In the preface to *Acta philosophorum*, Heumann himself suggested the parallels between the educational benefits of travelling and those of historical studies. If the history of philosophy undermined philosophical authorities, Heumann argued, travels through the confessional landscape of Europe was the best antidote to religious intolerance:

> Thus, we learn from the history of philosophy that philosophers are human and can make mistakes ... In this way we realize that eclectic philosophy is the best form of philosophy, yes I would even say, that no-one deserves the title philosopher, who is not an eclectic. He who is inexperienced in the history of philosophy, he binds his entire reason to the authority of one particular philosopher ... It is the same as with religion. He who never has set his foot

outside of Spain or Italy, he imagines that Lutherans are cursed heretics and already burning in hell, and he barely believes that they are real human beings. However, when he sees the world and through travels learns to know members of all kind of religions, then his eyes open. As he before believed that the Lutherans could make mistakes, he now starts to believe that also in the Roman Church the *Errare humanum est* is more than true (Heumann 1715-1723: I:20-21).

The Prince of Pedants
Another interesting example of the educational purpose of the journey is Heumann's encounter with Thomas Crenius in Leiden. Despite the severe warnings of the local Lutheran minister, quoted above, Heumann did not find an exciting freethinker in Crenius' house. Crenius, Heumann noted, had nothing "thorough" to say and "constantly contradicted himself" (Cassius 1768: 65). The philologist was nevertheless interesting as an example of a faulty, pedantic, and outdated academic and Heumann filled his journal with comments about Crenius' flaws. He was egocentric, arrogant, and convinced "that apart from himself there is no real erudite man in world." He mistakenly thought, "that he is the highest, yes, the only theologian" and "constantly boasted about himself." He preferred style and method to scholarship and arguments and demanded to be addressed as "Your Excellency." Crenius was, Heumann summarized, "the prince of that kind of scholars that the French call pedants" (65).

Heumann's remarks about the Lutheran philologist Crenius were harsher than his comments about any of his other Dutch encounters, including those with Catholics, Spinozists, and mystics. However, they were similar to descriptions of Crenius in other travel journals. In *Engeländischer Kirch- und Schulen-Staat*, Benthem warned against notes such as "that Gilbert Burnet during sermons gesticulates against the customs of the English and often takes his finger to his nose; that Isaac Vossius sleeps during the day and works at night; that Henry More smiles for every third word; that John Spencer doesn't have a beautiful face; that Edward Bernard eats more than usual before theater; and other similar useless things" (Benthem 1694: Vorbericht). Such matters, however, vividly interested the young scholars in the Thomasius' circle. They all made notes about the customs and manners of scholars and Gottlieb Stolle, Martin Mulsow argues, even used Thomasius' doctrine of temperaments as a starting point for his fieldwork among Dutch scholars (2007: 69). While such observations were irrelevant in the disembodied realm of arguments of Thomasius' Republic of Letters, and therefore also in the history of scholarship, they offered valuable lessons as to how one should navigate in the world.

Crenius was portrayed more frequently and in more detail than any other scholar in Holland or England. Some travellers only visited him to witness his impolite and pedantic behavior. Uffenbach, for example, opened his description of Crenius with the comment that "people had made a fool's portrait of [Crenius] in conversations ... However, we found him even odder than we had imagined" (von Uffenbach 1753-1754: III:465). Christoph August Lämmermann in 1710 defended Crenius, but remarked that: "Strangers and others who visit him speak much about his odd temperament" (Lämmermann 1792: 102). Almost all the travel journals noticed and condemned Crenius' pedantic style, his dirty clothing, and his arrogance. Even sympathetic visitors, such as Stolle and Lämmermann, could not ignore his appalling appearance and bad manners. Lämmermann noted that "he is a very badly dressed man" (102). Stolle declared that one could learn much from Crenius, "only one must have patience with him" (Stolle forthcoming: 525).

The educational purpose of the visit to Crenius was clearly different from the purpose of the conversations with underground freethinkers and sectarians. These dissenters seldom dressed *à la mode* and their opinions were also not in perfect accordance with the writings of Christian Thomasius. However, for the habituation of tolerance and open-mindedness, these breaches of the rules of decorum and enlightened reasoning were of minor importance. The dissenters were interesting because of their unusual ideas, not because of their appearances or knowledge of contemporary philosophy. Crenius, on the contrary, presented himself as a scholar and demanded that visitors should honor him as such. He insisted upon the significance of appearances in academic dialogue and thus became a target of the fierce critique of pedantry of the early German Enlightenment (also, Kosenina 2003: 55-84). He served as the model of an antiquated scholar and exemplified the academic vices of the past that a young German scholar should learn to shun if he wanted to become an enlightened philosopher.

The journals especially ridiculed Crenius' demand to be addressed "Your Excellency." This small vanity violated some of the central principles of scholarly communication of the German Enlightenment. During conversations in his private study or in his collegium, a scholar entered the disembodied realm of arguments and should relinquish all privileges of rank and estate. The force of arguments should replace the authority of books, clothes, and titles. Thomasius therefore also immediately prohibited new students at the University of Halle from addressing him as "Your Excellency." Only outside the study or the private lecture hall should the normal rules of decorum be enforced and obeyed (Eskildsen 2008). Crenius' fondness for titles revealed his inability to change personae and this weakness interested the

Thomasius' followers. Uffenbach, for example, purposefully challenged Crenius' vanity:

> He received us as a Stoic, and watched if we honored him adequately. In the beginning we diligently avoided to do so, to see if he would let us know, as he had others, that we must address him as Your Excellency. That happened immediately. He soon started to talk about his many titles, and how one had addressed him with this, another with that, while he each time said, 'he spoke to me: Your Excellency, what do you think of this book,' etc. After we had heard enough of this foolishness, we started diligently to call him Your Excellency, and he became much friendlier. Only there was no end to the bragging and boasting (von Uffenbach 1753-1754: III:465-466).

Uffenbach was not the only visitor who purposefully tested Crenius' vanity. Even Stolle noticed that if one failed to address Crenius as "Your Excellency," "then he soon comes up with a story about how an important prince or ambassador called him Excellency" (Stolle forthcoming: 528). As late as 1789, David Christoph Seybold's *Historisches Handbuch auf alle Tage im Jahre* published a short anecdote about Crenius' fondness of titles. The story recounted the experience of one German traveller, named Hauber, who visited Holland at the beginning of the century.[4] Like Uffenbach and Stolle, Hauber sensed that "Crenius liked no other title than Excellency ... However, [Hauber] on purpose did not address him so, to see how he would react" (Lämmermann 1792: 107-108).

From Fieldwork to Bildungsreise

Crenius' encounters with German travellers were not only encounters between different types of scholars, but also between two different concepts of the Republic of Letters. Crenius considered himself a representative and defender of a dying tradition that connected humanistic scholarship and Christian piety. "In Holland," he explained to Stolle, "one only appreciates paradoxical books," in particular Spinozistic books. The scholars in Germany, he lamented, "are truly donkeys, no professor is any good" (Stolle forthcoming: 472, 474-475). To the friendly Lämmermann he remarked "that barbarianism is again creeping into polite letters, thus only a few good men of letters can be encountered in Holland and Germany, very few in Italy, and none whatsoever

---

4   Possibly the later minister in the Sankt Petri Church in Copenhagen, Eberhard David Hauber, who went on his academic journey in 1720 (see Oeheme 1976: 12).

Figure 1. The fruits of academic travelling separated from the personal experiences of the journey and collected in the study (frontispiece in Uffenbach 1753-1754).

in France" (Lämmermann 1792: 103). Disrespect of past authorities and ignorance of ancient languages undermined the Latin *respublica literaria* that Erasmus of Rotterdam had envisioned in his *Antibarbarorum liber* two centuries earlier. To the travellers, however, Crenius' insistence upon the etiquette and values of the old *respublica literaria* violated enlightened standards for free academic dialogue, and thus the principles of Thomasius' new eclectic Republic of Letters.

During the first decades of the eighteenth century, German scholars increasingly questioned the ideals of the Latin *respublica literaria* and this critique also extended to the academic educational journey (see Eskildsen 2004). Another example of a young northern European scholar who sympathized with Thomasius and travelled to Holland during the early eighteenth century was the later Copenhagen professor Ludvig Holberg. Holberg visited Holland at the same time as Heumann, 1704-5, and returned again in 1714. Unfortunately his travel journals have not survived, but in his later writings Holberg often reflected on the significance of his journeys for his personal as well as scholarly development. By the 1740s, Holberg had convinced himself that travelling was of little or no academic significance:

On such journeys, one has the occasion to speak to one or another famous man. But the company of such men is not always easily accessible. For example, I visited Monsieur Le Clerc in Amsterdam a couple of times and was each time received lovingly. However, when I returned a third time, he knitted his brows. Thus, the so-called benefit only consists in the name, that one after returning can report that one has spoken with such a man. To enjoy this honor, one has squandered a large part of one's heritage and uselessly wasted a large part on mail-carriages. In libraries, one finds numerous rare writings that one cannot find in one's native country. That cannot be denied. However, for a tenth of the cost of the journey, one could become the actual owner of those writings that one otherwise only has the liberty to leaf through (Holberg 1865-1875: IV:91-92).

Holberg travelled constantly during his youth. He visited not only Holland, but also France and Italy, and lived for several years in England. Thus, his rejection of the value of travel should probably be read with some reservations. During the eighteenth century, academic travels remained not only a resource for the education of philosophers, but also a source of underground knowledge and an entry into the academic community. As late as 1752 the Jena professor Johann Andreas Fabricius recommended such journeys because "one here visits scholars, talk with them, sees their libraries, wins their friendship, and benefits from their oral teachings in [private] company, where these often are delivered more freely" (Fabricius 1752-1754: I:45). Professors at German universities continued lecturing on the proper rules for travelling to the end of the eighteenth century (Neutsch 1991).

Holberg's rejection of academic travelling, however, points to an important development in the history of scholarship. The ideals of eclectic and disinterested scholarship challenged the role of personal connections and mutual trust in scholarship. During the eighteenth century, academic specialization and the increasing importance of written communication and academic journals further undermined these early modern ideals. The emergence of new disciplinary communities fragmented the common European community of the Republic of Letters.

As scholars objectified knowledge in writing and within disciplines, the subjective experiences of the academic journey won new importance. Travel journals of European scholars increasingly focused on personal developments, impressions, and emotions, and consequently they found a new readership outside narrow academic circles (also, Kalb 1981, Brenner 1989, Sherman 1996: 149-268). Holberg's descriptions of his travels are clear examples of this development. In 1728, Holberg published the first volume of his memoirs and included a short account of his visits to Holland. This account reported nothing about his conversations with Dutch scholars. He described the naïveté and arrogance of his youth, dwelled on his illnesses and his lack of money, and charted his way to self-knowledge, but he ignored Dutch libraries, universities, and bookshops. The only trace of the old center of the European Republic of Letters was a short description of the statue of Erasmus in Rotterdam. Holberg noticed the schoolboys throwing stones at the statue and wondered if even the citizens of Rotterdam had forgotten the merits of the humanist (Holberg 1969-1971: XII:63).

References

Becker, P. J. 1980. "Bibliotheksreisen in Deutschland im 18. Jahrhundert." *Archiv für Geschichte des Buchwesens*, 21: 1361-1534.

Bekker, R. 2007. "Watches, Diary Writing, and the Search for Self-Knowledge in the Seventeenth-Century." In: P. H. Smith & B. Schmidt (eds.) *Making Knowledge in Early Modern Europe: Practices, Objects, Texts, 1400-1800*. Chicago: University of Chicago Press, 127-142.

Benthem, H. L. 1694. *Engeländischer Kirch- und Schulen-Staat*. Lüneburg.

Benthem, H. L. 1698. *Holländischer Kirch- und Schulen-Staat*, 2 vol. Merseburg.

[Benthem, H. L.]. 1700. *Pacifici Verini ohnmaßgebliches Bedencken, ob und wie die heutiges Tages gesuchte Vereinigung derer, welche die ohnveränderte Augspurgische Confession angenommen, mit den übrigen diene zum Wohlstande der Kirche Christi.* n. p.

Beyrer, K. 1985. *Die Postkutschenreise*. Tübingen: Tübinger Vereinigung für Volkskunde.

Bientjes, J. 1967: *Holland und die Holländer im Urteil deutscher Reisender 1400-1800*, Gronningen: J. B. Wolters.

de Blainville, M. 1743-1745. *Travels through Holland, Germany, Switzerland, and other parts of Europe; but especially Italy*. 3 vol. London.

Boedeker, H. E. 2002. "'Sehen, hören, sammeln und schreiben'. Gelehrte Reisen im Kommunikationssystem der Gelehrtenrepublik." *Paedagogica Historica*, 38(2-3): 505-532.

Brenner, P. J. 1989. *Der Reisebericht: Die Entwicklung einer Gattung in der deutschen Literatur*. Frankfurt am Main: Suhrkamp.

Brockliss, L. W. B. 2002. *Calvet's Web: Enlightenment and the Republic of Letters in Eighteenth-Century France*. Oxford: Oxford University Press.

van Bunge, W. 2003. *The Early Enlightenment in the Dutch Republic, 1650-1750*. Leiden: Brill.

Cassius, G. A. 1768. *Ausführliche Lebensbeschreibung des um die gelehrte Welt Hochverdienten D. Christoph August Heumanns*. Kassel.

Chartier, R. 2003. "Foucault's Chiasmas: Authorship between Science and Literature in the Seventeenth and Eighteenth Centuries." In: M. Biagioli & P. Galison (eds.) *Scientific Authorship: Credit and Intellectual Property in Science*. New York: Routledge, 13-32.

Curtius, C. (ed.) 1891. "Heinrich Christian Postel's und Jacob von Melle's Reise durch das nordwestliche Deutschland nach den Niederlanden und nach England. Aus einer Handschrift der Lübeckischen Stadtbibliothek." In: J. Schubring (ed.) *Einladung zu den auf den 19. und 20. März 1891 angeordneten öffentlichen*

Prüfungen und Redeübungen der Schüler des Katherineums zu Lübeck. Lübeck: Gebrüder Borchers, 1-48.

Deichsel, J. G. 1786-1787. "Reise durch Deutschland nach Holland und England in den Jahren 1717-1719." *Johann Bernoulli's Archiv zur neuern Geschichte, Geographie, Natur- und Menschenkenntniß*, 3(1786): 137-88, and 7(1787): 151-212.

Dibon, P. & F. Waquet 1984. *Johannes Fredericus Gronovius pèlerin de la République des Lettres: Recherches sur le voyage savant au XVIIe siècle.* Geneva: Droz.

[Erndl, C. H.] 1711. *De itinere suo Angelicano et Batavo annis 1706 et 1707 facto, relatio ad amicum.* Amsterdam.

Eskildsen, K. R. 2004. "How Germany Left the Republic of Letters." *Journal of the History of Ideas*, 65(3): 421-432.

Eskildsen, K. R. 2008. "Christian Thomasius, Invisible Philosophers, and Education for Enlightenment." *Intellectual History Review*, 18 (3): 319-336.

Fabricius, J. A. 1752-1754. *Abriss einer allgemeinen Historie der Gelehrsamkeit.* 3 vol. Leipzig.

Fostergill, R. A. 1974. *Private Chronicles: A Study of English Diaries.* London: Oxford University Press.

Franke, K. 1965. "Zacharias Conrad von Uffenbach als Handschriftensammler: Ein Beitrag zur Kulturgeschichte des 18. Jahrhunderts." *Börsenblatt für den Deutschen Buchhandel – Frankfurter Ausgabe*, 51(July 29): 1235-1338.

Fumaroli, M. 1994. "La conversation savant." In: H. Bots & F. Waquet (eds.) *Commercium Litterarium: La communication dans la République des Lettres.* Amsterdam: APA-Holland University Press, 67-80.

Goldgar, A. 1995. *Impolite Learning: Conduct and Community in the Republic of Letters.* New Haven, CT: Yale University Press.

Goodman, D. 1994. *The Republic of Letters: A Cultural History of the French Enlightenment.* Ithaca, NY: Cornell University Press.

Gundling, N. H. 1734-1736. *Vollständige Historie der Gelahrheit.* 5 vol. Frankfurt am Main.

Hartmann, A. 1991. "Reisen und Aufschreiben." In: H. Bausinger, K. Beyrer & G. Korff (eds.) *Reisekultur: Von der Pilgerfahrt zum modernen Tourismus.* Munich: C. H. Beck, 152-159.

Heumann, C. A. 1715-1723. *Acta Philosophorum, das ist: Gründl. Nachrichten Aus der Historia Philosophica, Nebst beygefügten Urtheilen von denen dahin gehörigen alten und neuen Büchern.* 18 vol. Halle.

Heumann, C. A. 1718. *Conspectus reipublicae litterariae sive via ad historiam litterariam iuventuti studiosae.* Hanover.

Heumann, C. A. 1764. *Erweiß daß die Lehre der Reformirten Kirche von dem Heil. Abendmahle die rechte und wahre sey*. Eisleben.

Holberg, L. 1865-1875. *Epistler: Udgivne med oplysende Anmærkninger*. 5 vol. Copenhagen.

Holberg, L. 1969-1971. *Værker i tolv Bind: Digteren – Historikeren – Juristen – Vismanden*. Copenhagen: Rosenkilde & Bagger.

Hull, W. I. 1933. *William Sewel of Amsterdam 1654-1720: The first Quaker Historian of Quakerism*. Philadelphia, PA: Paterson & White Company.

Israel, J. 1995. *The Dutch Republic: Its Rise, Greatness, and Fall, 1477-1806*. Oxford: Clarendon Press.

Israel, J. 2001. *Radical Enlightenment: Philosophy and the Making of Modernity 1650-1750*. Oxford: Oxford University Press.

Israel, J. 2006. *Enlightenment Contested: Philosophy, Modernity, and the Emancipation of Man 1670-1752*. Oxford: Oxford University Press.

Jacobsen Jensen, J. N. 1919-1939. *Reizigers te Amsterdam: Beschrijvende lijst van reizen in Nederland door vreemdelingen vóór 1850*. 2 vol. Amsterdam: Ipenbuur & Van Seldam.

Kalb, G. 1981. *Bildungsreise und literarischer Reisebericht: Studien zur englischen Reiseliteratur 1700-1850*. Nürnberg: Verlag Hans Carl.

Kannegieter, J. Z. 1971. *Geschiedenis van de vroegere quakergemeenschap te Amsterdam 1656 tot begin negentiende eeuw*. Amsterdam: Scheltema & Holkema NV.

Kniggen, H. 1727. "Lebens-Lauf des Wohlseeligen Herrn General-Superintendenten." In: H. L. Benthem (ed.) *Vorstellung und Betrachtung der Schrifften der alten Kirchen-Lehrer von der Wahrheit und Göttlichkeit Christlicher Religion*. Hamburg, n. p.

Kosenina, A. 2003. *Der gelehrte Narr: Gelehrtensatire seit der Aufklärung*. Göttingen: Wallstein.

Kulczykowski, M. (ed.) 1989. *Peregrinations academiques: IVème session scientifique internationale, Cracovie, 19-21 Mai 1983*. Krakow: Nakladem Uniwersytetu Jagiellonskiego.

Laeven, H. 2005. "Einleitung." In: J. B. Mencke. *Das Holländische Journal 1698-1699: Ms. Germ. oct. 82 der Staatsbibliothek Berlin*. Laeven, H. ed.. Hildesheim: Olms, 9-29.

Lämmermann, C. A. 1792. "Litterarische Anekdoten, die Elzevirische Buchdruckerey in Leiden, und die beiden dasigen Gelehrten, Herm. Boerhave und Thom. Crenius." *Historisch-litterarisch-bibliographisches Magazin*, 3(6): 99-109.

Lehmann-Brauns, S. 2004. *Weisheit der Weltgeschichte: Philosophiegeschichte zwischen Barock und Aufklärung*. Tübingen: Max Niemeyer Verlag.

Licoppe, C. 1995. *La Formation de la pratique scientifique: Le discours de l'expérience en France et en Angleterre 1630-1820*. Paris: Editions La Découverte.

Mager, I. 1987. "Die theologische Lehrfreiheit in Göttingen und ihre Grenzen: Der Abendmahlskonflikt um Christian August Heumann." In: B. Moeller (ed.) *Theologie in Göttingen: Eine Vorlesungsreihe*. Göttingen: Vandenhoek & Ruprecht, 41-57.

Marshall, J. 2006. *John Locke, Toleration and Early Enlightenment Culture: Religious Intolerance and Arguments for Religious Toleration in Early Modern and 'Early Enlightenment' Europe*. Cambridge: Cambridge University Press.

Mencke, J. B. 2005. *Das Holländische Journal 1698-1699: Ms. Germ. oct. 82 der Staatsbibliothek Berlin*. Hildesheim: Olms.

Miller, P. N. 2000. *Peiresc's Europe: Learning and Virtue in the Seventeenth Century*. New Haven, CT: Yale University Press.

Mulsow, M. 2002. *Moderne aus dem Untergrund: Radikale Frühaufklärung in Deutschland 1680-1720*. Hamburg: Meiner.

Mulsow, M. 2007. *Die unanständige Gelehrtenrepublik: Wissen, Libertinage und Kommunikation in der Frühen Neuzeit*. Stuttgart: J. B. Metzler: Stuttgart.

Murris, R. 1925. *La Hollande et les hollandaise au XVIIe et au XVIIIe siècles vus par les Français*. Paris: Campion.

Neutsch, C. 1991. "Die Kunst, seine Reisen wohl einzurichten: Gelehrte und Enzyklopädisten." In: H. Bausinger, K. Beyrer & G. Korff (eds.) *Reisekultur: Von der Pilgerfahrt zum modernen Tourismus*. Munich: C. H. Beck, 146-152.

Oehme, R. 1976. *Eberhard David Hauber 1695-1765: Ein schwäbisches Gelehrtenleben*. Stuttgart: W. Kohlhammer Verlag.

de Ridder-Symoens, H. 1996. "Mobilität." In: W. Rüegg (ed.) *Geschichte der Universität in Europa*. 4 vol. Munich: C. H. Beck, II, 335-59.

Roche, D. 2003. *Humeurs vagabondes: De la circulation des hommes et de l'utilité des voyages*. Paris: Fayard.

Sauerland, K. 1983. "Der Übergang der gelehrten zur aufklärerischen Reise im Deutschland des 18. Jahrhunderts." In: J. P. Stelka & J. Jungmayr (eds.) *Virtus et Fortuna: Zur Deutschen Literatur zwischen 1400 und 1720*. Frankfurt am Main: Peter Lang, 557-570.

Scattola, M. 2007. ">Historia literaria< als >Historia pragmatica<: Die pragmatische Bedeutung der Geschichtsschreibung im intellektuellen Unternehmen der Gelehrtengeschichte." In: F. Grunert & F. Vollhardt (eds.) *Historia literaria: Neuordnung des Wissens im 17. und 18. Jahrhundert*. Berlin: Akademie Verlag, 37-63.

Schneppen, H. 1960. *Niederländische Universitäten und deutsches Geistesleben: Von der Gründung der Universität Leiden bis ins späte 18. Jahrhundert.* Münster: Aschendorffsche Verlagsbuchhandlung.

Shapin, S. 1995. *A Social History of Truth: Civility and Science in Seventeenth-Century England.* Chicago: University of Chicago Press.

Shelford, A. G. 2007. *Transforming the Republic of Letters: Pierre-Daniel Huet and European Intellectual Life.* Rochester, NY: University of Rochester Press.

Sherman, S. 1996. *Telling Time: Clocks, Diaries, and English Diurnal Form 1660-1785.* Chicago: University of Chicago Press.

Siebers, W. 1991. "Ungleiche Lehrfahrten – Kavaliere und Gelehrte." In: H. Bausinger, K. Beyrer, and G. Korff (eds.) *Reisekultur: Von der Pilgerfahrt zum modernen Tourismus.* Munich: C. H. Beck, 47-57.

Sparn, W. 1988. "Philosophische Historie und dogmatische Heterodoxie. Der Fall des Exegeten Christoph August Heumann." In: H. G. Reventlow, W. Sparn & J. Woodbridge (eds.) *Historische Kritik und biblischer Kanon der deutschen Aufklärung.* Wiesbaden: Harrassowitz, 171-192.

Stagl, J. 1995. *A History of Curiosity: The Theory of Travel 1500-1800.* Chur: Harwood Academic Publishers.

Stolle, G. forthcoming. *Reise durch Deutschland und die Niederlande.* Cologne: Editions Marteau.

van Strien, C. D. 1993. *British Travellers in Holland during the Stuart Period: Edward Browne and John Locke in the United Provinces.* Leiden: E. J. Brill.

van Strien, K. 1998. *Touring the Low Countries. Accounts of British Travellers, 1660-1720.* Amsterdam: Amsterdam University Press.

Thomasius, C. 1705. *Außerlesene und in Deutsch noch nie gedruckte Schrifften.* 2 vols. Halle.

von Tschirnhauß auf Hackenau, W. B. 1727. *Getreuer Hofmeister auf Academien und Reisen.* Hanover.

von Uffenbach, Z. C. 1753-1754. *Merkwürdige Reisen durch Niedersachsen Holland und Engelland.* 3 vol. Frankfurt am Main.

Wielema, M. 2004. *The March of the Libertines: Spinozists and the Dutch Reformed Church 1660-1750.* Hilversum: Uitgeverij Verloren.

Daniel E. Clinkman

The Civil-Military Enlightenment in Britain
Links between the Royal Society of London and the British Military, 1761-1790

The long eighteenth century (1688-1815) was a time when early modern kingdoms and empires transitioned into modern states. This transition was characterized in part by increased regimentation of society – governments became more efficient compilers of information, collectors of taxes, and conscripts of soldiers. The collection and compilation of information was important because it made efficiency in other areas of government possible. The partnership between governments and information centers has become an important part of the modern state, with governments sponsoring research at universities and apparatchiks of political parties co-operating with affiliated think tanks.

An example of early co-operation between state agencies and nationally-oriented learned societies can be found in the relationship between the Royal Society of London and the British military in the decades between the Seven Years War and the French Revolutionary Wars. There had always been a loose affinity between the Royal Society and the military, as many military officers were Fellows, and the military and Royal Society had occasionally co-operated on matters of joint interest. This interaction increased in the 1760s through the 1780s with a series of joint survey missions. By the outbreak of war with France in 1793, the military had come to rely on the Royal Society for expert scientific advice, and the Royal Society had come to rely on the military for patronage for some of its highest profile, and most expensive, ventures.

Between 1761 and 1790, the time spanning from the first transit of Venus to the completion of William Roy's reports on the Greenwich survey, the profile of the military in the Royal Society rose substantially, particularly in the form of military contributions to and military-related articles in the *Philosophical Transactions*. As the Royal Society and military began co-operating on astronomical survey missions, voyages to far-flung destinations such as Batavia and Tahiti yielded unexpected discoveries in the fields of botany, chemistry, geography, and climatology. From 1761 to 1779, no less than eight Royal Navy vessels co-operated with the Royal Society to record celestial phenomena, chart the Pacific Ocean, and collect specimens of exotic life, including the human geography of the Pacific watershed (Banks 1962, Wales 1788). From 1761 to 1790, forty-seven articles on military subjects, of military authorship, or compiled with military assistance, by thirty-two different authors, were published in the *Philosophical Transactions*; of these, a naval captain, James Cook, two Physicians of the Army, Donald Monro and Charles Blagden, and one Army major-general, William Roy, contributed multiple articles; Cook and Roy led joint survey missions for the Royal Society and the military.[1]

The Royal Society's main military partners were the Royal Navy and the Board of Ordnance. The Royal Navy was interested in various areas of Royal Society expertise. Astronomical observations could assist the navy with navigation, geographic explorations could help chart the globe and provide the navy with the locations of new ports and natural resources, and research into physical sciences could yield discoveries in magnetology for compasses or biochemistry for improving shipboard health. The Board of Ordnance, and by extension the Army, had equally compelling interests. At various times the Board of Ordnance relied on Royal Society advice for safe storage of munitions in their depots, and for expert advice on surveying the British coasts and countryside.

Co-operation during the intervening three decades from the denouement of the Seven Years War to the commencement of the French Revolutionary Wars proceeded in four overlapping stages. The first stage consisted of initial co-operation between the Royal Society and the Royal Navy in preparation for the transits of Venus in 1761 and 1769. The second stage consisted of joint naval expeditions, notably those under Captain James Cook and Sir Joseph Banks. The third consisted of a period when the Royal Society began to offer consultations to the Board of Ordnance. The fourth was a period when co-operation between the Royal Society and Britain's land forces culminated in a joint survey mission of southern England involving the Royal Society, the Board of Ordnance, and the British Army.

Historiography

The Royal Society during the eighteenth century has been the subject of a single monograph, written by Charles Richard Weld and published in 1848 (Weld 1848). In her full-length study of the Society during the nineteenth century, Marie Boas Hall reflects on the lack of historiography relating to the eighteenth century and remarks that it "has been less studied, mainly because of an English conviction that English science did not thrive in this period" (Hall 1984: x). Thomas Sprat's *History of the*

---

1   See Appendix to this essay.

Royal Society covers the period of the Society's founding and early years but is not scholarly in the modern sense of the term; for that one can turn to Margery Purver's *The Royal Society: Concept and Creation* (Purver 1967). There have also been numerous studies of the nineteenth century Society. Hall's book, *All Scientists Now: The Royal Society in the Nineteenth Century* takes a non-sociological look at the Society from the death of Sir Joseph Banks to the end of the century (Hall 1984).

Hall's thesis is that it was only in the nineteenth century that the Royal Society professionalized itself, becoming "more influential (...) in regard to the Government than had been the case in earlier times" (Hall 1984: x). It is the thesis of this essay that Hall's characterization is overly broad and fails to appreciate the professionalism of eighteenth century Fellows and contributors to the *Philosophical Transactions*, as well as the important role the Society played as a government auxiliary. However, simply acknowledging the episodic assistance of the Royal Society to the government throughout the eighteenth century, and giving examples of official co-operation, does not shake Hall's appraisal of the Royal Society as being a "leisurely" practitioner of science (Hall 1984: 1).

It should be insisted upon that, while the majority of the Society's dues-paying members may have been men of leisure, many of its most active members were men of practical experience and professional affiliation – particularly the men who wore the uniforms of the British military. The physicians, engineers, and naval officers who contributed to the *Philosophical Transactions* and were inducted as Fellows were professional men, if not of science then of military service. But to dismiss the eighteenth century Fellows as unscientific and unprofessional is to take nineteenth century definitions of science and of professionalism and anachronistically transpose them into the eighteenth century Enlightenment. And by undermining a sense of the professionalism of these men, it is in turn easy to miss the importance they held collectively as an organization tasked with officially assisting the government.

John Gascoigne addresses the issue of science and British military and foreign policy both in his monograph *Science in the Service of Empire* and his article entitled "The Royal Society and the Emergence of Science as an Instrument of State Power" (Gascoigne 1998, 1999). Challenging the characterization of the Royal Society as leisurely, Gascoigne posits as the thesis of his article that the "apparent character of a gentlemen's club is transmuted when one considers that in the social and political context of the eighteenth century such an institutional milieu helped link the Royal Society to the workings of government." In other words, the Royal Society was just as professional as its political counterparts; the workings of the Society were suited to the time and place in which it existed. Gascoigne offers a special mention to the Society's relationship with the Admiralty and the navy dating back to the 1670s, through the Royal Observatory and the Royal Mathematical School (Gascoigne 1999: 172). He also relates the government's expanded relationship with the Royal Society as a result of the reforms of George III's reign aimed at increasing the country's military capacity against France (Gascoigne 1999: 173-174).

While studying the role of shipboard missions in natural philosophy, it is important to keep in mind Richard Sorrenson's work on "The Ship as a Scientific Instrument in the Eighteenth Century" (Sorrenson 1996a). Sorrenson holds that the ship offered a unique advantage over other forms of transportation and observation, such as pack animals, because it could give "a superior, self-contained, and protected view of the landscapes and civilizations" travelling natural philosophers encountered (Sorrenson 1996a: 222). Ships became the Royal Society's method of choice for exploration in the late eighteenth century, and explorers such as Sir Joseph Banks fully availed themselves of the security and mobility these platforms offered.

Military affairs themselves were becoming more scientific. Since the 1670s the Royal Society had assisted with training naval officers in mathematics. In addition, scientific practices were being incorporated into the field, such as the improvement of shipboard health in light of new advances in medical science (Lawrence 1995). The practice of science became part of the pursuit of empire, allowing governments to organize their imperial possessions and field military forces more effectively (Miller 1995). It was within this context that the Royal Society conducted a search for a vessel to conduct a team of astronomers in 1760.

The Royal Navy: The Transits of Venus and Expeditions to the South Seas
In 1760, while the Seven Years War was still being fought, the Royal Society began looking for a partner to facilitate observation of the impending transit of the planet Venus the following year. In order to collect the best possible information, the Royal Society intended to establish observation points around the globe, because good visibility would not be available from Europe (MM/10/106).[2] By compiling the information collected from these various points, the Royal Society could then determine the sizes and distances of various bodies in the solar system. Entreaties were made to private supporters. On July 22, 1760, the directors of the East India Company offered assistance to Royal Society observation teams bound for Bencoolen and St. Helena. The meeting between representatives of the Company and the Royal Society was recorded in a memorandum, in which the Royal Society noted that "The Company will give us all the assistance, and do our Astronomers all the Civilities

---

2   All entries beginning with MM refer to the Royal Society of London Manuscript Division.

in their Power; and seem extremely well pleased at our Resolution to pursue this inquiry" (MM/10/105).

However, the arrangements with the East India Company were not completed, and this development left the Royal Society scrambling for alternative arrangements and a new sponsor. At this point the Royal Navy began to take an active interest in the preparations. On July 30, 1760, only ten months before the transit was to take place, the Admiralty informed a Fellow, one Dr. Birch, that the Admiralty "have ordered a Ship to be fitted for the said Purposes" of carrying the Royal Society observation team to Batavia and the observation point (MM/10/110). On November 15, 1760 the Admiralty Office ordered the harbormaster at Plymouth to outfit the H.M.S. *Seahorse* with the Royal Society's observational instruments, and the vessel departed for Batavia in January (MM/10/119).

No sooner had the *Seahorse* left Plymouth than it was attacked by a French man-of-war patrolling the English Channel. The Seven Years War would not end for another two years, and while the British and French navies customarily treated one another's scientific vessels as neutrals, the French ship was evidently not aware of *Seahorse*'s scientific status and the ship, along with its scientists, was forced to return to port. The scientists, Charles Mason and Jeremiah Dixon, now "absolutely refused to proceed the voyage," according to an Admiralty representative on the scene, despite the Royal Navy's provision of a ship-of-the-line to escort them safely past the combat zone (MM/10/133). Threatened with legal action, Mason and Dixon denied that they had refused to board the *Seahorse* and sailed on the evening of February 3, 1761 (MM/10/134).

By now well behind schedule, the *Seahorse* could not make it to Batavia in time and instead stopped at another Dutch colony, the Cape of Good Hope. Here Mason and Dixon had hoped to use local laborers to construct their observatory, but the locals were inadequate to the task. The *Seahorse* now became not only a conveyance but an active participant in the mission. The ship's crew went to work in the locals' stead. Mason recorded that he "apply'd to Captain Grant (whose favor and good nature I have always experience'd) for his Carpenter, who very freely ordered them to assist, and I hope it will soon be compleated [sic]" (MM/10/135).

The *Seahorse* expedition was not able to reach its intended destination, but the partial success that was achieved at the Cape of Good Hope was due in large part to the willingness of the civilian and military components of the mission to work together. The Royal Society relied on the navy for transport, escort, and manpower; in exchange, the navy received astronomical data that would help them navigate around the world. This small episode of fruitful co-operation set a precedent for more ambitious expeditions and surveys to follow.

Eight years later, Venus made another transit between Earth and the Sun, and co-operation between the Royal Society and the Royal Navy resumed with the collaboration between Sir Joseph Banks of the Royal Society and James Cook of the Royal Navy. Taking the bark *Endeavour*, Cook and Banks circumnavigated the world via the South Seas. This was a complex operation involving a Royal Navy vessel along with eleven Royal Society personnel. While the scientific results of the South Seas voyage are well known, it takes on an even greater significance when considered as the first large-scale field expedition involving both the military and a learned society. In contrast to 1761, when Mason and Dixon had been passengers on board the *Seahorse*, the *Endeavour* was an integrated ship. While the *Seahorse* had been a means of transportation, *Endeavour* was a vessel of exploration, and Cook was a scientific captain. Upon *Endeavour*'s return to England, Cook wrote several papers about the voyage for the *Philosophical Transactions* (*Phil. Trans.* LXI/XLIII-XLV). In his turn, Banks had made many ecological discoveries in the Pacific Islands, and the wealth of knowledge he returned with set him on the path to the Presidency of the Royal Society, a position he would use to encourage future joint expeditions.

This voyage was a career-maker for both Cook and Banks, and a second expedition was prepared for 1772. However, while collaboration between the Navy and the civilians had worked on the first voyage, Cook baulked at Banks' desire to modify the ship H.M.S. *Resolution* to hold additional scientists and their paraphernalia. *Resolution* departed without Banks' contingent, but Cook and his ship's surgeon, William Anderson, still submitted three reports to the Royal Society upon his return in 1776 (*Phil. Trans.* LXVI/XXII, XXVI, XXXVI).

The Board of Ordnance: The Purfleet Committee and Roy's Geographic Surveys

The failure to co-operate on the mission of the *Resolution* was not a fatal setback to the burgeoning civil-military efforts. In 1772, the Board of Ordnance determined that its ammunition magazines at Purfleet were in danger from lightning strikes, and resolved to build safeguards but were first "desirous of knowing the sentiments of the Royal Society upon the propriety thereof before any steps are taken" (MM/3/71). A few weeks later Sir Charles Frederick, the Surveyor-General of the Ordnance as well as an F.R.S., wrote to the Royal Society to add that "we shall request this favor that the [recommendations] may be executed under your direction" (MM/3/72). Not only was the Royal Society to offer advice, it was to oversee the construction of the new lightning rods.

Accordingly, a committee was formed consisting of Benjamin Franklin, Benjamin Wilson, and three other Fellows with expertise in electricity. The committee wrote a report for the Board of Ordnance that largely relied on Franklin's prior expertise in developing the lightning rod. Wilson dissented from the main report, arguing that the committee's suggested use of pointed rods would actually increase the risk to the magazines. Wilson, who had been awarded a Copley Medal and was an authority on electricity, challenged the committee's analysis and proposed an alternative set up of the lightning conductors. In three meetings in August 1772, Wilson repeatedly objected to the committee findings and Franklin, taken aback, suggested referring the matter to the council of the Royal Society when the report was read aloud (MM/11/41). The matter was not resolved – Wilson maintained his opposition, and the findings of the committee were printed in the *Philosophical Transactions* in three parts. The first was the main report signed by Franklin and the other Fellows; the second was Wilson's minority report, signed by himself alone and addressed not to the Royal Society but personally to Charles Frederick; and thirdly a rebuttal by the committee refuting Wilson's objections in a letter addressed to the Secretary of the Royal Society (*Phil. Trans.* LXIII/VIII-X).

This incident, while to an extent a comedy of errors and personalities with Franklin and Wilson engaging in a clash of egos as well as ideas, remains significant for its structural aspects. The Board of Ordnance recognized that, when faced with the particular problem of the lightning risk at Purfleet, it could get a better solution from outside consultants than from its own internal engineers. Just as the Navy considered the Royal Society's astronomical teams to be of superior expertise to their own personnel, the Board of Ordnance, and by extension the Army, knew that civilian specialists assisting the military could augment their existing capabilities. Interestingly, the lightning rods at Purfleet, while built according to Franklin's specifications and personally inspected by the committee in 1773, failed in a lightning storm in 1777; yet the Board of Ordnance once again consulted the Royal Society, which dispatched a new committee and prepared a fresh report in June 1777 (MM/4/83-84, MM/3/21). Despite the failure of the initial design, the Board of Ordnance clearly benefited from having access to such a concentration of experts in one society, able and willing to assist the military enterprise, and recognized the Royal Society's value by returning to them for the second appraisal in 1777 (see also Schiffer 2003: 194-200).[3]

The Royal Society was equally aware that it stood to gain from advising the Board of Ordnance: both in terms of influence and, as an organization reliant on

---

3   For a fuller account of this episode, see Schiffer (2003), in particular pp. 194-200 regarding Wilson's failed scientific vendetta and his attempts to disprove Franklin's designs. Also see *Phil. Trans.* LXVIII/XXXVI, XXXVII and XLIV for the official attempts to refute Franklin in the *Philosophical Transactions*.

internal funding from its fellows, in terms of valuable and prestigious financial patronage. The Royal Society was motivated by science, but it was also motivated by the money and access to resources necessary to conduct science. For example, government funding for observing the transits of Venus quintupled between 1761 and 1769 as the Royal Society partnered more closely with the military (McClellan 1985: 216). No-one understood the importance of this relationship better than Joseph Banks, who became President of the Royal Society in 1778.

Banks was a keen networker and inter-institutional player who understood and cultivated the webs of connections radiating throughout London. During his four-decade tenure as President, he worked hard to maintain inter-institutional links both within Britain and with overseas partners. Most importantly, he pursued increased contact between the Royal Society and the government (Gascoigne 1998: 22). An opportunity to make the Royal Society both militarily and diplomatically useful became available after the conclusion of the War of American Independence.

In late 1783 the French, represented by the geographer Cassini de Thury, requested a joint survey of the English Channel in the interests of science and good diplomatic relations. The request was forwarded to Banks by the foreign ministry, who as president of the Royal Society gave the assignment to his most accomplished mapmaker and trusted friend, Major General William Roy, F.R.S. Roy had risen through the military ranks since joining an Ordnance survey team in Scotland after the 1745 Jacobite uprising. He had gained recognition as a mapmaker and was asked to join the team of military engineers conducting the Military Survey. Roy was one of a handful of civilians on the team, which was broken into eight squads of eight men each, seven soldiers and one surveyor (Gardiner 1977: 439-441). Throughout the eight-year survey he never took military rank, maintaining his status as a civilian in a military world. Nonetheless, on the outbreak of war with France in 1756, the year following the survey's completion, he was given formal rank in the Corps of Engineers and commissioned as an officer in a line regiment serving with the British expeditionary force in Germany, and a year later was commissioned as an Ensign of Engineers (Gardiner 1977: 443). By the 1770s, Roy was a member of both the Royal Society of London and the Royal Society of Antiquaries, and had published his first entry in the *Philosophical Transactions* in 1777 (*Phil. Trans.* LXVII/XXXIV). Roy was now assigned by Banks to accomplish what Cook had done with the Navy – he would lead a joint survey mission that would yield military, political, diplomatic, and scientific benefits for both the military and civilian organizations.

Roy immediately seized upon the survey, designed to trigonometrically and astronomically link the Paris and Greenwich observatories, as a platform from which to launch a national survey, which he had been advocating for two decades (Gardiner

1977: 447). After having repeatedly failed in obtaining funding for his national survey, Roy was able to convince King George III himself to finance the smaller joint operation with the French, crediting him in the *Philosophical Transactions* as a "generous and beneficent Monarch, whose knowledge and love of the sciences are sufficiently evinced by the protection which HE constantly affords them, and under whose auspices they are seen daily to flourish" (*Phil. Trans.* LXXV/XXIV: 389) The King's contribution, and Roy's fundraising skill, was all the more impressive considering the great rarity with which the monarch supported the Royal Society (Sorrenson 1996b: 36).

Roy pursued the Paris-Greenwich survey work aggressively, his zeal attracting both military and civilian volunteers as well as prompting feuds with his instrument maker, Jesse Ramsden, who was tardy in delivering his orders to the field. Roy was conscious that not only his personal reputation, but that of the Royal Society and of the country was on the line, writing that good management and conduct of the enterprise would be "to the credit of the Nation in general, and of this Society in particular" (*Phil. Trans.* LXXV/XXIV: 390). He also knew that a successful survey in co-operation with the French would give him the political capital needed to expand the survey to the entire island (Gardiner 1977: 447).

As Roy worked through the summer with his joint crews of civilian and military engineers, Banks himself came down from London on July 15, 1784, bringing with him a trio of civilian volunteers as well as an important intangible element: publicity. For the rest of the summer, Banks played host to a stream of visitors taking an interest in the project, including the survey's financier, the King, whom Roy briefed and reported to the Royal Society that the work "met with HIS gracious approbation" (*Phil. Trans.* LXXV/XXIV: 425-426, 456). This placed the survey squarely within the context of London's polite society. On August 30, in the presence of a mixed team of civilians and officers, Roy completed the initial baselines of the survey (458-459).

In 1785, Roy presented an account of his work, entitled "An Account of the Measurement of a Base on Hounslowheath" and the following year the Royal Society awarded him the Copley Medal for the accuracy and speed of his achievements. For much of the rest of the decade he worked on the triangulation measurements, always at a pace slower than he would have liked, publishing his progress in subsequent editions of the *Philosophical Transactions*. Roy feuded with Jesse Ramsden over the increasingly late delivery of the theodolite necessary to conduct the trigonometric measurements, an affair that grew more and more ugly. Banks led his civilian friends in a vigorous defense of Roy's honor and conduct – all concerned being well aware that the tarnishing of Roy's reputation in the eyes of the public could jeopardize future cartographic missions (Widmalm 1990: 197-199).

With his friends successfully defending him and his reputation intact, and with the theodolite still in Ramsden's workshop, Roy used the time on his hands to great profit by lobbying for subsequent surveys. Roy was aware that Britain lagged behind France in mapping its territory, by now considering even his own survey of Scotland to be inadequate. To close this map-gap, Roy lobbied Charles Frederick for a national survey, and the Court of Directors of the East India Company for a survey of British possessions in Bengal. He met with resistance. Many members of Parliament were concerned that the use of the military to survey the British interior could threaten their liberties and property rights (Widmalm 1990: 202-203). No doubt they were also concerned that a more accurate sense of the size and dimensions of the country would raise their land taxes. Roy would eventually succeed with the founding of the Ordnance Survey in 1791, but his final success was posthumous, coming a year after his death.

Independent Naval Expeditions Reporting Back to the Royal Society
In contrast with the collaborative voyage of the *Endeavour*, the Royal Navy sent out many ships of its own without Royal Society participation on missions that were, nonetheless, reported back and published in the *Philosophical Transactions*. Often an officer on the voyage was a Fellow of the Royal Society, or became one upon his return home based on the quality of his contributions. In any event, the contributions of Royal Navy officers helped to build a valuable knowledge base shared by the Royal Society and the Royal Navy, enriching both organizations' capabilities.

Between the transits of Venus, the Royal Navy ordered Commodore John Byron to take a small squadron consisting of the H.M.S. *Dolphin*, H.M.S. *Tamer*, and H.M.S. *Florida* on an expedition. Byron's squadron sailed through the Straits of Magellan, and encountered a tribe of giant men who were eight feet in height. One of Byron's officers, Charles Clarke, wrote an account of this tribe and submitted it to the Royal Society for publication. The benefit was twofold: it would alert the Royal Navy to the presence of this tribe in case of future contact, and make a contribution to the field of human geography (*Phil. Trans.* LVII/VII).

Upon the expedition's return to England, the *Dolphin* almost immediately put to sea again for a circumnavigation under the command of Captain Samuel Wallis. The *Dolphin* was accompanied to the Straits of Magellan by another exploratory vessel, H.M.S. *Swallow*, under Captain Philip Carteret. There the *Dolphin* and the *Swallow* were separated in the fog, and Wallis took *Dolphin* west while Carteret changed course and took *Swallow* east. Carteret reported his exploits off the southern African coast to the Royal Society upon his return to England in 1770 (*Phil Trans*. LX/III). When the *Dolphin* returned to England after completing the circumnavigation in 1772, some of Wallis' observations taken in the South Seas were included in a communication regarding solar eclipses from Mr. George Wichell, F.R.S. and Master of the Royal Academy at Portsmouth (*Phil Trans.* LXII/IV).

While Wallis and Carteret were away to the south, another Royal Navy vessel, H.M.S. *Emerald*, was on a hydrographical survey mission in the Arctic Ocean. The captain of the *Emerald*, Charles Douglas, was also a Fellow of the Royal Society and reported back his findings, which included not only temperature readings of the waters but reports of the sighting of a Kraken (*Phil. Trans.* LX/VI). In 1774, H.M. sloop *Otter* was off the northern coast of the Canadian maritime provinces at Labrador, and the ship's lieutenant, Roger Curtis, wrote a memorable account for the *Philosophical Transactions*: "The barrenness of the country explains why it has been so seldom frequented. Here avarice has but little to feed on." The rest of Curtis' account is an extensive description of the geography of the Labradoran coast (*Phil Trans.* LXIV/XL). In a return to Labrador in 1776, Lieutenant Richard Pickersgill, commanding H.M. brig *Lion*, made several observations on calculating longitude and also noted variations in his compass readings, having apparently stumbled upon the magnetic north pole (*Phil. Trans.* LXVIII/L). These findings were reported to the Astronomer Royal, Nevil Maskelyne, who in turn published them in the *Philosophical Transactions*.

These small and isolated examples illustrate a larger point: even in cases where there was no formal co-operation between the Royal Navy and the Royal Society, communication still existed to the benefit of both parties. The Royal Society eagerly accepted field observations from naval officers, and in turn the navy's officers were better informed about the places they were going to and the people and things they might take with them or find there.

Other Individual Military Contributions to the Philosophical Transactions

Observation of an anomaly or new discovery and consequent filing of a report to the Secretary or President of the Royal Society, followed by publication in the *Philosophical Transactions*, was a common practice amongst all Fellows in the eighteenth century. While many entries to the *Philosophical Transactions* concern replicable experiments or careful observation, often field entries are rough sketches of strange new forms of life or physical phenomena. Military officers were among the best-equipped people to submit entries on such curiosities, given the combination of their field experience and the greater concentration of maladies that seem to have followed military units in the eighteenth century, be they afloat or on land.

Of course, naval officers were not the only observers of new phenomena. Military medical men, at sea and on land, were frequent contributors. Strikingly, these medicos did not confine themselves to their professional area of expertise. Donald Monro, Physician to the Army in the years 1771-1772, authored two entries in the *Philosophical Transactions* of those years on the subjects of fossilized salts and mineral waters (*Phil. Trans.* LXI/LII and LXII/III). In the early 1780s, another Physician to the Army and future Secretary of the Royal Society, Charles Blagden, contributed three entries on the waters of the Gulf Stream, the "congelation of quicksilver," and an observation of a meteor shower (*Phil. Trans.* LXXI/XVIII, LXXIII/XXI, and LXXIV/XVIII).

Scientifically inclined military officers were an important group; most of the *Philosophical Transactions* entries by military officers of this time period came from officers who had no formal connection to the Royal Society. Some would file a single report with the Secretary and then disappear from the records; others would use their contribution as a stepping-stone to election to a fellowship. Contributions ranged from that of Captain Alexander Rose of the 52nd Regiment of Foot, who while serving in Quebec in 1765 kept a meteorological diary, to that of Lieutenant James Glenie of the Royal Regiment of Artillery, who in 1777 submitted a paper on pure mathematical theory. Rose had no further contact with the Royal Society; Glenie was elected a fellow in 1779. Contributions were as varied as the men who wrote them, and often had no relationship to the officer's military vocation.

A small number of contributions were directed to the Royal Society from officers in foreign militaries. Interestingly, these contributions came from officers of France and Spain, two powers hostile to Britain throughout this time period. In 1766, J. O. Justamond, Surgeon to the First Troop of Horse Grenadier Guards, wrote from France to share the results of a bone surgery (*Phil. Trans.* LVI/XXXIII). In 1774, Lieutenant Joseph Varelaz of the Spanish Navy submitted an observation of Saturn's ring, and in 1779, even as Britain and Spain were at war, Admiral Don Antonio Ulloa of Spain offered observations of an annular eclipse of the sun taken from the deck of his flagship the *Espagne* en route from the Azores to Cape St. Vincent (*Phil. Trans.* LXIV/XIII and LXIX/XI). These outlying reports are entirely within the character of similar reports from British officers. What makes them exceptional is that they represent a backdoor to the enemy during times of both cold peace and hot war. Together with the practice of treating science vessels as neutrals and Roy's joint survey mission, the co-operation between the military and the Royal Society offered not only scientific and political benefits but diplomatic ones as well.

Self-Interest and the Pursuit of Knowledge

All parties involved in these joint surveys benefited from them, though the Royal Society and its military partners often wanted different things from a mission. The military's goals were clearly practical; it was interested in scientific advances that would allow it to fight wars more effectively. The motivations of the Royal Society tended to be more complex, striking a balance between the desire to accumulate philosophical knowledge, to be patriotic supporters of government, and to improve society through the application of useful knowledge.

The efforts of the Royal Society as an institution, and its Fellows as individuals, took place within a culture of "improving" knowledge in Enlightenment Britain. As explained by David Hume, the goal of Enlightenment knowledge was to bring the benefits of rigorous philosophical enquiry to the reading public (Hume 1996: 1). Knowledge was to be accessible and presented in such a way that the reader could usefully employ its lessons, be they of moral or natural philosophy, in his social interactions. The Royal Society enacted this ethos on an institutional scale by putting its scientific expertise to work in a way that served a useful need and in making that knowledge publicly available through the printing of the *Philosophical Transactions* (see also Golinski 1992).

In this setting, usefulness translated into patriotism. By supplying the military with useful knowledge that would allow it to conduct its operations more effectively, the Royal Society raised its own stature above that of other learned societies, and credibly cast itself as a truly national organization. As the most useful and therefore most patriotic learned society, the Royal Society was elevated above competing organizations such as the Royal Society of Antiquaries and the Royal Society of Arts. In the case of William Roy, he saw no distinction between his various activities, considering them all to fall under the aegis of British patriotism. Introducing his contribution to the *Philosophical Transactions* of 1785, Roy declared that "accurate surveys of a country are universally admitted to be works of great public utility" that were of use both for civil improvement and military campaigning (*Phil Trans*. LXXV/XXIV: 385). He saw harmony between the monarchy, which provided his Greenwich survey team with funds; the Royal Society, which provided him with institutional support; the Admiralty, which provided him with supplies; and the Corps of Engineers, which provided him with troops. With prestige came popularity, and with popularity it became possible to recruit richer and better-connected Fellows, thereby increasing the Royal Society's existing assets and further making them indispensable as the genesis of the British military-intellectual complex.

Finally, co-operating with the military did yield philosophical knowledge, but this seems to have been an afterthought both for the Fellows of the time as well as historians who have subsequently studied them. As Hall reminds us in the introduction to her book, the eighteenth century is often regarded as a barren land between the age of Newton and the more refined science of the nineteenth century.

Yet Sir Joseph Banks remains the Society's longest serving and most active President, and it was the empirical groundwork laid in the eighteenth century that allowed the advanced work of subsequent decades. The observations collected by Banks, Cook, Roy, and others were valuable for their usefulness but also for extending natural philosophy in its own right.

The benefits to the military are equally straightforward. The expertise of Fellows yielded important advances in the military's awareness of the geography of Britain and its ability to efficiently conduct operations such as powder storage. The Royal Navy benefited from advances in hydrography and in the expansion of the knowledge of human geography, especially since the Pacific Ocean and South Seas were largely devoid of European populations, meaning that in the event of emergency British vessels would need to turn to native assistance and knowledge of their friendliness as well as language and cultural mores would be of material help in deciding the outcome.

Individuals within the Royal Society and the military benefited along with their organizations. Enlightenment Britain had many progressive qualities but also some conservative ones and that included the desire of military officers to be socially accepted as gentlemen in polite circles. Election to the prestigious Royal Society of London served as a validation of a military officer's acceptability, and could rescue him from dismissal as a sort of tradesman who worked with his hands. The combined notoriety from a successful military operation, social connectivity, and the production of philosophical knowledge could make a career, as James Cook and William Roy discovered.

Social fame was particularly important for key individuals who positioned themselves as mediators between the civilian and military worlds. There had always been some crossover, but in an increasingly institutionalized modern state, Banks, Cook, and Roy stood out as men who understood how organizations worked and how the savvy playing of those organizations with and off of each other could result in achieving personal objectives. Banks desired fame, Cook desired promotion, and Roy desired to be useful. All three achieved their objectives by finding ways to position themselves in the gaps between British institutions and serving as nexuses to achieve the goals of all concerned.

Conclusion:
The Legacy of Co-operation and the Military-Intellectual Complex
By the 1790s, these episodic cases of military-Royal Society co-operation had created a framework for regular institutional co-operation during the Revolutionary and Napoleonic Wars, which subsequently stretched through the nineteenth and

twentieth centuries to the present day. With the Board of Ordnance and the Royal Society, William Roy successfully created the Ordnance Survey, which became a permanent mapping agency of the British government that continues its work today. The Board of Ordnance also frequently consulted the Royal Society on issues such as trigonometry, gunpowder processing, and fire suppressant casings throughout the Napoleonic Wars. In turn, the Royal Society would appoint liaison committees from amongst its Fellows to work with the military on each problem. By the 1820s, the Board of Ordnance worked with Captain Henry Kater to complete Roy's mapping of Scotland through a full trigonometric survey, and the Board of Longitude co-operated with the Royal Society in establishing a stellar observatory in South Africa (Hall 1984: 12-14). During the Napoleonic Wars, Banks worked to assist the Royal Navy through active co-operation with the Board of Longitude (Gascoigne 1998: 28-30).

The joint field research done by the military and Royal Society yielded short-term military, political, and scientific benefits; the long-term result was one of civil-military co-operation and the establishment of overlapping, permanent links between the Royal Society and its military counterparts. The Royal Society had established an important precedent during the decades prior to the war by ensuring that, despite its status as a private, independent society of both amateur and professional scientists, the Royal Society of London stood ready and willing to assist the state when it was in need. In exchange, the society expected financial and social rewards for its efforts; the advice offered was of a professional caliber, utilizing the best expertise resident within the Society's membership.

The co-option of the Royal Society into the state's military apparatus was at odds with its role as a civilian association and forced a choice between patriotism and cosmopolitanism. On the one hand, the Royal Society carried on fruitful correspondences with foreign counterparts and inducted foreign members, and scientific vessels were treated by governments as neutrals on the high seas. Yet the Royal Society also helped maintain the Royal Observatory for the benefit of the Royal Navy and the merchant marine. When faced with a choice between acting as a private learned society or as a national academy on the lines of the Continental model, the Royal Society chose to act as a functional appendage of government, laying the groundwork for the Society's governmental advisory role of the nineteenth century and setting a precedent for the fully-matured military-intellectual complex of the present day.

References

Banks, J. 1962. *The Endeavour Journal of Joseph Banks, 1768-1771*. 2 vols. Sydney: Halstead Press.

Beaglehole, J. C. 1962. *The Endeavour Journal of Joseph Banks 1768-1771*. 2 vols. Sydney: Halstead Press.

Gardiner, R. A. 1977. "William Roy, surveyor and antiquary." *The Geographical Journal*, 1433: 439-450.

Gascoigne, J. 1998. *Science in the Service of Empire*. Cambridge: Cambridge University Press.

Gascoigne, J. 1999. "The royal society and the emergence of science as an instrument of state policy." *British Journal for the History of Science*, 32(113, Pt 2): 171-184.

Golinski, J. 1992. *Science as Public Culture: Chemistry and enlightenment in Britain, 1760-1820*. Cambridge: Cambridge University Press.

Hall, M. B. 1984. *All Scientists Now: The Royal society in the Nineteenth Century*. Cambridge: Cambridge University Press.

Hume, D. 1996. "Of Essay Writing." In: S. Copley & A. Edgar (eds.) *Selected essays*. Oxford: Oxford University Press, 1-4.

Lawrence, C. 1995. "Disciplining disease: scurvy, the navy and imperial expansion, 1750-1825." In: D. Miller & P. H. Reill (eds.) *Visions of Empire: Voyages, Botany and Representations of Nature*. Cambridge: Cambridge University Press.

McClellan, J. E. 1985. *Science reorganized: Scientific Societies in the Eighteenth Century*. New York: Columbia University Press.

Miller, D. 1995. "Joseph Banks, empire, and 'centers of calculation' in late Hanoverian London." In: D. Miller & P. H. Reill (eds.) *Visions of Empire: Voyages, Botany and representations of nature*. Cambridge: Cambridge University Press.

Purver, M. 1967. *The Royal Society: Concept and Creation*. Cambridge, Mass.: MIT Press.

Roy, William. *The Great Map: The Military Survey of Scotland, 1747-55*. Scotland: Birlinn in association with the National Library of Scotland and the British Library.

Royal Society of London Manuscript Division – Folios 3, 4, 10, 11.

Schiffer, M. B. 2003. *Draw the Lightning Down: Benjamin Franklin and Electrical Technology in the Age of Enlightenment*. Los Angeles: University of California Press.

Sorrenson, R. 1996a. "The ship as a scientific instrument in the eighteenth century." *Osiris, 11*: 221-236.

Sorrenson, R. 1996b. "Towards a history of the royal society in the eighteenth century." *Notes and Records of the Royal Society of London*, 50(1): 29-46.

Various Authors 1761-1790. *Philosophical Transactions, Giving some Account of the Present Undertakings, Studies and Labours of the Ingenious in Many considerable Parts of the World.* 28 vols. London: L. Davis and C. Rhymers, Printers to the Royal Society, Against Grays-Inn Gate, in Holbourn.

Wales, W. 1788. *Astronomical Observations, Made in the Voyages which were Undertaken by Order of His Present Majesty, for Making Discoveries in the Southern Hemisphere, And Successively Performed by Commodore Byron, Captain Wallis, Captain Carteret, and Captain Cook, in the Dolphin, Tamer, Swallow, and Endeavour. Drawn up and Published by Order of the Commissioners of Longitude, from the Journals which were Kept by the Several Commanders, and from the Papers of Mr. Charles Green, formerly Assistant at the Royal Observatory.* London.

Weld, W.R. 1848. *A History of the Royal Society: With Memoirs of the Presidents, Compiled from Authentic Documents.* London: J. W. Parker.

Widmalm, S. 1990. "Accuracy, rhetoric, and technology: The Paris–Greenwich triangulation, 1784-1788." In: T. Frangsmyr, J. Heilbron & R. E. Rider (eds.) *The Quantifying Spirit in the Eighteenth Century.* Berkeley & Oxford: University of California Press, 179-206.

Appendix: List of articles in Phil. Trans.

Volume Number	Year	Author(s)	Rank and Position	Title
LII	1761	Charles Mason, Jeremiah Dixon	Fellows aboard H.M.S. *Seahorse*	"LX. *Observations made at the Cape of* Good Hope, *by Mr.* Charles Mason *and Mr.* Dixon, *reduced to apparent Time by Mr.* Mason. *With an Appendix.*"
LII	1761	Charles Mason, Jeremiah Dixon	Fellows aboard H.M.S. *Seahorse*	"LXI. *Latitude of the Observatory at the Cape of* Good Hope, *reduced from the Observations of different Stars; by Mr.* Charles Mason."
LII	1761	William Hirst	Chaplain, Royal Navy	"LXII. *An Account of the Transit of* Venus *over the Sun, on the 6th of* June 1761, *at* Madras; *by the Rev. Mr.* William Hirst, *Chaplain of one of His Majesty's Ship in the* East Indies : *Contained in a Letter wrote by Him to the Right Honorablethe Earl of* Macclesfield, *President of the Royal Society. Dated Ft.* St. George, *1 July 1761.*"
LVI	1766	J. O. Justamond	Surgeon to the First Troop of Horse Grenadier Guards (France)	"XXXIII. *An Account of the Extraction of three Inches and ten Lines of the Bone of the upper Arm, which was followed by a Regeneration of the bony matter; with a Description of a Machine made use of to keep the upper and lower Pieces of the Bone at their proper Distances, during the Time that the Regeneration was taking Place; and which may also be of Service in fractures happening near the head of that Bone. By Mr.* Le Gat, *Professor of Anatomy and Surgery at* Rouen, *Member of several Academies, and F.R.S. Translated from the* French *by* J. O. Justamond, *Surgeon to the First Troop of Horse Grenadier Guards.*"
LVI	1766	Alexander Rose	Captain, 52nd Regiment	"XXXVI. *Abstract of a Journal of the Weather in* Quebec, *between the 1st of* April 1765, *and 30th of* April 1766. *By Cap.* Alex. Rose, *of the 52d Regiment; communicated by the Rev.* P. Murdoch, *D.D.F.R.S.*"
LVII	1767	Charles Clarke	Officer, H.M.S. *Dolphin*	"VIII. *An Account of the very tall Men, seen near the Streights of* Magellan, *in the Year* 1764, *by the Equipage of the* Dolphin, *Man of War, under the Command of the Honourable Commodore* Byron; *in a Letter from Mr.* Charles Clarke, *Officer on board the said Ship, to M.* Maty, *M.D. Sec. R.S.*"
LIX	1769	Jardine	Lieutenant, R.N.	"XLV. *Observations of the Transit of* Venus, *and other Astronomical Observations, made at* Gibraltar; *contained in a Letter to the Astronomer Royal from Lieutenant* Jardine."

Volume Number	Year	Author(s)	Rank and Position	Title
LX	1770	Philip Carteret	Captain, H.M.S. *Swallow*	"III. *A Letter on a* Camelopardalis *found about the* Cape of Good Hope, *from Capt.* Carteret *to* Mathew Maty, *M.D. Sec. R.S.*"
LX	1770	Charles Douglas	Captain, H.M.S. *Emerald*	"VI. *An Account of the Result of some Attempts made to ascertain the temperature of the Sea in great Depths, near the Coasts of* Lapland *and* Norway; *as also some Anecdotes, collected in the former: By* Charles Douglas, *Esquire, F.R.S. then Captain of his Majesty's Ship the* Emerald, Anno 1769."
LXI	1771	[Charles Green and] James Cook	Lieutenant, H.M.S. *Endeavour*	"XLIII. *Observations made, by Appointment of the Royal Society, at* King George's Island *in the* South Sea; *by Mr.* Charles Green, *formerly Assistant at the Royal Observatory at* Greenwich, *and Lieut.* James Cook, *of his Majesty's Ship the* Endeavour."
LXI	1771	James Cook	Lieutenant, H.M.S. *Endeavour*	"XLIV. *Variation of the Compass, as observed on board the* Endeavour Bark, *in a Voyage round the World. Communicated by Lieut.* James Cook, *Commander of the said Bark.*"
LXI	1771	James Cook	Lieutenant, H.M.S. *Endeavour*	"XLV. *Transitus* Veneris & Mercurii *in euroum Exitu e Disco* Solis, *4to Mensis* Junii & *10mo* Novembris, 1769, *observatus. Communicated by Capt.* James Cook."
LXI	1771	Donald Monro	Physician to the Army	"LII. *An Account of a pure native crystallized Natron, or fossil alkaline Salt, which is found in the Country of* Tripoli *in* Barbary: *By* Donald Monro, *M.D. Physician to the Army, and to* St. George's *Hospital, Fellow of the Royal College of Physicians, and of the Royal Society.*"
LXII	1772	Donald Monro	Physician to the Army	"III. An Account of the sulphureous mineral Waters of Castle-Loed and Fairburn in the County of Ross, and of the salt purging Water of Pitkeathly, in the County of Perth in Scotland: By Donald Monro, *M.D. Physician to the Army, and to* St. George's *Hospital, Fellow of the Royal College of Physicians, and of the Royal Society.*"

Volume Number	Year	Author(s)	Rank and Position	Title
LXII	1772	[George Wichell and] Samuel Wallis	Captain, H.M.S. *Dolphin*	"IV. *Extract of a Letter from Mr.* George Wichell, *F.R.S. and Master of the Royal Academy at* Portsmouth, *to* Charles Morton, *M.D. Sec.R.S. including some Account of a Solar Eclipse observed at* George's Island, *by Captain* Wallis; *and several Astronomical Observations made at* Portsmouth."
LXII	1772	Charles Newland	Captain, H.M.S. *Kelsall*	"X. A Letter to the Rev. Mr. Maskelyne, Astronomer Royal, F.R.S. accompanying a new Chart of the Red Sea, with two draughts of the Roads of Mocha and Judda, and several Observations made during a voyage on that Sea, by Capt. Charles Newland."
LXII	1772	James Cook	Lieutenant, H.M.S. *Endeavour*	"XXV. *An Account of the Flowing of the Tides in the* South Sea, *as observed on board His Majesty's Bark the* Endeavour, *by Lieut.* J. Cook, *Commander, in a Letter to* Nevil Maskelyne, *Astronomer Royal, and F.R.S.*"
LXIII	1773-1774	Benjamin Franklin, H. Cavendish, William Watson, J. Robertson	Liaison committee between Royal Society and Board of Ordnance	"VIII. *A Report of the Committee appointed by the Royal Society, to consider a Method for securing the Powder Magazines at Purfleet.*"
LXIII	1773-1774	Benjamin Wilson	F.R.S.	"IX. *Observations upon Lightning, and the Method of securing Buildings from its Effects: In a Letter to Sir* Charles Frederick, *Surveyor-General of His Majesty's Ordnance, and F.R.S. By* Benjamin Wilson, *F.R.S. & Ac.R.Ups.Soc.*"
LXIII	1773-1774	Benjamin Franklin, H. Cavendish, William Watson, J. Robertson	Liaison committee between Royal Society and Board of Ordnance	"X. A Letter to Sir John Pringle, Bart.Pr.R.S. on pointed Conductors."
LXIV	1774	Joseph Varelaz	Lieutenant, Spanish Navy	"XIII. *The Disposition of* Saturn's Ring, *observed by* Joseph Varelaz, *Lieutenant of the Royal Navy of the King of* Spain, *and Professor of Mathematics, in the Academy of* Guard-Marine *at* Cadiz. *In a Letter to Dr.* Morton."

Volume Number	Year	Author(s)	Rank and Position	Title
LXIV	1774	George Sproule	Ensign, 59th Regiment	"XXII. *Observations of the Immersions and Emersions of the Satellites of* Jupiter, *taken in the Year 1768, by Ensign* George Sproule, *of his Majesty's 59th Regiment, on the South Point of the Entrance of* Gaspee Bason, *which bears from* Cape Ferrilong, *or the Cape forming the Bay to the Northward, N. 68 ¼ W. by the true Meridian, distant 12 ¼ Marine Miles. Communicated by the Astronomer Royal.*"
LXIV	1774	Roger Curtis	Lieutenant H.M. Sloop *Otter*	"XL. *Particulars of the country of* Labradore, *extracted from the Papers of Lieutenant* Roger Curtis, *of His Majesty's Sloop the* Otter, *with a Plane-Chart of the Coast. Communicated by the Honourable* Daines Barrington."
LXVI	1776	William Clayton	Lieutenant, R.N.	"V. *An Account of the Falkland Islands. By* William Clayton, *Esq. of His Majesty's Navy.*"
LXVI	1776	James Cook	Captain, H.M.S. *Resolution*	"XXII. *The Method taken for preserving the Health of the Crew of His Majesty's Ship the* Resolution *during her late Voyage round the World. By Captain* James Cook, *F.R.S. Addressed to Sir* John Pringle, *Bart. P.R.S.*"
LXVI	1776	Alexander Small	Surgeon to the Train of Artillery at Minorca	"XXV. *Extract of a Letter from Mr.* Alexander Small, *Surgeon to the Train of Artillery at* Minorca, *to Sir* John Pringle, *Bart. P.R.S. Dated St. Philip's, Aug. 8, 1775.*"
LXVI	1776	James Cook	Captain, H.M.S. *Resolution*	"XXVI. *Of the Tides in the* South Seas. *By* Captain James Cook, F.R.S."
LXVI	1776	William Anderson	Surgeon, H.M.S. *Resolution*	"XXXVI. *An Account of some poisonous Fish in the* South Seas. *In a Letter to Sir* John Pringle, *Bart. P.R.S. from Mr.* William Anderson, *late Surgeon's Mate on board His Majesty's Ship the* Resolution, *now Surgeon of that Ship.*"
LXVII	1777	James Glenie	Lieutenant, Royal Regiment of Artillery	"XXIII. *The general Mathematical Laws which regulate and extended Proportion universally; or, a Method of comparing Magnitudes of any kind together, in all the possible Degrees of Increase and Decrease. By* James Glenie, *A.M. and Lieutenant in the Royal Regiment of Artillery.*"
LXVII	1777	William Roy	Colonel, British Army	"XXXIV. Experiments and Observations made in Britain, in order to obtain a Rule for measuring Heights with the Barometer."

Volume Number	Year	Author(s)	Rank and Position	Title
LXVIII	1778	Charles Hatton	Military Academy at Woolwich	"III. *The Force of fired Gunpowder, and the initial Velocities of Cannon Balls, determined by Experiments; from which is also deduced the Relation of the initial Velocity to the Weight of the Shot and the Quantity of Powder. By Mr.* Charles Hatton, *of the Military Academy at* Woolwich. *Communicated by* Samuel Horsley, *LL.D. Sec.R.S.*"
LXVIII	1778	Sir George Shuckburgh	F.R.S.	"XXXII. *Comparison between Sir* George Shuckburgh *and Colonel Roy's Rules for the Measurement of Heights with the Barometer; in a Letter to Col. Roy, F.R.S. from Sir* George Shuckburgh."
LXVIII	1778	Dr. Musgrave	F.R.S.	"XXXVI. *Reasons for differing from the Report of the Committee appointed to consider Mr.* Wilson's *Experiments; including Remarks on some Experiments exhibited by Mr.* Nairne."
LXVIII	1778	Mr. Edward Nairne	F.R.S.	"XXXVII. *Experiments on Electricity, being an Attempt to shew the Advantage of elevated pointed Conductors.*"
LXVIII	1778	Benjamin Wilson	F.R.S.	"XLIV. *New Experiments upon the* Leyden *Phial, respecting the Terminations of Conductors.*"
LXVIII	1778	Richard Pickersgill	Lieutenant, H.M. Brig *Lion*	"L. *Track of His Majesty's armed Brig* Lion, *from England to* Davis's Straights *and* Labradore, *with the Observations for determining the Longitude by Sun and Moon and Error of common reckoning; also the Variation of the Compass and Dip of the Needle, as observed during the said Voyages in 1776. By Lieutenant* Richard Pickersgill, *late Commander of the said Vessel; communicated by* Nevil Maskelyne, *D.D.F.R.S. and Astronomer Royal.*"
LXIX	1779	Don Antonio Ulloa	Admiral, Spanish Navy	"XI. *Observations on the total (with Duration) and annular Eclipse of the Sun, taken on the 24th of June, 1778 on Board the* Espagne, *being the Admiral's Ship of the Fleet of* New Spain, *in the Passage from the* Azores *towards* Cape St. Vincent's. *By Don* Antonio Ulloa, *F.R.S. Commander of the said Squadron; communicated by* Samuel Horsley, *LL.D. F.R.S.*"
LXX	1780	William Roxburgh	Assistant-Surgeon, Fort St. George	"XIV. *A Continuation of a Meteorological Diary, kept at Fort* St. George, *on the Coast of* Coromandel. *By Mr.* William Roxburgh, *Assistant-Surgeon to the Hospital at the said Fort; communicated by* Joseph Banks, *Esq. P.R.S.*"

Volume Number	Year	Author(s)	Rank and Position	Title
LXXI	1781	Charles Blagden	Physician to the Army	"XVIII. *On the Heat of the Water in the Gulf-Stream. By* Charles Blagden, M.D. *Physician to the Army, F.R.S.*"
LXXIII	1783	Charles Blagden	Physician to the Army	"XXI. *History of the Congelation of Quicksilver. By* Charles Blagden, M.D. F.R.S. *Physician to the Army.*"
LXXIV	1784	Charles Blagden	Physician to the Army	"XVIII. *An Account of some late fiery Meteors; with Observations. In a Letter from* Charles Blagden, M.D. *Physician to the Army, Sec. R.S. to Sir* Joseph Banks, *Bart. P.R.S.*"
LXXV	1785	William Roy	Major General, British Army	"XXIV. An Account of the Measurement of a Base on Hounslowheath."
LXXVI	1786	Richard McCausland	Surgeon, 8th Regiment of Foot	"XI. *Particulars relative to the Nature and Customs of the Indians of North-America. By Mr.* Richard McCausland, *Surgeon to the King's or Eight Regiment of Foot. Communicated by* Joseph Plantz, *Sec. R.S.*"
LXXVI	1786	Sir Benjamin Thompson	Colonel, British Army	"XIV. *New Experiments upon Heat. By Colonel Sir* Benjamin Thompson, *Knt. F.R.S. In a Letter to Sir* Joseph Banks, *Bart. P.R.S.*"
LXXVII	1787	William Roy	Major General, British Army	"XIX. *An Account of the Mode proposed to be followed in determining the relative Situation of the Royal Observatories of* Greenwich *and* Paris. *By Major-General* William Roy, *F.R.S. and A.S.*"
LXXVII	1787	N/A	N/A	Appendix: "*A Supplement to Major-General* Roy's *Account of the Mode proposed to be followed in determining the relative Situation of the Royal Observatories of* Greenwich *and* Paris."
LXXX	1790	William Roy	Major General, British Army	"XII. *An Account of the Trigonometrical Operation, whereby the Distance between the Meridians of the Royal Observatories of* Greenwich *and* Paris *has been determined. By Major-General* William Roy, *F.R.S. and A.S.*"

Anke Fischer-Kattner

Entangled Experiences, Disentangling Disciplines
Antoine and Arnauld d'Abbadie's Voyages in Ethiopia

After his return from the northeastern African interior to France in 1848, Antoine d'Abbadie painted a distinct picture of the figure of a great voyager in the clean copy of his journal:

> A great voyager is not an ordinary man, even after one has read and gotten to know him. In his account, there are quite a few hidden characteristics which, in removing but a corner of the veil that covers him, leave him in the solitude of mystery. Not even the longest conversation could bring back to his home country these ever renewed impressions, these insecurities, and these everyday reflections. Which of you would understand his long anxiety in unknown lands? His pain is greater than that of a young lover taken away from the object of his cares because he does not have any distractions other than memories of the past and hope for the future. And when he shakes his white hair while recounting his travels, is he not like the portrait of a dead person?[1]

Abbadie's voyager appears as an intriguing personality, physically and mentally shaped by his experience in the field. It is impossible for him to convey the variety of this experience to his stay-at-home countrymen. No narrative account can give them a true insight into the strange world that has touched his mind and heart. While he was putting these thoughts into well-rounded prose, Antoine was himself already working on narratives and publications of his and his brother's journey into the Ethiopian highlands. So what were the pictures of these two travellers like? What image of themselves were they about to present to the European public? Today, the internet provides us with printed portraits of the brothers, unfortunately

---

[1] Bibliothèque Nationale Française (subsequently BNF), n. a. f. 21 301 (Copie du volume V des manuscrits faits par An.e d'Abbadie en Ethiopie), 126, no. 373. All English translations of French archival source material cited in the text are made by the author.

not of sufficient quality to be reprinted here, but easily accessed on any computer.[2] The contrast between the two portraits that have come to be the public visual representations of the travellers is striking.

Antoine, depicted as a young man, does not conform to his own presentation of the seasoned explorer. One might see the serious look of the learned man in his face, a detail that might be taken to hint that this man was to publish numerous learned books. Nothing in the portrait, however, indicates that it was this man's traveling experience that would form the basis for his becoming president of the Academy of Sciences in 1882. Antoine's portrait is an engraving that corresponds to a painted likeness in Abbadia castle on the coast of Acquitaine. Antoine had this building designed to combine the features of a *savant*'s study with the pseudo-historic grandeur of an aristocratic dwelling, and later bequeathed it to the French Academy of Sciences as the basis of a special foundation. Although the castle's interior decoration is intimately related to Abbadie's exotic experiences, the portrait of its owner does not show any trace of these.

The public picture of his brother Arnauld can be regarded as a complete contrast. The photograph seems to match Antoine's description of the great traveller perfectly: he looks aged, but still excites an impression of bodily strength. He is wearing exotic dress, supposedly the costume of the country he had lived in for so long. His face is averted from the viewer, so his gaze seems to be turned back towards past adventures rather than towards his admirers at home, the readers of the publications in which his picture appears. But maybe these striking differences only result from the practicalities of comparing portraits taken at different stages of a man's life? Maybe they are simply due to the developments of portraiture that occurred between the youth of the elder and the old age of the younger brother? Perhaps they are, but I will argue that they can also be taken to exemplify the diverging strategies of self-presentation employed by the Abbadie brothers.

Narratives of Self and Knowledge from the Field

The strategies employed by Antoine and Arnauld d'Abbadie to present a well-defined *self* to the world, particularly to the world of readers interested in travel writing and scientific knowledge in mid-nineteenth-century Europe, were intimately related to the processes in which they constructed narratives and knowledge from their

---

2   A simple search engine query for Antoine and Arnauld d'Abbadie can demonstrate the wide dissemination of the two images. Easy access to both portraits is provided by Wikipedia: Antoine, retrieved August 9, 2011 from http://commons.wikimedia.org/wiki/File:Anton_abbadia.png, and Arnauld, retrieved August 9, 2011 from http://en.wikipedia.org/wiki/File:Catholic_Encyclopedia_-_Arnauld_d%27Abbadie.jpg.

experiences *in the field*. Generally, any sort of empirical knowledge has to be put into the form of language; it must be given a certain narrative structure in order to become a social instead of just an individual fact. The philosophical problem of the connection between words and world has been treated in various contemporary disciplines: not only philosophers of language and philosophically-minded linguists, but also anthropologists and even philosophers and historians of science have dealt with the question of whether human knowledge of a (positivist) reality could exist without being steeped in language. As it turns out, even scientific description clearly follows narrative conventions: its authors employ linguistic tricks to convince their readers that the text represents a genuine experience.

Science studies have demonstrated the importance of this and other factors in the construction of true scientific knowledge. A striking example is given by Bruno Latour, who used a detailed linguistic analysis of a scientific text about Pasteur's famous experiment on lactic acid fermentation to show the multiple influences on the narrative of discovery. Besides the writing scientist, the most influential factors are the social world around him and the unexpected activities of the supposedly passive *object of study* (Latour 1999: 115). Latour argues that this process is one of social and textual construction, but he opposes this term's connotations of subjectivity and – following from this – lack of reality. The text and the discovery of a new entity it produces become part of reality, of a reality that is constantly being recreated by the production of new knowledge and thus of new entities or "propositions" within their particular "spatiotemporal envelope" (Latour 1999: 156). The process of knowledge production thus becomes one of co-operation and mutual enabling of humans and non-humans, working together in events that change reality (Latour 1999: 126, 131-132).

Latour's complex model of scientific referentiality as a system of different "flows" or "chains" of circulating reference (Latour 1999: 99, 69) is not based solely on scientific activities in the laboratory; it also relates to the confrontation between humans and non-humans *in the field*, in this case the Brazilian rainforest of Boa Vista (Latour 1999: 30). He tries to break down the cognitive barriers between the (controlled) laboratory and the (uncontrolled) field, between (active) men and (passive) things, between the ideas of construction and re-presentation of reality. The constant interactions of all these aspects can be seen as a fitting description of the entangled activities of scientifically oriented travellers such as Antoine and Arnaud d'Abbadie. But for a detailed historical analysis of the narrative strategies of their manuscript and published records one has to take into account the fact that they were not only dealing with non-humans, but for the most part also with unknown humans.

Latour himself is very careful not to suggest that his analytical approach to the activities of (natural) science might simply be transposed to texts describing social relations or particular interactions between people (Latour 1999: 78). However, many of his concepts and models might help to describe the processes at work in the sciences of mankind and culture.

Some striking parallels can be drawn between Latour's approach and the important methodological debates that evolved around the narrative dimension of the field science of ethnography from the late 1970s. Clifford Geertz has pointed out that all textual attempts at representing unfamiliar people and their societies or cultures have a literary quality. Geertz argues that the most convincing examples of this kind of representation obtain their authority through well-chosen literary stylistics, which helps to create an impression of genuine experiences and insights (Geertz 1990).

In the early 1980s, James Clifford and George E. Marcus united several practitioners of the study of culture from various disciplines for a series of workshops. In the resulting publication, all the contributors question the "ideology" of uncritical ethnographic representations of foreign realities and explain that their "focus on text making and rhetoric serves to highlight the constructed, artificial nature of cultural accounts" (Clifford 1986a: 2). All the book's contributors argue for an acceptance of the fact that any description of a foreign country or people is shaped by the ethnographer's rhetoric and the socio-cultural patterns s/he reproduces in writing. Ultimately, the activity of writing culture "is always caught up in the invention, not the representation of cultures" (Clifford 1986a: 2). At the same time, any ethnographic text is open to numerous interpretations by its readers (whichever culture they might belong to). However, this conceptual shift from the idea of "truthful representation" to "contested interpretations" does not necessarily diminish the value of descriptions as long as their authors are aware of their own narrative construction and allow for its openness (Clifford 1986b: 119-120).

Is the "crisis in anthropology" (Clifford 1986a: 3) that is expressed in these attitudes really only a result of a "post-scientific world"? A world in which "post-modern ethnography" introduces polyphony and experience into descriptive texts where before there was only "the illusion of a transcendental," of the securities of science, as Stephen A. Tyler argues (Tyler 1986)? From a point of view which might be called that of the *cultural historian of science*, if it were a less awkward expression, this is not the case. Even if Latour's assertion that "We have never been modern" (Latour 1991) sounds ahistorical too, his provocative stance provides an important insight: the modernist position of positivism might have been very dominant from the mid-nineteenth century, especially in popular forms of scientific thought, but it never stood completely unquestioned.

The period of early "scientific" European travel into the African interior (roughly between the mid-eighteenth and mid-nineteenth century) is an extremely interesting case supporting this point: in the wake of the so-called second age of exploration (exemplified by the sea voyages of Cook and Bougainville, in which Enlightenment interests in natural history and the varieties of mankind played an important role) the African continent, well known in its outlines since the fifteenth century, became the object of a new interest. The famous French geographer D'Anville visually marked the uncertainties of knowledge about the interior of Africa on his map of 1749 by leaving large stretches blank or inserting texts that contrasted diverging traditional opinions. Information that was based on hearsay or collected by uneducated or specifically interested people (merchants, soldiers, or the early catholic missionaries were often regarded as such) no longer appeared sufficient. The *philosophical traveller*, well-read and instructed in recent scientific as well as historical theory and therefore competent to make observations and describe his experience in an appropriate form, made his appearance in Africa.

Thus the mid-eighteenth-century French traveller Adanson, though he was employed by a trading company, presented his trips from the coastal trading centers into the West African interior as opportunities for unbiased observation. The introduction to the English edition of his report did not even mention commercial interests, but emphasized his philosophical mind (Adanson 1795: vii, x). The primary objective of his travels was supposed to be the production of well-founded knowledge for his readers back in Europe. He was thus expected to contribute to the completion of the understanding of the world towards which the proponents of contemporary scientific systems, e. g. those of Linné's rules for plant classification, were working towards (Müller-Wille 2001). However, these "African travellers" could not yet position themselves in securely bounded scientific disciplines and rely on particular methodologies. Their era was rather one of trials and errors, one in which different epistemological standpoints co-existed and confronted each other. At this time the differentiation that was to mark the second half of the nineteenth century, of broad subjects like natural philosophy into specific disciplines with institutional and academic foundations, was by no means a clearly discernible goal.[3]

---

3   The peculiarities of this era of African travel is the subject of my PhD dissertation, in which nine different examples of European travelogues about the African interior will be taken as examples of common socio-cultural and individual strategies of processing experience into textual knowledge. Within the general framework of this period of non-linear development, I will try to point out features that are common to all of these texts, but I will also show that they differ according to the backgrounds of the author-travellers, to the regions they traveled in, and to the times of their voyages.

The textual productions of Antoine and Arnauld d'Abbadie stand between this era of experimentation in knowledge production and the formation of rather clearly delineated academic disciplines. They were struggling with the multi-faceted, *entangled* experience of travel,[4] not yet to be grasped conceptually as *fieldwork*, though practically such, and with the need to fit this into different, socially given frames of textual representation. Because they stayed in their field for about twelve years, during which time these socio-cultural molds for science and travel were entering a period of accelerating change, they had to adapt themselves and their texts to a transforming social and scientific environment at home. I will argue that their strategy was one of *disentangling* their work: they were fitting what they had seen and done into distinct frameworks as required by the formation of specialized scientific disciplines in the mid-nineteenth century.

This process might already have been foreshadowed in their youth and earliest occupations. It definitely began while they were in the field, as it started to shape their experiences and actions to some extent. Finally, it required textual adaption (of language and genre) afterwards. Disentangled disciplines and clarity were increasingly asked for, and the detailed descriptions of personal insecurity and interactive complexity that had been woven into travelogues with a scientific bent since the late Enlightenment had to cede to the trend. That the brothers held a fragile intermediate position between an earlier tradition and a new development in the perception of scientific study during their travels becomes evident in the stories of their lives. Therefore, any analysis of the d'Abbadie's travels and writings has to be embedded both in the background from which they set out and in the further development of their biographies.

Entangled Experiences

Arnauld and Antoine were the sons of Michel d'Abbadie, a native of Arrast in the Basque country, who had emigrated from France to Ireland during the French Revolution. He married the Irishwoman Eliza Thompson and had six children with her. Their oldest son was Antoine, born in Dublin in 1810; his younger brother Arnauld Michel followed in 1815. In 1820, Michel took his family back to France.

---

4   My use of this expression has been to some extent inspired by the methodological suggestion of "entangled histories" presented in Germany by Sebastian Conrad and Shalini Randeria (Conrad & Randeria 2002). The methods I use in this case study, however, do not strictly adhere to their concept, which is concerned with relations between different societies as a whole and seeks to go beyond the concepts of cultural transfer and traditional national comparison. My own approach is certainly shaped by an awareness of such a background, but seeks to point out individual, though socio-culturally embedded strategies of historical figures who were confronted with the experience of entangled strangeness and familiarity.

They first moved to Toulouse, where the brothers received their basic education, and in 1828, they settled down in Paris.⁵

Apparently, Antoine had been longing to be an explorer for some time: he devised a plan for an expedition into Africa in 1829 and organized his studies accordingly. Even though he was a law major, he heard lectures on astronomy, physics, geology, mineralogy, and zoology. While Arnauld went to the new French colony of Algeria as a soldier in 1833, Antoine compiled an Arabic-French dictionary. He was even said to have changed his lifestyle to prepare his body for a difficult journey in unknown lands: he exercised, practiced swimming, and lived on a diet consisting only of vegetables, eggs, and milk. In 1836, Antoine got his first chance to prove himself suitable as a scientific voyager on a trip to Brazil, when he was commissioned by the famous rival scientists and political antagonists Arago and Becquerel (Levitt 2009: 129-160) for the Académie Royale des Sciences. Antoine was asked to measure variations in terrestrial magnetism over a few months, and he accepted the offer. However, his main field of interest remained the African region known as Abyssinia, which had been so fascinatingly described in a travelogue by the late-eighteenth-century Scottish explorer James Bruce (Bruce 1790).

Antoine was determined to direct his efforts towards the African continent, and he would take his edition of Bruce's *Travels* – the second or third, by Alexander Murray, who had added a detailed biography of Bruce to the text – with him.⁶ In October 1837, he met his brother Arnauld in Cairo to begin a journey into the interior of northeastern Africa. From February 17, 1838, when they arrived at the port city of Massawa on the Red Sea, the two brothers were going to live and travel, both jointly and separately, in the area that was still usually known as Abyssinia to Europeans. However, they later preferred to call the empire they had visited Ethiopia, thus shunning the injurious Arabic term which the natives did not like to hear applied to their country (Abbadie 1980: 79). This striking change in the terminology the brothers used can be seen as an indication of the transformative power of their experience.

Ethiopia is usually linguistically explained as a term derived from the ancient Greek name for black Africans, referring to their supposedly sunburnt faces. Did the Abbadie brothers just take up older Eurocentric frameworks of description? No; rather, they pointed out that this was the term the inhabitants of the region preferred

---

5   A concise history of the Abbadie family is given by Tubiana, 1980.

6   In 1840, Antoine had listed different things that he was to take into Abyssinia in his journal, among them a "Life of Bruce" that would at times be referred to later: BNF, n. a. f. 21 300 (manuscript journal by Antoine d'Abbadie, probably copied from original notes; microfilm 8397), 4/5.

for themselves (Abbadie 1980: 80). The name of Ethiopia was a traditional indigenous expression for the Christian state, referring to Ityopis, a mythical ancestor from the oral traditions of the Axumite Empire (Munro-Hay & Nosnitsin 2007). So the *other* side of the entangled experience was not simply erased from the brothers' presentation of their *modern selves* after their return. The links between entangled experiences and disentangled disciplines in Antoine's and Arnauld's writings have apparently not been completely severed. In order to find out what these links were, one has to gain an overview of the chronology of those activities that can be established from the source materials. I will therefore give a sketch of the rather complicated succession of trips taken by the brothers in these years. Antoine himself provides a basis for this in two slightly differing versions of a *Récit abrégé* of their travels, one written in 1851 for a French minister, the other probably for the Geographical Society.[7]

Initially, their group included the Catholic cleric Father Sapeto and the Englishman Richards besides several European and native servants. They all left the course of the Nile on December 25, 1837 to cross the desert to the Red Sea. On its shores, they boarded a ship to Massawa, the usual port chosen by those who wanted to enter the Christian empire of the highlands. There, however, they split up: Arnauld, his English friend and Father Sapeto were to proceed to the territory of the Dedjazmatch Oubie in order to convince this powerful prince to give his assent to the brothers' travel plans.[8] Antoine was to stay in Massawa with the baggage and the rest of the group. The brothers joined up, but made their way separately to Gondar, the capital of the Christian empire, founded in the seventeenth century by the emperor Fasil (Abir 1975: 557).

Because Antoine found his scientific instruments inadequate for giving accurate measurements to map the country while on the road, the brothers decided to split up again. While Antoine returned to France in 1839 in order to procure better instruments, Arnauld was to travel to Saka, the capital of Innarya, in the south of the Ethiopian lands, where they supposed the source of the Nile was. He also intended to continue with the scientific observations. Arnauld himself admits that his simple plan to travel through this completely unknown country within a year, after which he was to meet Antoine in Massawa, was slightly naive (Abbadie 1980: 43-46). In fact, he first resided in Gondar a little longer, which enabled him to give lively descriptions of the daily life of the townspeople as well as of his interviews with the imperial elite in his publication (Abbadie 1980: 169-197). Arnauld then accompanied the son of

---

[7] BNF, n. a. f. 21 302 (papers of A[ntoine] d'Abbadie IV, "Géographie de l'Ethiopie"), 384-430.

[8] All Ethiopian titles and names will be given as they were spelled in Arnauld's publication except when a different transliteration is cited directly from manuscript sources.

Guoscho, the governor (or Dedjazmatch) of the province of Gojam, to his father's home. The French traveller got more and more involved in the life of Guoscho's camp followers and even accompanied the governor on military expeditions.

His life at Guoscho's court in Gojam was "une revelation" for Arnauld (Abbadie 1980: 383). Looking back on this time after his return to Europe, Arnauld described how his energetic intentions of exploration and geographic science were completely changed by the experience of companionship with the Ethiopian soldier-elite:

> The urbanity, the Christian spirit, and some antique and chivalric *je-ne-sais quoi* that reigned at his court had made me wish to know it better; I started to learn Amarigna [one of the Ethiopian languages], and the campaign I had just undertaken with the Gojamite army had made me determined to give a new direction to my studies and to postpone my voyage to Innarya to some later time. The geography of Gojam, of Damote, and of Agaw-Médir was still unknown, it's true; the information regarding that great water-course of Innarya was still to be verified, a question that had made such a strong impression on my brother; but, since his departure, time had gone by without giving me the chance to execute our program. I knew my brother would come back soon, and that he would – with a competence so much superior to mine – take up again the geographic works that I had interrupted so suddenly for our campaign in Liben. In all cases, the exceptional position that I owed to the goodness of the Dedjadj Guoscho made me hopeful, if I continued to live at his court, of being able to facilitate and render less perilous the explorations that my brother might try to undertake in the Galla's country in the event that his further investigations were to confirm his belief that the waters which moisten their country contributed to form the White Nile. The Dedjadj Guoscho held up friendly relations with the king of Innarya, and his influence extended to the intermediate Galla peoples. These considerations determined me to devote myself without reserve to the new life that I led in Gojam. To my initial indifference for the Christian populations of Ethiopia had succeeded this affectionate interest that it is necessary to feel in order to understand the people (Abbadie 1980: 383/384).

Arnauld felt that his life became entangled with those of the people he was living with. The scientific plans he had made with his brother no longer seemed as important as before once the traveller became increasingly absorbed in the joys and sorrows of native life. And Arnauld even put forward his conviction that the changes he had undergone enhanced his understanding of the Ethiopian Christians. But the Frenchman did not go completely native. He never allowed himself to get so absorbed as to forget the plan to meet his brother in Massawa.

The brothers met again, as they had planned, in February 1840. They proceeded to Adoa, where Antoine undertook some geographical surveying. Again the Dedjazmatch Oubie, still one of the most powerful men in the country, refused to allow them to travel in his territory. Arnauld decided to join Guoscho and his son Birro, who were said to have rebelled against the Ras, the highest officer of the empire. Antoine went back to Massawa in order to explore its surroundings but soon wounded his eye in a slightly embarrassing accident with his own gun. In order to find the help of "European medicine," he travelled to Aden and on to Cairo, where he was treated. When he had recovered, he wanted to enter Ethiopia again by the port of Berberah, where Arnauld joined him. They were convinced that their plan was counteracted by the British agent in Aden. The British also prevented the brothers' entering the region from Toudjourrah. Nevertheless, they were able to collect commercial and geographical information about both areas. Arnauld's published narrative of 1868 ended at this point. The volumes about the rest of their travels were never published.

Once more the brothers separated, this time to the east of the Red Sea, on the Arabian coast. While Arnauld was preparing another trip into the Ethiopian highlands at the port city of Jiddah, Antoine travelled to Hodeydah. At the conclusion of their individual travels in Arabia, the brothers went back to Massawa, which was to become their point of departure for the interior once more in August 1841. Arnauld went straight back to the highlands and Gojam, while Antoine first collected the vocabulary of the Saho, the nomadic inhabitants of the desert lowlands near the Red Sea. Afterwards the older brother set out for Gondar, where he stayed until the beginning of 1843. In the capital city he studied the ancient religious language of Abyssinia, Ge'ez, with indigenous professors. From his base in this town, he also made several excursions into the surrounding country, e. g. to Lake Tana and to the famous rock churches of Lalibela.

In the spring of 1843, Antoine moved on to join his brother in Gojam. Meanwhile, however, the political situation had become extremely precarious for Arnauld's protector Guoscho. Arnauld decided to stay with the prince, whom he had begun to regard as his feudal lord, as he expressed it in his dictation for further volumes of his publication: Guoscho was "mon maître"[9]. Thus Antoine was on his own when he undertook a trip to the southern countries of Innarya and Kafa

---

[9] BNF, n. a. f. 21 299 (Papers of Arnauld d'Abbadie, vol. 1: "Recit des Voyages en Ethiopie faits par Arnauld d'Abbadie (copie originale)"), 362.

in 1843. He had some information that led him to believe that the source of the White Nile might be found in these areas (see manuscript source quoted above). But neither the caravan of Oromo (or Galla as they were called by the indigenous Christians and the European travellers) traders he joined nor the rulers of the two countries he visited were willing to let a stranger gather too much local geographical knowledge. Antoine got entangled in the world of politics and superstitions at the courts of Innarya and Kafa. Only a threat from Arnauld, who was by now known as a favorite of the powerful Guoscho, could extricate him from his difficult situation. Antoine returned to Gojam in December 1843 without having reached the source of the Nile. But the brothers were not going to give up on this particular goal, which fascinated intellectual elites all over Europe and the tradition of which went back to expeditions by such important figures of ancient history as Alexander the Great and Nero (Romm 1992: 149-155).

Although the famous Scottish traveller James Bruce's claim to have reached the sources of the Nile in the province of Maitsha had been proven to be erroneous (he had only reached the source of the so-called Blue Nile),[10] the ancient tradition that these sources were to be found in the Ethiopian highlands still held its place among geographers. The Abbadies hoped to succeed where their predecessor Bruce had failed. Arnauld's political influence was waning as his protector Guoscho had been defeated and taken prisoner by his opponent, Ras Ali. The role of a geographical traveller must have gained a new attraction for the younger Abbadie in this situation. His brother Antoine returned to Gondar in December 1844 in order to obtain the money and presents that were necessary for a new trip to the south. He then travelled part of the way with the army of the Ras Ali, thus securing this powerful man's acquiescence for their trip, and arrived safely in Innarya.

But Arnauld was held up. The chiefs of the region had allegedly been brought to hate all "Whites" by two Englishmen, Plowden and Bell, who were also on their way to the sources of the Nile: they had killed two men and had – according to Antoine's reports – generally behaved in an improper and stupid way. Arnauld now had to avoid their tracks. This deviation meant that he only arrived in Inarya in December 1845. The brothers were sure that the Oromo natives of the region believed in a river-god, so they convinced them that they also wanted to pay their respect to the godly sources. Thus, Arnauld and Antoine were brought to what they thought to be the source of the White Nile on January 19, 1846. For their voyage back to Gojam,

---

10   Bruce himself admitted that the Blue Nile could never water Egypt if it were not for the constant stream of the White River (Bruce 1790, vol. 3: 666), but he persisted in his claim to have reached the sources of the Nile, as is evident from the title of his publication.

the brothers split up again. Arnauld reached Gojam in June; Antoine followed in December.

Meanwhile, the brothers' family, who had not had any news from them for some time, had become worried for their lives. Their youngest brother, Charles, arrived in Massawa in 1848 in order to enquire after them. He proceeded to the ancient city of Axum in the Ethiopian highlands and corresponded with his brothers. Antoine, troubled by ophthalmic complaints, left Ethiopia first, arriving back in France in May 1849. Arnauld stayed longer in Africa to finish researches about Abyssinian law and to nurse Charles, who was suffering from an illness, in Cairo. Finally, after Arnauld had bought some Arabic breeding horses, he also returned to Paris in October 1850.

Disentangling Disciplines
During their time in Ethiopia, the brothers – like all travellers in unfamiliar areas – had to deal with extremely complex situations. Particularly when they interacted with native people or attempted to make scientific measurements and observations, they were confronted with a host of uncontrollable influences. At first, this complexity must have appeared as an unwelcome obstacle. But eventually, experience began to change their attitude, as Arnauld stated in retrospect in his published account (see manuscript source quoted in note 8). As the travellers changed, new questions and new perspectives on the land and people they encountered arose. Arnauld's reflections on his own and his intentions' mutability might serve to reinforce Homi Bhabha's assertion of a "contramodernity" in the nineteenth century (Bhabha 1994: 173). They demonstrate that even in the era of the rise of imperialism, situations of contact brought up supposedly postmodern topics like "aporia, ambivalence, indeterminacy, the question of discursive closure, the threat to agency, the status of intentionality, the challenge to totalizing concepts, to name but a few" (Bhabha 1994: 173).

But Arnauld's statement also undermines a conviction Bhabha shares with many other postmodernist critics: their proposition that *hybridity* was characteristic of a colonial discourse, of stereotypical differentiation in situations of European dominance, implies that these were the only possible configurations of interaction (or at least those that mainly existed) in the age of European expansion since the fifteenth century (see, for example, Cohen 1980: xv). For a very general description of the origins of imperialist injustice and of long traditions of Eurocentric arrogance, this assumption has been morally important and politically helpful. For an analysis of the particular situation of pre-colonial travellers it is not. The French Abbadie brothers were not in a position of power in Ethiopia. They were frequently rather helpless, as their instruments and mental apparatus proved inadequate for the country. They had to conform to the wishes of indigenous rulers, and they both found themselves equally entangled in a web woven from both their own expectations and unexpected local agencies. They knew they were not in control of their situation. On their first trip to Gondar in 1838 they had been held up by a local lord, and this incident provided an opportunity for reflection on their goals and the means to achieve them:

> During our forced stay on the plain of Igr-Zabo, we had had much leisure for reflection; experience was already modifying our preconceived opinions; the initial chaos was beginning to calm down, and our voyage appeared to us under new aspects. From Massawa to Gondar, we had meticulously measured the country with the compass, but the magnetic attraction caused by the ferruginous nature of the soil introduced into this work insecurity, which voyagers would do well to think about more. […]; Antoine was to return to France in order to buy new instruments, and it was decided that during the absence of my brother I would at least go to Saka, the capital of Innarya; and in order to make better use of my voyage, I exercised myself under his direction in making the necessary astronomical observations for determining the position of any location, as well as the meteorological observations so they would be continued until his return (Abbadie 1980: 42-43).

Arnauld's report also indicates that he and his brother had at first maintained the idea of replacing each other in their knowledge-gathering activities. Arnauld even made an effort to acquire the necessary scientific skills to continue their joint project. Experience of the alien country and culture, scientific measurements, and the insecurities of interaction with the natives all formed a complex whole which both brothers had to deal with.

In the same vein, Antoine pointed out in his work on the geography of Ethiopia that any exclusive concentration on "the exact sciences" while studying a foreign country was insufficient. It had to be complemented by the study of man; a discipline Arnauld had declared his special area of interest in his publication.[11] Nevertheless, Antoine wrote:

> Moreover, in spite of the great attraction of the exact sciences, for which I have always had a passion, the perspective of visiting – only as a geographer or

---

11 Arnauld explained the different motivations of the members of their initial groups in his statement: "My brother was sustained by the love of science [including, as opposed to English usage, knowledge in general], the Lazarist priest by religious enthusiasm and I by the desire to study unknown peoples" (Abbadie 1980: 13).

naturalist – little or unknown areas, seemed less attractive to me than the study of languages, religions, political and legislative constitutions, and literature, which seemed to offer particularities worthy of interest of these southern regions [...] Since then I let myself be won over by the thought that the most elevated field of study man can turn his attention to is that of his kin (Abbadie 1873: i).

With this assertion of the importance of the study of mankind, Antoine tried to preserve a link between his intellectual position and the traditions of Late Enlightenment philosophy and science, in which anthropology dominated as a focal point of reference for all scientific efforts. In the late eighteenth century these had not yet been distinctly separated into cultural and natural sciences, and savants, not yet specialized professionals, often contributed to and were interested in extremely diverse topics within this broad field (Moravia 1973, Garber 1999: 138-140, Meyer 2003, Dietz & Nutz 2005). However, this entanglement no longer shaped the intellectual environment in which Antoine worked after his return to Europe.

The last two lengthy statements by Arnauld and Antoine, as quoted above, appeared in totally different kinds of specialized publications. Arnauld's quotation was taken from an adventurous travelogue, Antoine's from a scientific work of geography that provides a technical explanation of his surveying method and tries to prove its correctness in many tables with endless columns of numbers and complicated calculations. These different uses the brothers made of their experiences in the foreign country were not just the result of *ex post* reflection. Maybe they were due to personal preferences they had started to develop before their trip. During their voyage, the brothers were definitely taking up current European ideas of scientific specialization.

Thus, Antoine – while on his way back into Ethiopia in 1840 – was invited by a member of the Paris Société de Géographie to become a specialist on the Galla people, as other parts of northeastern Africa had already been described by other famous travellers.[12] Antoine's prospective work is put into a context of what might be called specialized *area studies* on particular African regions: "This is not to say: neglect Abyssinia, surely not; [...] But I would like to see your name become for the Gallas what the name of Bruce is for Abyssinia, that of Bowdich for the Ashanti, that of Clapperton for Sakkatou, &c."[13] Not only Antoine's expected publication, but

12 A Member of the Société de Géographie [D'Auzac?] to Antoine, Paris, December 1, 1840: BNF, n. a. f. 23 853 (Papiers d'Abbadie VI, Correspondance d'Antoine et Arnaud d'Abbadie sur leurs voyages en Abyssinie (Microfilm 25 528)), 75-76.

13 BNF, n. a. f. 23 853 (Papiers d'Abbadie VI, Correspondance d'Antoine et Arnaud d'Abbadie sur leurs voyages en Abyssinie (Microfilm 25 528)), 75v-76r.

also the works of previous famous African travellers were thereby fit into a model of specialist disciplines that did not exist when they were writing their publications. Likewise, several letters by Jomard,[14] the president of the Société de Géographie, prove that Antoine was keeping in touch with the scientific establishment at home. He was thus constantly reminded of the necessities of specialization that increasingly shaped scientific careers in Europe, even while he was faced with the complex realities of interaction during his field work.

The brothers were not insensitive to these developments. Whenever they were separated and lived or travelled individually among the Ethiopians, Antoine and Arnauld kept up a lively correspondence. In these letters, the brothers can be said to have started to disentangle separate fields of interest: while Antoine was back in Paris in 1839, for example, Arnauld, who had stayed behind in Gondar, realised that his prolonged experience of local life distinguished him from his brother. In a letter to Antoine, he gave a lengthy account of an episode of conflict with a member of the native elite and its final resolution, and then he concluded:

> I will tell you in detail this and several other episodes of the same kind, which are very instructive about the customs of the country. What I regret every day is that before departing for Europe, you didn't stay here for longer. This is a bad country if one does not know it, but it changes a lot during a stay. Not that it seems better, but nearly everything that seemed an obstacle to you is smoothed out. There is a delicate choice to make between the customs one adopts and one rejects. I have often been off the mark, but when I was successful, the impression I produced was always favorable for me.[15]

Even though a sort of Orientalist prejudice against the country still pervades Arnauld's language, he described to his brother the advantages of a longer stay in Ethiopia. Prolonged direct contact allowed him a more intimate experience of the land, its people, and local customs, and this experience – translated into appropriate behavior if possible – could remove many an obstacle for a traveller and observer. But was Arnauld still a traveller when he started to become more and more involved in local politics and warfare? Antoine himself sometimes seemed to doubt this. In a letter that was probably written in 1842, he urgently tried to persuade Arnauld to join him for some more travelling: "It is indispensable that you come as soon as

---

14   Jomard to Antoine, several letters from the years 1841-1848: BNF, n. a. f. 22 430 (Correspondance d'Antoine d'Abbadie I (A-L)), 268-281.

15   BNF, n. a. f. 23 848 (Papiers d'Abbadie I, Correspondance famille), 132r.

possible. I have pulled you into travels as you will pull me into politics."[16] Apparently, each of the brothers was carving out his personal field of activity.

These initial traces of disentanglement refer to the brothers' practice of lengthy separations and to their different individual lifestyles in Ethiopia. While Arnauld was fascinated by what Europeans regarded as the *feudal* way of life led by the Ethiopian warrior elite, Antoine fashioned himself as a travelling man of knowledge, similar to those learned strangers who had from antiquity been highly respected among the Ethiopians, the head of whose church, for example, was a cleric chosen by the bishop of Alexandria. This role could explain and legitimize his interest in churches and manuscripts to his native hosts, who often sought his advice or tested his knowledge.[17] However, his geographical measuring and his efforts at keeping aloof from political conflicts must have set him apart from native customs, in contrast with his brother. Antoine's role was probably closer to that of other contemporary European scientific travellers in Ethiopia, like the French cartographers Ferret and Galinier, the German Botanist Schimper, or Antoine's personal enemy, the British geographer Beke. Like all of these men, Antoine had to rely on the support of powerful natives, so Arnauld's elevated political position was favourable for both of them. Obviously, the Abbadie brothers divided up their fields of interest to a certain extent. In practice, however, they still had to co-operate for their joint goal of providing Europe with information about Abyssinia. Their different roles in Abyssinian life contributed to the entangled practice of exploration.

The persisting practical entanglement of their activities and interests is illustrated by the simple fact that they kept up their correspondence whenever they separated. Moreover, they established a sort of base camp in the capital city of Gondar, where their Basque servant Domingo Lorda maintained their house and interests until his death in January 1845. He received frequent written instructions by both brothers. His own letters, written not only in his sometimes shaky French, but also in the Basque and Ethiopian languages,[18] described his activities for the travellers: he dealt with officials and conducted business for the brothers. In particular, he acted as a commercial agent for the acquisition and transport of their collections of natural

---

16 Antoine to Arnauld, Gondar, Thursday, October 27, [1842], BNF, n. a. f. 23 848 (Papiers d'Abbadie I, Correspondance famille), 40r.

17 For example, Antoine was tested as a knowledgeable man and felt he passed during his stay in Ras Ali's camp. Another time, he felt deeply touched by some Ethiopian monks, who asked him for religious instruction. BNF, n. a. f. 21 300 (manuscript journal by Antoine d'Abbadie, probably copied from original notes; microfilm 8397), 129-130, 144.

18 For letters by Domingo to Antoine and Arnauld, see BNF, n. a. f. 22 430 (Correspondance d'Antoine d'Abbadie I (A-L)), 368-435.

historical specimens, art, and manuscripts. He frequently stressed his emotional attachment to his employers, but he was noticeably proud of his own success in finding his way in this strange environment and at familiarizing himself with Ethiopian customs and lifestyles. The brothers often had conflicts with this self-confident servant, but their recurring complaints to each other about his lack of loyalty only give further proof of the importance of his role.[19] In spite of all their troubles, they kept him as cultural and commercial mediator in their joint service. Domingo's activities tied one highly important knot in the entangled network of the Abbadies' interactions during their work in the field.

In the brothers' subsequent specialized publications, however, the traces of this practical entanglement were erased to considerable extent. The pivotal role played by Domingo Lorda appears nowhere; he is hardly ever mentioned. Even in Arnauld's detailed narrative account, he appears as one of many servants whose activities are chiefly determined by their masters' will (e.g. Abbadie 1980: 1, 4/5, 25, 220, 476, 542-543). It is only in the originally unpublished parts of Arnauld's text that a longer passage on Domingo appears: it refers to his death and the lengthy conflict Arnauld has with the Ethiopian queen-mother about his servant's inheritance.[20] Arnauld, the ethnographic adventurer, is the main actor of his narrative. This was not simply the result of personal vanity. Antoine's scientific works never mention Domingo Lorda either. The disentangling of disciplines demanded clarity; the entangled complexities of fieldwork had to give way in narratives, as well as in the developing self-perceptions of the fieldworker-authors, to the distinct agendas of individual actors. The different lifestyles and interests that had been suggested by the brothers' lives up to 1837, and which were becoming increasingly marked during the last years of their voyage, were a first sign of this new tendency. This basic differentiation, already socially sanctioned for their European personalities, could develop further according to the

---

19 Arnauld expressed his ambiguous views on Domingo, who had stayed with him while Antoine had travelled back to France in 1838, in a letter to Antoine written from Gondar, as early as February 3, 1839: "Domingo has been a zealous majordomo to me, the only fault I can find with him is a tendency to act up; this derives from the attention he is the object of in his capacity as a white. I eat alone; he sits at the table of the domestics. He speaks tolerable Amharic, reads a little and is learning to write; by the way, he is aware of the resources of Gondar, and he might help you by pointing out some usages. While bad at the beginning, his health has been re-establishing for some time, he will, I believe, be very useful to you here" (BNF, n. a. f. 23 848 (Papiers d'Abbadie I, Correspondance famille), 133v.). Domingo's growing pretensions as a "white man" endangered his submission as a servant, but his cultural knowledge was probably very useful to Antoine on his return.

20 BNF, n. a. f. 21 299 (Papiers d'A. d'Abbadie, vol. 1: Récit des Voyages en Ethiopie faits par Arnauld d'Abbadie (copie originale)), 402-437.

general trend of scientific culture and its application in travel writing. Differentiation marked the brothers' textual self-presentations as well as their later lives.

After their return to France, Antoine established himself as a famous scientist. He corresponded with the scientific elite throughout Europe. He published several specialized works on the geography and linguistics of Ethiopia (Abbadie 1859, 1866, 1873), and he led a heated debate about the location of the sources of the Nile with his favorite opponent from his travels in Abyssinia, Beke (for an overview over this debate, see Beke 1851). He had already received the gold medal of the Paris Société de Géographie in 1839. After his return, he was president of this society in both 1869 and 1883. He was founding president of the Société de Linguistique in 1864 and 1865. In 1852 he became a corresponding member of the prestigious Académie des Sciences and was elected to full membership in 1867. In 1878, he was nominated member of the Bureau des Longitudes. As an honored scientist, he participated in the Brussels anti-slavery conference of 1889/90 and was a member of a number of foreign scientific societies (Abbadie 1980: vi).

Arnauld, in contrast, returned to Abyssinia in 1852, but internal strife in the country made it too dangerous for him to remain there.[21] He returned to France again and dictated a lively and colorful account of his travels to a secretary for publication in 1868. He thereby disentangled his experience to a certain extent from the primarily scientific interests of his brother: he apparently tried to answer a new popular demand for entertaining books of adventure (Allen 1976: 138) and for information about Ethiopia, where the British general Napier was fighting against the new emperor Theodoros at the time (Abebe 1998: 99-101). The roles of adventurer-voyager and scientist-voyager were disentangled and separated between the brothers. Such a clarification of their roles enabled individual travellers to integrate themselves into differentiated European fields of knowledge. Thus, they were ultimately able to sell their different forms of experiential knowledge in specialized, disentangled portions, even though the original entanglement occasionally shone through.

Conclusion

The gradual loss of the universalistic ethos of Enlightenment travel went hand in hand with feelings of definite European superiority and, ultimately, imperialist ambitions. Affection for the object of one's interest – as Arnauld claimed to justify his authority on the country and its inhabitants (Abbadie 1980: 384) – was no longer regarded as

---

21 Arnauld's difficult position after the death of his protector Guoscho was described to Antoine by a correspondent in Cairo: Etienne[?]Barthélemy to Antoine, Cairo, April 17, 1853 (BNF, n. a. f. 23 850 (Papiers d'Abbadie III, Lettres adressés aux membres de la famille d'Abbadie, Anstey-Vissefier, XVIIIe – XIXe siècles), 29-31.)

an acceptable category of science. Similarly, scientific accounts of particular aspects of a foreign country no longer made very fascinating reading for a broad audience, as the travel accounts of the late eighteenth and early nineteenth century had – in spite of their learned pretensions. Antoine and Arnauld had disentangled the figures of the travelling adventurer and scientist, but both still referred to a partly shared experience, to a joint voyage that was still connected to older traditions of travel in the African interior. At the same time, their efforts to disentangle these shared experiences, to make them available to modern scientific disciplines and popular writing, point to new tendencies that were to transform travel writing in the era of high imperialism.

In a short summary of their travels written for a French administrator in 1851, the shadow of a third figure appeared: It was that of the nationalist-voyager, whose importance was to increase immensely in the second half of the century. His travelogues would take up the adventurous dimension of travel and travel-writing, but they would rather deny complexity, replacing it with disentangled power and the conviction of national and racial superiority. Historical examples of this figure are well known as explorers: the former missionary David Livingstone travelled as an agent of the British Government on his second and third trips. His public image was mainly formed by the journalist Henry Morton Stanley, who far surpassed his model when he violently "opened up" the Congo basin for the Belgian king and ultimately European imperialistic activity in general. Antoine and Arnauld seemed to foreshadow this development when they worked on the following terms to explain their motives: "We were of that sort of young people [changed to 'men'] of which France has so many, who want to dedicate their fortune and energy to extend [added in pencil: "our relations with the exterior at the same time as"] the vast horizon of our [changed to "the"] sciences."[22] Being manly and extending the state's foreign relations now became more important. Science might still have been a legitimate motive, but it was no longer worthy of the first person plural possessive pronoun *our*. Science ultimately had to move to the background of national interest in exploration, as disentangled disciplines took over from the entangled experience and insecurity of earlier travelling.

Arnauld's effort to preserve traces of the connection between adventure, cultural adaptation, and ethnographic science in his publication was not completed until the new edition of his work was finished in 1999. It is significant that his publication had been forgotten until new developments within the discipline of anthropology or ethnology re-established the reflection on the complex connections between

---

22   BNF, n. a. f. 21 302 (Papiers d'A. d'Abbadie IV, "Géographie de l'Ethiopie"), 384.

experience and writing, between constructing the self and the other, between convincing science and good literature. From this new standpoint, the in-between position of Antoine and Arnauld d'Abbadie during their time in the field and afterwards becomes a fascinating example of the crossing of different chronologies in the particular time and space of mid-nineteenth-century travel.

In the end, however, they might not only be interesting as an exemplary case study. Arnauld stated that during his first stay in Gondar, instructed by his daily experience and a venerable native mentor, his attitude towards the Ethiopians changed: they no longer appeared to him as typical examples of barbarian or biblical manners, but as a fascinating people among whom he could meet interesting individuals if he learned their language and adopted their dress as well as their intricate customs (Abbadie 1980: 173). Likewise, Antoine and Arnauld might well stand for different socio-cultural tendencies among their contemporaries, but their individual approach to Ethiopia and the people among whom they travelled and lived there also makes them a deeply fascinating object of study.

References

D'Abbadie, Ant. 1859. *Catalogue raisonné de manuscrits éthiopiens appartenant à Antoine d'Abbadie*. Paris: Imprimerie Impériale.

D'Abbadie, Ant. 1866. *L'Arabie, ses habitants, leur état social et religieux, à propos de la relation du voyage de M. Palgrave*. Paris: Challamel.

D'Abbadie, Ant. 1873. *Géodésie d'Éthiopie ou Triangulation d'une partie de la Haute-Éthiopie, exécutée selon des méthodes nouvelles*. Paris: Gauthier-Villars 1873.

D'Abbadie, Ar. 1980. *Douze ans de séjour dans la Haute-Éthiopie (Abyssinie)*. Vol. 1 (originally Paris: Hachette 1868). Edited by J.-M. Allier. Studi e testi 286. Vatican City: Biblioteca Apostolina Vaticana.

D'Abbadie, Ar. 1999. *Douze ans de séjour dans la Haute-Éthiopie (Abyssinie)*. Vol. 4. Edited by J.-M. Allier. Studi e testi 286. Vatican City: Biblioteca Apostolina Vaticana.

Abebe, B. 1998. *Histoire de l'Ethiopie, d'Axoum à la révolution (c. IIIe siècle avant notre ère – 1974)*. Addis Abeba: Center Français des Études Éthiopiennes, Paris: Maisonneuve & Larose.

Abir, M. 1975. "Ethiopia and the Horn of Africa." In: R. Gray (ed.) *The Cambridge History of Africa. Volume 4: From c. 1600 to c. 1790*. Cambridge: Cambridge University Press, 537-577.

Adanson, M. 1759. *A Voyage to Senegal, the Isle of Goree, and the River Gambia*. London: Nourse & Johnston.

Allen, D. E. 1976. *The Naturalist in Britain. A Social History*. London: Penguin Books.

Beke, C. T. 1851. *An Enquiry into M. Antoine d'Abbadie's Journey to Kaffa, in the Years 1843 and 1844, to Discover the Source of the Nile*. 2nd edition. London: James Madden.

Bhabha, H. K. 1994. *The Location of Culture*. London and New York: Routledge.

Bruce, J. 1790. *Travels to Discover the Source of the Nile, in the Years 1768, 1769, 1770, 1771, 1772, and 1773*. 5 vols. Edinburgh: Ruthven, Robinson & Robinson.

Clifford, J. 1986a. "Introduction: Partial Truths." In: J. Clifford & G. E. Marcus (eds.) *Writing Culture. The Poetics and Politics of Ethnography*. Berkeley: University of California Press, 1-26.

Clifford, J. 1986b. "On Ethnographic Allegory." In: J. Clifford & G. E. Marcus (eds.) *Writing Culture. The Poetics and Politics of Ethnography*. Berkeley: University of California Press, 98-121.

Cohen, W. B. 1980. *The French Encounter with Africans. White Response to Blacks, 1530-1880*. Bloomington and London: Indiana University Press.

Conrad, S. & S. Randeria 2002. "Einleitung. Geteilte Geschichten – Europa in einer postkolonialen Welt." In: S. Conrad & S. Randeria (eds.) *Jenseits des Eurozentrismus. Postkoloniale Perspektiven in den Geschichts- und Kulturwissenschaften.* Frankfurt: Campus Verlag, 9-49.

Dietz, B. & T. Nutz 2005. "Naturgeschichte des Menschen als Wissensformation des späten 18. Jahrhunderts. Orte, Objekte, Verfahren." *Zeitschrift für historische Forschung* (1): 45-70.

Garber, J. 1999. "Selbstreferenz und Objektivität: Organisationsmodelle von Menschheits- und Weltgeschichte in der deutschen Spätaufklärung." In: H. E. Bödeker (ed.) *Wissenschaft als kulturelle Praxis, 1750-1900.* Göttingen: Vandenhoeck & Ruprecht, 137-185

Geertz, C. 1990. *Die künstlichen Wilden. Anthropologen als Schriftsteller.* Munich and Vienna: Hanser.

Latour, B. 1991. *Nous n'avons jamais été modernes. Essai d'anthropologie symétrique.* Paris: La Découverte.

Latour, B. 1999. *Pandora's Hope. Essays on the Reality of Science Studies.* Cambridge, MA, and London: Harvard University Press.

Levitt, T. 2009. *The Shadow of Enlightenment. Optical and Political Transparency in France, 1789-1848.* Oxford: Oxford University Press.

Meyer, A. 2003. "The experience of human diversity and the search for unity: Concepts of mankind in the late Enlightenment." *Cromohs,* 8: 1-15.

Moravia, S. 1973. *Beobachtende Vernunft. La scienza dell'uomo nel Settecento.* German edition. Munich: Hanser.

Müller-Wille, S. 2001. "Carl von Linnés Herbarschrank: Zur epistemischen Funktion eines Sammlungsmöbels." In: A. te Heesen & E. C. Spary (eds.) *Sammeln als Wissen. Das Sammeln und seine wissenschaftsgeschichtliche Bedeutung.* Göttingen: Wallstein, 22-38.

Munro-Hay, S. & D. Nosnitsin 2007. "Article 'Ityiopis'." In: S. Uhlig (ed.) *Encyclopaedia Aethiopica.* Vol. 3 (He-N). Wiesbaden: Harrassowitz Verlag, 245-246.

Tubiana, J. 1980. "Préface." In: Ar. D'Abbadie 1999. *Douze ans de séjour dans la Haute-Éthiopie (Abyssinie).* Vol. 4. Edited by Jeanne-Marie Allier. Studi e testi 286. Vatican City: Biblioteca Apostolina Vaticana, v-xx.

Romm, J. S. 1992. *The Edges of the Earth in Ancient Thought. Geography, Exploration, and Fiction.* Princeton, NJ: Princeton University Press.

Tyler, S. A. 1986. "Post-Modern Ethnography: From Document of the Occult to Occult Document." In: J. Clifford & G. E. Marcus (eds.) *Writing Culture. The Poetics and Politics of Ethnography.* Berkeley: University of California Press, 122-140.

Casper Andersen

Explorer-Engineers Take the Field
Imperial Engineers, Africa, and the Late Victorian Public

From the last quarter of the nineteenth century, tropical Africa became a new field of operation for British engineers. Over the ensuing decades this professional group played a significant role in disturbing the political status quo as they equipped the expanding British possessions in Africa with extensive infrastructural systems geared for the machinery of economic and political imperialism (Headrick 1981, Kubicek 1999). In doing so, engineers left a noticeable stamp on the colonial sphere. Influences, however, also went in the other direction – from the imperial fields to the metropolitan scene, where the increasing importance of the empire had a profound impact on developments in the British engineering profession (Thompson 2007: 22-26).

These feedback processes were particularly strongly felt within the public domain. Imperial projects gave engineers new opportunities to approach the British public by associating their work with the vibrant late Victorian cultures of empire and exploration and this in turn brought about changes in public perceptions of what it meant to be a British engineer.

This chapter will examine this issue by analyzing the public activities of three influential imperial engineers. Elements of self-representation are involved here and as we shall see engineers – like men of science – were adept at fashioning their public images in order to serve their own needs and ends (Biagioli 1993, Marsden & Smith 2007: 242-243). Specifically, this article argues that the way British engineers presented themselves to the public differed greatly according to the standing of the engineers within the complex social and professional hierarchy that was in place by the time British engineers began operating in Africa – a hierarchy consisting of assistant engineers and chief engineers, with independent consulting engineers at the top (Griffith & Porter: 1988). The position of individual engineers within this professional hierarchy profoundly influenced the motives as well as the means they had for engaging with the public. As we shall see, some engineers presented themselves as "explorer-engineers" in the African field, while others approached the public as gentleman experts on colonial development and policy making. In this process a number of enduring images emerged of the engineers as well as of the African fields they claimed to be taming.

Importantly, a number of cultural images of engineers had already developed throughout the nineteenth century, making the "heroic engineer" a well-established icon in Victorian culture (Chew & Wilson 1994). In particular, the inventors and engineers whose names were associated with the transformation of Britain into the leading manufacturing nation and "workshop of the world" became the subjects of numerous portraits and eulogies, as for example in Samuel Smiles' immensely popular books on the lives of eminent engineers published in the 1850s and 1860s (Smiles 1997). These *Smilesian* engineering heroes were portrayed as self-reliant, hardworking, and individualistic men of humble background who – unspoiled by formal, theoretical education and book learning – had a developed talent for devising practical solutions. Challenging and overcoming physical barriers in the British Isles as well as the social barriers that stood in their way, the engineers of the Smilesian mold were accredited with having produced faster means of transport, cheaper goods, employment, and better living conditions for their fellow countrymen. The heroic engineers triumphed through their stamina, high moral integrity, noble ideals, and solid character and were described as the embodiment of the massively popular Victorian philosophy of self-help (Jarvis 1997).

Traditionally, historians have argued that the popularity of the engineering heroes reached its zenith in the mid-Victorian period – the Great Exhibition in London 1851 marking a convenient watershed – and from then on faded in tune with what in an influential historiography has been labeled "the decline of the industrial spirit" (Rolt 1970, Wiener 1981). More recently, several historians have challenged this view and demonstrated that inventors and engineers were the subject of complex cultural interpretations and appropriations that do not fit dichotomous notions of ascent and decline (Cannadine 2004, Macleod 2007). Most forcefully, Christine Macleod, in a study impressive in both scope and detail, has shown that engineers and inventors were celebrated far beyond the pages of Samuel Smiles and that they remained important, if ambiguous, cultural icons even after 1860. Macleod also demonstrates that towards the end of the century images of inventors and engineers toiling in their workshops or in the British countryside had trouble competing for public attention with other iconic figures, most notably explorers and soldiers. She notes, however, that "although engineers active after 1860 failed to capture the imagination of contemporary biographers or historians, they did enjoy some imperial limelight" (Macleod 2007: 378). Building on this revisionist historiography, this article argues not only that this assertion holds true, but also that it was a much more widespread and important phenomenon in late Victorian culture than has hitherto been recognized. Indeed, as we shall see, it is justifiable to speak of the emergence of a distinct cultural persona combining the characteristics of the Smilesian engineer with those of the explorer.

The Explorer-Engineer

With the possible exception of soldiers, nobody in the second half of the nineteenth century could compete for public attention and admiration with the explorers (Mackenzie 1994, 2000, Riffenburgh 1994). Yet the category of the explorer was ambiguous. In the second half of the nineteenth century the cultural icon of "the explorer" underwent constant negotiations and appropriations and this process has continued throughout the twentieth century (Driver 2001: 199-219). Several groups associated themselves with exploration and, in particular, laid claims to following in the footsteps of Victorian icons such as David Livingstone and the more controversial figure Henry Morton Stanley (Hopkins 2008). Field scientists, doctors, travel writers, tourists, and others would – often in subtle and ambiguous ways – relate their activities to those of the explorers who had "trotted the field" before them.

Engineers constituted another group laying claim to being the heirs of the explorers. Central to this view was the idea that engineers were representative of the next phase: where explorers had taken control of the frontier by mapping it, engineers were now bending it to commercial exploitation and political subjugation by establishing infrastructure. The fact that engineers from the third quarter of the nineteenth century were hard on the heels of explorers and mapmakers added much to the persuasiveness of this idea; settling railway trajectories or surveying potential mining areas often meant travelling along routes and through regions where few (white) people had gone before. Thus, there were significant overlaps in the field activities of engineers and the famed explorers – and also accompanying ideological overlaps. Livingstone's credo of Civilization, Christianity, and Commerce was conceived to be open to scientists, soldiers, merchants, and indeed engineers who could describe themselves as taking part in this mission by emphasizing a fourth C of civil engineering: in this view infrastructure would create political stability, strengthen the position of missionaries, encourage investment and commerce, and thus allow the African continent with its allegedly backward inhabitants to take its place as a supplier of raw produce in a world economy based in London. The practical and ideological overlaps became even more obvious in the more violent phase of exploration which, as Felix Driver has convincingly argued, gradually became dominant in the last quarter of the century. It is worth recalling that Stanley, the controversial "geography militant" icon of this period, earned his nickname *Bula Matari* – the breaker of rocks – from blasting away the rocks blocking the road that he had been employed by King Leopold to construct as a forerunner of the railway connecting the Atlantic Coast with the navigable sections of the Congo River (Stanley 1885: 182-185). Stanley was an explorer who also took on the role of the road builder. Engineers made the same journey but from the opposite direction to Stanley. By 1880 the ideological as well as the practical distance between path finder and railway builder was shorter and more blurred than ever: these were ideal conditions for explorer-engineers to flourish.

One consequence of this was the creation in British culture of an identifiable image of the engineer in Africa, an image that was fed through a number of channels and by a number of agents: by observant editors and publishers of popular magazines and technical journals (Andersen forthcoming), or by prolific authors such as Rudyard Kipling and the young Winston Churchill who expressed a widespread sentiment when he claimed that:

> Civilization must be armed with machinery if she is to subdue these wild regions [of East Africa] to her authority. Iron road, not jogging porters; tireless engines, not weary men; cheap power, not cheap labor; steam and skill; not sweat and fumbling: there lies the only way to tame the jungle – more jungles than one (Churchill 1908: 51).

Even more important for the dissemination of visions of engineering in Africa were journalistic writers such as Frederick A. Talbot, whose handsomely illustrated and popular book series on *Railway Wonders of the World* and *The Railway Conquest of the World* in chapters on "the romance of construction" and "the adventurous life of the railway surveyor" pictured engineers as conquerors of the globe, annihilators of space and time, civilizers of "savage natives" – as "the advance guard of civilization" (Talbot 1911: 2, 9-10).

Little is known about the scale and changing faces of the phenomenon of the explorer-engineer, which occupied a place at the intersections of engineering and empire over a period spanning at least fifty years. However, as a cultural icon it was strong enough to constitute an essential cultural backcloth against which engineers could present themselves and their profession in the public spheres. Indeed, engineers actively and increasingly associated themselves with the vibrant Victorian and Edwardian cultures of empire and exploration in a number of ways. Autobiographies were one channel they used. The first example was probably John Hawkshaw's *Some Reminiscences from South America* of 1838, which described his experiences as a young mining engineer in Bolivia (Hawkshaw 1838). Hawkshaw's book predated the era of the "new imperialism" but it was a genre that remained popular for over a century which often commented on the role and character of the engineer venturing into the imperial field. In his autobiography, *The Adventures of a Civil Engineer. Fifty Years on Five Continents*, published in London in 1909, Charles O. Burge provided a typical character sketch of the frontier-going British engineer:

> Here [Decca Province in India], I made my first acquaintance with the typical constructing engineer abroad, and after meeting him since in many other lands, I may say that he is *sui generic* among his craft. He differs in many respects from the type I had left at home. Sunburnt, bearded, with the pipe ever in his mouth, a daring rider, full of energy, exhaustless in resource when difficulties arise, hospitable to the last degree, and full of queer anecdote, he has little tolerance for fussy namby-pambyism in his superiors or his comrades, and expects his men to work as hard as himself, with due allowance for the various disabilities of the races with which he has to deal (Burge 1909: 86).

According to Burge the engineer of the imperial field was resourceful and accustomed to dealing with tropical perils and people of different races – *sui generic*, in particular through his stamina and independence when faced with these challenges.

Large-scale engineering projects in the tropics were collective efforts involving hundreds and occasionally thousands of people, but the writings highlighted the contribution of the "lone heroic engineer." Scholars who have analyzed the writings of Victorian explorers have emphasized that the collective nature of exploration was usually downplayed and in particular that the often crucial contributions of indigenous people almost disappeared from the published narratives (Pratt 1992, Kennedy 2007). This was also the case in the writings of explorer-engineers; when they triumphed they usually did so in spite of their workforce rather than in collaboration with it. Indeed, the engineers' writings generally reflected the racial and cultural prejudice of a time in which indigenous peoples in Africa and elsewhere were viewed as very inferior in particular with regard to their abilities in handling "Western" technologies (Adas 1989).

This aspect featured in a range of ways. For example Percy Girouard, a military engineer, colonial administrator, and prolific writer who first made his name as "Kitchener's railway man" during the Anglo-Sudanese war (1896-1898), expressed in his autobiography very little patience with the Egyptian and Sudanese workforce he felt he had to make do with:

> Never had a motlier crew been brought together in the guise of a Railway Battalion […]. If the best theodolite goes smash upon the ground, if the locomotive runs without a gauge glass and burns a fire box, if the precious belongings are lost and the white mans' Western ire is raised – "Why oh why Excellency be perturbed? Ten thousand years ago it was ordained I should do these things this day" […] The Egyptian fellahin, his nearest prototype in fatalism is the Chinese, his modern prototype the Russian moujik […][1]

For Girouard there was a clear link between the inability of the Egyptians to appreciate even the simplest "Western" tools and the invariable superstition and fatalism they suffered from. Frederic Shelford, a consulting engineer to whom we shall return shortly, expressed a similar view but also warned a "pioneering engineer" not to

> imagine a native carrier is a mere beast of burden, without soul or feeling. He is after all, a human being, and there is a distinct difference between the services rendered by a well-treated native and an ill-treated one. It is easy, by gentle but firm dealing, to obtain the affection of one's carriers (Shelford 1908).

While it is important not to overlook the fact that Victorian engineers also expressed strong opinions about the inferiority of the white unskilled workers employed in domestic construction work – "the notoriously drunk Irish navvy" – there is no question that the perceived lack of ability of "the natives" added to the drama of construction projects and by implication to the achievements of the explorer-engineer.

A connected and equally important feature is that the writings of engineers *sensationalized* the surrounding environments in which the construction projects were set. *Sensationalizing* is a "tag" introduced by the historian Anne Crozier in a study of the writings of colonial doctors in Africa to denote that the characteristics of the tropical field were exaggerated with the purpose of creating an ideological effect (Crozier 2007, Driver & Martins 2007). Sensationalizing the African field by highlighting its abnormality, "tropicality," and danger was a way for engineers to stress individual heroism and skill, and it enabled them to operate above usual social, moral, and professional roles. The African field was one of wild animals, deadly diseases, and hostile natives that only the *sui generic* explorer-engineer was fit to

---

[1] Girouard, Percy [Undated Manuscript]: *Autobiography*. Rhodes House Archive, MSS. Afr. S. (subsequently RHA-MSS-AFR) 1865, 16-17.

handle. The sensationalizing of the environment was another feature that set apart the explorer-engineers from earlier "heroic engineers," whose achievements were set in a domestic context.

John Henry Patterson – The Assistant Resident Engineer
Several of these aspects come together in the case of John Henry Patterson, an Anglo-Irish military engineer who won fame as the slayer of *The Man-Eaters of Tsavo*. In 1898-1899 Patterson was employed as assistant resident engineer on the Uganda Railway constructed by the British government between Mombasa and Lake Victoria for military strategic purposes. Against normal procedures this railway project was funded directly by the British exchequer, and with no clear economic incentive to back the strategic rationale the Uganda railway was by far the most controversial and publicly debated railway project instigated during the Scramble for Africa (Hill 1957). The story of Patterson and "the man-eating lions" of Tsavo is well-known. From March to December 1898 two male lions attacked and killed twenty-seven Indian workers and an unknown number of Africans who were constructing a railway bridge by the river Tsavo. Following an intense and hazardous hunt the lions were eventually shot and killed by Patterson (Patterson, 1907, Miller 1977: 242-299).

Patterson had first published the story of the man-eaters in early 1899 in two short articles in the popular journal *The Field* and in 1907 he expanded his story of the lion hunt into a book entitled *The Man-Eaters of Tsavo and other East African Adventures*. The book is an illustrative example of how railway construction was placed in the context of sensationalized Africa. The stories of game hunting and close encounters with lions, of "superstitious Indian coolies," and of "lazy natives" run parallel with descriptions of plate laying and bridge construction. The narrative was supported by the book's many illustrations of railway work placed alongside trophy shoots of lions, rhinoceros, and dancing "natives." According to the book's 1907 preface the enthusiastic big game hunter Theodore Roosevelt considered "the incident of the Uganda man-eating lions [...] the most remarkable account of which we have any record" (Patterson 1907: p. xii), and one of its English reviewers claimed that "in all books of adventure written since the days of Herodotus it would be hard to parallel such a story."[2] Patterson's book was an immediate success and has remained so ever since, having gone through at least 12 editions by 2009.

It deserves to be put on record that lions had no significant impact on the construction of the Uganda Railway (or on any other railway in Africa for that matter – though lions were of course a serious matter for the individual workers that

---

2   "Press cuttings." Whitehouse Papers, RHA-MSS-AFR, 1046 (11).

ran the risk of attack). The works on the Uganda Railway that were disturbed by the lions during two short intervals in 1898 were carried out to construct a permanent bridge at Tsavo in replacement of a temporary bridge that already existed. The main workforce was occupied at the railhead, which at the time was pushed forward from a point forty miles further inland (Hill 1957: 131-142). However, while the Tsavo lions had minimal impact on the ground in East Africa, they certainly conquered the British public, where the story of the two lions gained almost instant fame. Over the following years Patterson and his lions were subjects of articles in youth magazines, travel literature, and newspapers, and have since appeared in novels, tourist brochures, and in three separate motion pictures including the 1996 Hollywood blockbuster *The Ghost and the Darkness*.

Very little is known of Patterson's life prior to his employment on the Uganda Railway, by which time he was in his early thirties. Purportedly the son of an Irish clergyman, Patterson had been trained as a military engineer in India where he had also married a young teacher in 1895. It was from India that he was transferred to East Africa. The diary he kept while stationed in Africa reveals that the salary he made as assistant resident engineer on the railway enabled him to send home £20 a month to his wife and newborn child.[3] For Patterson, however, his encounter with the Tsavo lions in 1898 led to dramatic changes in his engineering life and career. The fame he won as the slayer of the lions proved to be his entry ticket into the high society whirl. Upon his return to Britain (after soldiering in the South African War 1900-1901), the military engineer had become known as one of the great game hunters of his day and enjoyed the company and friendship of notables such as the actor Sir Henry Irving and the daughter of the explorer Sir Samuel Baker. A few years later Patterson published his full-length, sensational bestseller on his encounters with the Tsavo lions and his other East African adventures. This was undoubtedly a much welcomed financial opportunity for a royal engineer approaching middle age with a family to provide for. The year the book was published the British government appointed Patterson to the high position of *Chief Game Warden* in the East African Protectorate. He later returned to soldiering and became a leading and controversial figure in the Zionist Movement (Brian 2008).

George Whitehouse – The Chief Engineer

Associating their work with sensationalized African fields opened new opportunities for engineers in the public realm. However, not all engineers actively seized upon these opportunities. In this respect George Whitehouse, who as Chief Engineer to

---

[3] Diary entry March 4, 1898 in J. H. Patterson, "Diary 1898-1899." RHA-MSS-AFR, r. 93.

the Uganda Railway was Patterson's boss in East Africa, provides a telling contrast to his subordinate's public vigor. Whitehouse displayed very little interest in making use of the opportunities for public exposure that he undoubtedly had as chief engineer to the controversial Uganda Railway during the seven years it took to construct. Whitehouse did not publish any books or articles on his East African experiences and he never lectured in the learned associations of which he was a member such as *The Society of Arts*, *The Royal Geographical Society*, and the *Institution of Civil Engineers*, in spite of the fact that he went to their meetings when he was in London.[4] The few interviews he gave to agents from Reuter's were sobering in comparison with contemporary statements about the Uganda Railway.[5] Whitehouse appears not to have had a taste for public exposure and as an employee in the government's service with an annual salary of £1,200 he may also have lacked sufficient incentive for seeking it (North 2005: xv).

Others, however, were more than willing to do this for him. In 1903 *The Graphic* – at the time the most successful direct competitor to *Illustrated London News* – attributed the entire railway to the chief engineer when the popular journal carried a short biographical sketch of Whitehouse under the headline "The Constructor of the Uganda Railway."[6] *The Evening News* explained that Whitehouse had been chosen for the post as Engineer-In-Chief of that important undertaking largely on account of his wide experience in railway construction:

> He had already built lines in Peru, Mexico, India and Natal, but when he reached Mombassa in 1895 it looked to most people that he had undertaken a perfectly hopeless task. The difficulties which lay before him were stupendous, and it was only his indomitable energy and pluck that carried him through.[7]

From there the article attributed the entire construction of the railway – and a great deal more – to Whitehouse:

> The first problem to solve was that of labor. The Swahili, or coast men, are too lazy to work, the Wakamba, further in land are agriculturists by deputy, all the

---

4   Diary entries May, 13-18, 1903. Whitehouse Papers, RHA-MSS-AFR, 1046 (6).

5   See "Press cuttings." Whitehouse Papers, RHA-MSS-AFR, 1046 (11). Sources used in this article to reconstruct the press coverage of Whitehouse's role in the construction of the Uganda railway are to be found in this collection of press cuttings.

6   "Press cuttings." Whitehouse Papers, RHA-MSS-AFR, 1046 (11).

7   "Press cuttings." Whitehouse Papers, RHA-MSS-AFR, 1046 (11).

hard work being done by the women, the Masai warriors would not dream of demeaning themselves by manual labor, and so forth, so Sir George brought his labor from India to the tune of 15,000 men, built corn mills to feed them, condensing plants to give them drink, and hospitals for the sick, and finally completed his task with wonderfully small loss of life considering the conditions. Sir George indeed appears to work on a Kitchener plan, and is undoubtedly an organiser of first rank.[8]

The strong personification of the Uganda railway in its chief engineer was fed through many sources. A noteworthy incidence occurred when George Whitehouse's brother, Benjamin Whitehouse, lectured in the *Society of Arts* in January 1902 (Whitehouse 1902). Benjamin Whitehouse was Commander in the Royal Navy and had in 1898-1900 been placed in charge of a survey of the British section of Lake Victoria.[9] In the *Society of Arts* he spoke under the headline "To the Victoria Nyanza by the Uganda railway" and credited not only the railway to George Whitehouse: he also highlighted his brother's role in the survey he himself had been in charge of by claiming that "the British half of Victoria Nyanza with all the islands known to exist in it had been surveyed under the superintendence of the chief engineer." In the ensuing discussion on Benjamin Whitehouse's lecture, the London-based engineer Sir Guildford Molesworth – who had visited Whitehouse in East Africa in 1899 as Inspector-in-Chief to the Uganda Railway– insisted that:

> Commander Whitehouse had been too humble in speaking about the wonderful way in which his brother had carried out the important work of the Uganda railway. The work of constructing the railway presented unique difficulties, which no other railway had to encounter. The difficulties were extraordinary. There was unknown country, then the newness of the staff, the want of water, the absence of all means of animal transport owing to the mortality caused by tsetse fly, the having to establish everything, the policy of having to push forward the railway at any cost and at any sacrifice, so as to get through somehow or other. Then there had been the difficulty of lions, the difficulties of jiggers, which had often caused men to loose their toes [...] The way in which these enormous difficulties had been surmounted by Whitehouse and his staff was deserving of the very greatest credit (Whitehouse 1902: 241).

8   "Press cuttings." Whitehouse Papers, RHA-MSS-AFR, 1046 (11).

9   Whitehouse, B. "Account of Survey of the Northern Section of Lake Victoria." RHA-MSS-AFR, 1294.

Molesworth's sensationalized descriptions were reproduced in several of the next morning's papers. In its coverage of Benjamin Whitehouse's lecture *The Morning Post* mainly focused on the fact that the session had been chaired by Henry Morton Stanley on the very day of the explorer's 61st birthday.[10] *The Times*, however, commented on the words of praise directed at the chief engineer of the Uganda Railway and noted that it echoed the eulogy that the founder of the *Royal Niger Company*, George Goldie Taubman, had written on George Whitehouse in the paper a few weeks earlier.[11] In it George Goldie, who had met Whitehouse in Uganda in 1901, also attributed the completion of the railway to "the energy, organization genius, and fertile inventiveness of its chief engineer."[12]

The different levels of public activity displayed by Whitehouse and Patterson bring out an important point. Sometimes it was the case that engineers actively took on the role of the explorer-engineer, while in other instances a set of characteristics were assigned to them by (identifiable) agents in their environment. Like other groups, engineers did not fully control how they were presented and perceived in public spheres. Indeed, while Whitehouse – unlike Patterson – displayed little interest in being cast as a lion fighting, railway constructing civilizer of Africa, it could sometimes be difficult to escape a sensationalizing public; when in 1903 *The Liverpool Daily Chronicle* published one of the "un-sensationalizing" interviews Whitehouse gave to Reuters it was placed alongside a humorous cartoon depicting a snake confronting an overweight passenger on the notorious Uganda Railway.[13]

Frederic Shelford – The Consulting Engineer
By the closing decades of the nineteenth century, "the explorer-engineer" had become a cultural icon that embodied a recognizable set of characteristics and character traits associated with engineering in the colonial field, and with which engineers actively associated. In assessing the significance of this it is imperative to recognize that there was no single and uniform class of engineers to engage with cultures of empire and exploration. The position of assistant resident engineer Patterson differed from that of chief engineer Whitehouse; the distance to a third category of engineers was even greater. These were the London-based consulting engineers who occupied positions at the very top of the engineering profession. Consulting engineers were

10 "Press cuttings." Whitehouse Papers, RHA-MSS-AFR, 1046 (11).
11 "Press cuttings." Whitehouse Papers, RHA-MSS-AFR, 1046 (11).
12 "Press cuttings." Whitehouse Papers, RHA-MSS-AFR, 1046 (11).
13 "Press cuttings." Whitehouse Papers, RHA-MSS-AFR, 1046 (11).

Figure 1. Rare glimpse of Frederic Shelford in Africa ("The Sierra Leone Railway." The Engineer, May 8, 1899).

independently practicing engineers, a small elite segment within the profession who could lay claim to the status of "independent professionals." London consultants were often affluent and among those few professionals who could aspire to what *Queen's Magazine* called the "Upper Ten Thousand"; that is, the roughly 10,000 people in late Victorian society who sustained an annual income of over £10,000 (Perkin 1988: 61-88).

Consulting engineers also had strong imperial agendas. From offices congregated in central London, leading consultants designed engineering works across the globe and in particular within the territories of the empire. The consultants served as intermediaries between the imperial Metropole and the colonial projects devising technical solutions, drawing up specifications, procuring financial solutions, and lobbying the political establishment. Displaying gentlemanly postures of decorum, trustworthiness, and respectability, leading late Victorian consultants such as Sir Benjamin Baker and Sir Douglas Fox were well-connected with cultural, political, and financial elites. Importantly, their positions in London and their illustrious names made available to the consultants public channels that other engineers did not have – ranging from highbrow journals and technical publications to the halls and rostrums of learned societies and professional engineering associations (Andersen 2009).

London's consulting engineers made extensive use of these opportunities in the public sphere, where they were among the vigorous debaters of the role of imperial engineers in Africa. A case in point was Frederic Shelford, a consulting engineer who through his diverse public engagements combined the characteristics of a field-going explorer-engineer with those of a metropolitan gentleman engineer. Shelford had joined the business of his aging father in 1899 to form the consultancy *Shelford & Son*. The company was mainly engaged in railway construction in the British colonies in West Africa, where the consultants were hired by the Crown Agents of the Colonial Office in London to design and oversee railway construction (Shelford 1909).

As David Sunderland has demonstrated in detail, the West African railways constructed by *Shelford & Son* for the Crown Agents were subject to much controversy, complaint, and critique (Sunderland 2002). Indeed, several contemporaries independently of each other criticized Frederic Shelford's West African railway projects for exceeding (very generous) timeframes, for being overpriced by comparison with railways constructed under roughly similar conditions, and for being of very inferior quality once finished. Those who believed that Shelford only kept getting assignments with the Crown Agents because he was married to the daughter of the Permanent Under-Secretary for the Colonies, Sir Montague Ommanney, could certainly argue their case very convincingly (Sunderland 2004).

In spite of the poor reputation of *Shelford & Son* – or perhaps, rather, because of it – Frederic Shelford was eagerly seeking to establish himself as an authority on questions pertaining to engineering, imperialism, and African fieldwork. In the first decades of the twentieth century his name frequently appeared in the technical and non-technical literature on African railway development. Shelford's first public appearance in this regard occurred in July 1899 in the sensationalist weekly *The Graphic*. The story behind the article was that Shelford had led a survey expedition in unmapped territory between the coastal settlement of Sekondi and the Ashanti capital Kumasi in the Gold Coast with the aim of settling the trajectory for a future railway. *The Graphic* described the trials that the three white men of the expedition had surmounted and the accompanying illustrations were based on a number of sketches that Shelford had produced during the expedition and afterwards presented to the editors of the journal in London.[14] One sketch pictured Shelford and the white men swimming across a turbulent river, another showed "native" porters fleeing into the jungle, and a third sketch illustrated the safe arrival of the expedition at Kumasi. Such active use of visual representations was characteristic of Shelford's public appearances in the years that followed. He contributed more sketches and photographs to *The Graphic*, and in the numerous articles on the West African railways that were published in the technical journals it was common to see Shelford by a bridge, a railway viaduct, or depicted floating in a ropeway above the West African rainforest.

Normally journalists were, however, more likely to find Shelford in London than in the African field. He was the first person interviewed for a series of articles that the journal *African Engineering* began in 1905 under the headline "Talks with Engineers." The series was – as the journal's editor Stafford Ransome declared – specifically "devoted to engineers engaged in opening new regions [of Africa] for business" (African Engineering 1905: 137). Shelford was interviewed in Westminster, where the interviewer had called upon the engineer at work in his office in Great George Street. In the interview Shelford explained that his interest in West Africa had first been aroused from reading *With Edged Tools* – Henry Seton Merriman's bestselling West African novel from 1894 – and he made clear how the railway lines that *Shelford & Son* constructed were rapidly opening for exploitation the vast agricultural and mineral potential of the West African region.

The latter point always featured prominently in Shelford's public writings and statements. When he was interviewed in his office by *The Financier* in 1901, "Mr. Shelford was especially enthusiastic over the properties of *Ashanti Goldfields Corporation*, the resources of which he described as simply wonderful." He expressed similar enthusiasm with regard to other areas in West Africa in *The Times, Financial News, The Mining Journal, African World*, and numerous other newspapers and journals.[15] These public statements ranged from short extracts from the interviews that Shelford gave to Reuter's to long reports from proceedings in learned societies and banquet dinners where Shelford addressed fellow experts and potential investors with African interests (Shelford 1898, Shelford 1900).

---

14 A comprehensive collection of media appearances have survived in Frederic Shelford Papers, "Press Cuttings." RHA-MSS-AFR, 2193, Box 1.

15 Frederic Shelford Papers, "Press Cuttings." RHA-MSS-AFR, 2193, Box 1.

In the latter case, the *African Society* constituted Shelford's most important institutional platform. Established in 1901 in honor of the memory and ideas of explorer and travel writer Mary Kingley this society's objective was "the enhancement of knowledge, pertaining to the law and customs" of West African societies (Fage 1995). Shelford was a founding member and in 1910 became Vice President of the Society.

At the meetings of the *African Society*, Shelford addressed his audiences as an expert on what today would be called development economics and politics. He was an outspoken champion of Chamberlainian ideas of a "constructive imperialism," which stressed Britain's obligations "to cultivate the undeveloped estates" of the empire for the prosperity of Britain and to the advantage of the inhabitants of the colonies (Porter 2004: 198-202). In this view, private trade and industry were the driving forces of economic development, but for these civilizing influences to flourish infrastructural systems needed to be established, when necessary, through public investments in harbors, telegraph lines, and railway lines such as those designed by *Shelford & Son* (Shelford 1907). According to Shelford, railways were the backbone of colonial development. When he spoke in the *African Society* on "Sierra Leone in the Making," he claimed that the great dividing line in the history of the colony was the introduction of the railway in the late 1890s. Prior to this the colony had been subjected to "the efforts of the missionaries to impart some measure of Christianity into these pagan savages," but Shelford claimed that in spite of their efforts the colony was still "struggling as a second-class British possession in which no one took any interest" (Shelford 1929). In his analysis this only changed fundamentally when the British turned to the engineers who commenced equipping the colony with a proper infrastructure in the shape of a narrow gauge railway.

For Shelford, it was largely a question of replacing Livingstone's C of Christianity with the more efficient C of Civil engineering. It is worth mentioning that far from all consultant engineers drew this contrast; most instead argued that Christianity and engineering should march hand in hand. For example, the consulting engineer Francis Fox, whose company were consulting engineers to the British South Africa Company for more than three decades, in one of his autobiographies asserted that:

> Engineers in the execution of great works visit all parts of the world, and I would ask them to uphold the truth. It is, I am sure, unnecessary to appeal to them to uphold our national honor, or to conduct themselves as gentlemen, for no English man worthy of the name would be found wanting in these virtues; but I would ask them to go much farther than this – to use their influence to suppress drink, to protect the honor of women and the innocence of children, to maintain

the observance of Sunday or the Lord's day, to encourage all true missionary work, and to sympathise with the missionaries themselves, who in consequence of many and great difficulties, are frequently liable to be depressed and cast down (Fox 1904: 192).

Like many late Victorian engineers, Francis Fox was deeply religious. Work in his office began with a communal morning prayer and for him it was decidedly not a question of civil engineering replacing Christianity as the tool of civilization (Fox 1924: 125-145).

Shelford also used the rostrum of the *Africa Society* to vehemently oppose what he called "inaccurate articles" and "perfectly absurd speeches" that criticized the West African railways constructed under his company's supervision (Shelford 1902). He insisted that it was a critique leveled by people who had little or no experience of the extreme conditions pertaining to engineering works in Africa. But, as he also insisted before a crowd of potential investors at the *Liverpool Chamber of Commerce* in 1900, he spoke as a man who had "devoted both body and mind to the subject of West African railways for several years past" and he was therefore able to present the correct (and positive) assessment of the work done by British engineers in the region (Shelford 1900: 5).

Shelford claimed authority not only with regard to theoretical ideas of colonial development based on his experiences as an explorer-engineer; he also successfully established himself as an expert on what he in 1907 labeled "Pioneer Engineering" in a series of articles he wrote for the technical journal *The Engineer* (Shelford 1908: 469-471, 495-496, 528-529, 549-550). He later expanded the articles into a book with the more broadly appealing title *Pioneering* (Shelford 1909). The articles and the book described the equipment needed for engineering expeditions and survey parties, giving advice on a host of subjects ranging from the treatment of "natives" to methods for running a traverse in dense tropical forests. These publications, Shelford explained, were not aimed at "the wealthy explorer who can spend what he likes" and who is looking for "mere adventures," but:

> the pioneer who, in however humble a capacity, is doing good work, in introducing into undeveloped countries the advantages of civilization and the benefits of twentieth-century talent and invention (Shelford 1908: 469).

Shelford also here referred to his explorative endeavors, claiming that his "hints to the traveller" were based on hard-won exploration experiences. He was a person who had "swallowed a good deal of quinine," who knew from experience that "Berkefeld's pump filters tended to clog up," and he noted that "many a time has a soldier, or a civilian official examined my kit, and begged for an article or two from my equipment which has after many years' experience been brought to somewhat near perfection" (Shelford 1908: 470, 495-496).

If long and frequent stays in West Africa were a prerequisite for knowing the needs of African colonies as well as those of "the pioneer engineer," Shelford does not qualify as an ideal expert. As far as can be established, he only visited West Africa three times prior to 1910 and each time spent less than two months there. Like most consultants his real fields of operation were the engineering offices in Westminster, the government offices in and around Whitehall, the lecture theatres of learned societies, the dinners of commercial bodies, and the conversation rooms of gentleman clubs. In fact, one had to be quite lucky to catch Shelford in Africa floating in a ropeway above the jungle of Sierra Leone wearing a gentleman's tie. Yet Shelford had the access, the need, and the wish to make public use of his limited experiences of the African field in pursuit of his agendas.

In the late Victorian period, consulting engineers such as Shelford were among the most vigorous and visible engineers in the British public sphere. The consultants' public activities were in many cases linked to their commercial interests. Unlike state employed engineers – such as George Whitehouse – consultants practiced as independent engineers and often engaged the public to win support for new projects or to fend off voices critical of their ongoing works. Serving as intermediaries between the colonial engineering fields and the imperial center, London-based consultants enjoyed privileged access to a range of public platforms that allowed them to highlight to wider audiences their particular contribution to the successful execution of projects – and by implication their contribution to the advancement and consolidation of British imperial presence in Africa. These consultants took part in gentlemanly conversations where the role of the British engineering profession was seen as entailing broader and more abstract imperial obligations. This was brought out, for example, in the discussions of the relation between engineering and Christianity in Britain's "civilizing mission." Underlying these discussions was an assured belief in the importance of the engineer, a belief expressed by Sir John Coode – consulting engineer to several harbor works in South and West Africa – when he took the chair as President of the Institution of Civil Engineers in London in 1889:

> So long as the present dispensation may last, so long will there be a continuous progress in the science and practice of every branch of labor in the field

appertaining to the Civil Engineer. Neither to the Engineer, nor indeed to any other disciple of natural science, would it seem to have been announced – I say it with all reverence – "Thus far shalt thou go, but no farther" (Coode 1889: 39).

Moving in the highest metropolitan circles, consulting engineers engaged in public discourses on empire that added even more layers to the ideas of imperial engineers, on top of those connected with the sensationalized African field of the explorer-engineer.

Conclusion: Engineering the Public Field

This article has demonstrated that imperial engineers engaged the British public in a number of ways and for a number of reasons, depending not least on their position within the profession. As a result they constructed a range of images of the engineer in the African fields. Often a familiar and important contrast was drawn between on the one hand progressive, white imperial engineers and on the other a sensationalized, yet passive and dark African continent. To paraphrase Edward Said, it was clearly the case that the engineers felt that they knew and understood Africa better than the Africans (Said 1978). Yet the case of the engineers also goes beyond the construction of "otherness" along racial lines. There was great diversity in the public activities and postures of engineers, from the manly explorer-engineers roaming the sensationalized African fields to gentleman engineers in London discussing the role of the profession in an imperializing world. David Cannadine has argued that class-based hierarchies rooted in Britain's domestic culture were a decisive factor in molding the ideological outlook of British imperialism (Cannadine 2005). This certainly holds true in the case of engineers. Their positions within the social and professional hierarchy were fundamental to the ways in which they pursued the opportunities offered by imperial vocabularies.

What consequences did this growing association with empire have for how engineers were viewed in wider Victorian culture? The analysis suggests that British engineers gained significant public respect through their work in the empire. Indeed, while it makes no sense to speak of engineers experiencing anything like the late Victorian obsession with explorers and soldiers, there is no question that the "heroic engineer" was given a new lease of life that this cultural persona would not have had without the element of drama provided by exploration and imperial rivalry. Sensationalized landscapes and peoples clearly set the explorer-engineer apart from the fading Smilesian engineer, whose achievements were set in the domestic scene. Furthermore, the Smiliesian image of the self-reliant and individualistic engineer may have been more convincing and easier to uphold in relation to projects in the colonial world at a time when engineering work in the British Isles increasingly came to be associated with specialized teamwork, bureaucratic organizations, and employment in industry and laboratories. The idea of the manly, independent engineer trained through practice and the trials of life had found a temporary refuge in the imperial fields.

References

Adas, M. 1989. *Machines as the Measure of Men. Science, Technology, and Ideologies of Western Dominance*. Ithaca: Cornell University Press.

African Engineering 1905. "Talks with engineers." 1(3): 134-137.

Andersen, C. (forthcoming). "Engineering journals and Imperial Africa 1860-1914." In: C. Andersen & B. Marsden (eds.) *Literary Engineering – Print Culture and engineers in the nineteenth century*.

Andersen, C. 2009. *Consulting engineers, Imperialism and Africa 1880-1914*. Unpublished PhD Thesis, Aarhus University.

Biagioli, M. 1993. *Galileo. Courtier. The Practice of Science in the Culture of Absolutism*. Chicago: Chicago University Press.

Brian, D. 2008. *The Seven Lives of Colonel Patterson. How an Irish Lion Hunter Led the Jewish Legion to Victory*. New York: Syracuse University Press.

Burge, C. O. 1909. *The Adventures of a Civil Engineer. Fifty Years on Five Continents*. London: Alston Rivers.

Cannadine, D. 2002. *Ornamentalism. How the British Saw their Empire*. Oxford: Oxford University Press.

Cannadine, D. 2004. "Engineering History or the History of Engineering." *Transactions of the Newcomen Society*, 74: 163-180.

Chew, K. & A. Wilson 1994. *Victorian Science and Engineering Portrayed in The Illustrated London News*. Stroud: Alan Sutton Publishing.

Churchill, W. 1908. *My African Journey*. London: Hodder & Stoughton.

Clifton, G. & D. Porter 1988. "Patronage, Professional Values, and Victorian Publics: Engineering and Contracting the Thames Embankment." *Victorian Studies*, 31(3): 319-349.

Coode, J. 1889. "Presidential Address." *Proceedings of the Institution of Civil Engineers*, 99: 1-40.

Crozier, A. 2007. "Sensationalising Africa: British Medical Impressions of sub-Saharan Africa 1890-1939." *Journal of Imperial and Commonwealth History*, 35(3): 393-415.

Driver, F. 2001. *Geography Militant, Cultures of Exploration and Empire*. Oxford: Wiley Blackwell.

Driver, F. & L. Martins (eds.) 2005. *Tropical Visions in an Age of Empire*. Chicago: Chicago University Press.

Fage, J. D. 1995. "When the African Society Was Founded. Who were the Africanists?" *African Affairs*, 94(376): 369-381.

Fox, F. 1904. *River, Road, and Rail: Some Engineering Reminiscences*. London: John Murray.

Fox, F. 1924. *Sixty-three Years of Engineering, scientific and Social Work*. London: John Murray.

Hawkshaw, J. 1838. *Some Reminiscences from South America*. London: John Murray.

Headrick, D. R. 1981. *The Tools of Empire. Technology and European Imperialism in the Nineteenth Century*. New York: Oxford University Press.

Hill, M. F. 1957. *Permanent Way*. Nairobi: East African Railways and Harbours LTD.

Hopkins, A. G. 2008. "Explorers' Tales: Stanley Presumes – Again." *Journal of Imperial and Commonwealth History*, 364: 669-684.

Jarvis, A. 1997. *Samuel Smiles and the Construction of Victorian Values*. Stroud: Alan Sutton Publishing.

Kennedy, D. 2007. "British Exploration in the Nineteenth Century: A Historiographical Survey." *History Compass*, 56: 1879-1900.

Kubicek, R. 1999. "British Expansion, Empire and Technological Change." In: R. Porter (ed.) *The Oxford History of the British Empire. Vol. III. The Nineteenth Century*. Oxford: Oxford University Press, 258-277.

MacKenzie, J. M. 1994. "Heroic Myths of Empire." In: J. M. MacKenzie (ed.) *Popular Imperialism and the Military, 1850-1950*. Manchester: Manchester University Press, 109-139.

MacKenzie, J. M. 2000. "The iconography of the exemplary life: the case of David Livingstone." In: G. Cubitt & A. Warren (eds.) *Heroic Reputations and Exemplary Lives*. Manchester: Manchester University Press, 84-104.

Macleod, C. 2007. *Heroes of Invention: Technology, Liberalism and British Identity, 1750-1914*. Cambridge: Cambridge University Press.

Marsden, B. & C. Smith 2007 [2005]. *Engineering Empire: A Cultural History of Technology in Nineteenth-Century Britain*. Hampshire: Palgrave Macmillan.

Miller, C. 1977. *The Lunatic Express*. London: Penguin.

North, S. J. 2005. *Europeans in British Administrated East Africa. A Biographical Listing*. London: Stephen J. North.

Patterson, J. H. 1907. *The Man-Eaters of Tsavo and other East African Adventures*. Surrey: Surrey Press.

Perkin, H. 1989. *The Rise of Professional Society. England since 1880*. London: Routledge.

Porter, B. 2004. *The Lion's Share: a Short History of British Imperialism, 1850-2004*. London: Longman.

Pratt, L. 1992. *Imperial Eyes: Travel Writing and Transculturation*. London: Routledge.

Riffenburgh, B. 1994. *The Myth of the Explorer: The Press, Sensationalism, and Geographical Discovery*. Oxford: Oxford Paperbacks.

Rolt, L. T. C. 1970. *Victorian Engineering*. London: Penguin.

Said, E. W. 1978. *Orientalism*. New York: Pantheon Books.

Smiles, S. 1997 [1857-1905]. *The Collected Works of Samuel Smiles*. London: Routledge/Thoemmes Press.

Shelford, A. E. 1909. *The Life of Sir William Shelford, KCMG, Chevalier of the Order of the Crown of Italy, Member of Council of the Institution Of Civil Engineers*. London: Printed for Private Circulation.

Shelford, F. 1898. "Railway surveying in Tropical Forests." *Proceedings of the Institution of Civil Engineers*, 133: 339-350.

Shelford, F. 1900. *Address on West African Railways, The African Trade Section of the Incorporated Chamber of Commerce Liverpool*. Liverpool: Chamber of Commerce Liverpool.

Shelford, F. 1902. "On West African Railways." *Journal of the Royal African Society*, 1(3): 339-351.

Shelford, F. 1907. "Ten Years of Progress in West Africa." *Journal of the Royal African Society*, 6(24): 341-349.

Shelford, F. 1908. "Pioneer Engineering I-IV." *The Engineer*, 104: 469-471, 495-496, 528-529, 549-550.

Shelford F. 1909. *Pioneering London*. London: E. & F.N. Spon.

Shelford, F. 1929. "Sierra Leone in the Making." *Journal of the Royal African Society*, 28(11): 235-240.

Stanley, H. M. 1885. *The Congo and the Founding of its Free State. A Story of Work and Exploration*. London: Sampson Low.

Sunderland, D. 2002. "The departmental system of railway construction in British West Africa, 1895-1906." *Journal of Transport History*, 23(2): 87-112.

Sunderland, D. 2004. *Managing the British Empire: the Crown Agents, 1833-1914*. London: Royal Historical Society.

Talbot, F. A. 1911. *The Railway Conquest of the World*. Philadelphia: J.B. Lippincott Company.

Thompson, A. 2007. *The Empire Strikes Back? The Impact of Imperialism on Britain from the Mid-Nineteenth Century*. Harlow: Pearson Education Limited.

Whitehouse, B. 1902. "To the Victoria Nyanza by the Uganda railway." *Journal of the Society of Arts*, 38(2): 229-41.

Wiener, M. 1981. *English Culture and the Decline of the Industrial Spirit 1850-1980*. Cambridge: Cambridge University Press.

Palle O. Christiansen

From Collection to Fieldwork
The Field Research of Danish Folklorist Evald Tang Kristensen, 1870-1890

The beginnings of modern fieldwork in the nineteenth century gave several cultural disciplines new insights into the world of reality. Some folklorists, dialectologists, and ethnographers gained new and different knowledge of the people whose mental culture they studied, and this resulted in more diverse and accurate knowledge than previous collections had been able to offer. But armchair ethnography never quite lost its adherents, and the academic milieu was not always able to question previous practice and integrate the new insights into their activities. For that reason several early field researchers had no direct successors, and their methodological insights were forgotten. This was the case with the Danish folklorist Evald Tang Kristensen and the collecting work he did amongst the people of Denmark.

Tang Kristensen (1843-1929) is considered to be the individual in the western world who personally collected the most non-material folk culture, especially in the form of ballads, tales, and legends. On his death he left the result of no less than 50 years collecting oral tradition, mainly from the Jutland peasantry, both as written texts and as original recordings that have been preserved in their entirety until today. Much of the material was recorded in what he called "the cottages."

Tang Kristensen's collecting and working methods are considered to be of a high standard compared with the way ethnography and folklore studies were normally carried out in the nineteenth century. Oddly enough his activities in the field have never been studied in detail, and this is the background for the present article.[1] I will also give an account of the way his practice generated insights that in some areas overstepped the boundaries of previous notions of what the new field of study called popular culture consisted of in the nineteenth century. Finally I would like to sum up his guidelines for the behavior of the researcher among the people – which are on the whole still relevant – and to compare his advice with others' statements about his

---

1   See however Kofod 1983: 113ff, Holbek 1987: 70ff. For references to sources I use, see the abbreviations in Kristensen 1981: 145ff.

activities in the field. The information mainly comes from personal letters and notes, from introductions and postscripts to the source editions, and from his memoirs.

Fieldwork as outdoor research

As a result of modern anthropology in particular, the term fieldwork has come to describe almost any kind of questioning and observation among living people, formerly mainly in "exotic" surroundings. But fieldwork as a collecting technique goes back a long way. When botanists, linguists, geologists, construction researchers and archaeologists left their desks in the eighteenth and nineteenth centuries to study and gather material "on location," they were early fieldworkers. This kind of personal outdoor research contrasted not only with the work of other scholars and scientists in their studies or in academic libraries, but also with the gathering of information from travellers, Christian missionaries, and organized informants – that is, people who more or less systematically sent the researchers information on rare flowers, inscriptions, customs, and the like.

The crucial thing in scholarly fieldwork is that collection and research are not separated. This means that the collector himself knows the disciplinary questions that the collected material is expected to help to answer; otherwise the field researcher cannot engage in a deliberate and intuitive search. This is why Tang Kristensen, who was trained as a schoolteacher, also had to be trained in philology and folklore by the university reader and later professor Svend Grundtvig (1824-83), so that he would be qualified to work in the field. From 1869 and for some years to come this took the form of intensive correspondence and academic studies alternating with the practical work he did out in the field.

In the twentieth century ethnographic, linguistic, and folkloristic fieldwork developed such that it no longer made sense to compare it to the botanist's or the geologist's more impersonal hunt for plants or unknown rock types. This more socially oriented fieldwork also stands out today as being a type of work carried out among people, by people (cf. Olwig 2002: 121). Geographical *surveys* also involve outdoor research, but the so-called "intensive fieldwork" of the humanities is characterized by closer relations with people in the given localities. This is where Tang Kristensen became a pioneer in Danish folklore and national ethnography. At first he is unlikely to have been aware of this himself, and still more strangely, his experience had no consequences in the expert milieu.

Since the 1840s the philologist Svend Grundtvig, a son of the poet and moral philosopher N.F.S. Grundtvig, had been Denmark's leading figure in the collection and publication of folklore, especially in the field of ballad and folk tale research. His aim was both scholarly and nationalist, since it was thought that the

Figure 1. Tang Kristensen at the beginning of the 1870s, around 30 years old. This was the time when he began his long-lasting fieldwork and had published his first book (Kristensen 1924).

oral tradition of the countryside included old – indeed very old – elements and in concentrated form represented the cultural specificity of the nation vis-à-vis other nations. As in most European countries, folklore had been collected and material had been published since about 1800; but in Denmark Svend Grundtvig became the first systematic organizer. Personally he collected hardly anything, but worked through a nationwide network of over 300 collectors, who occasionally sent him material.[2] At the time this was normal collecting practice, and in parallel with this scholars throughout most of Europe engaged in correspondence with one another.

On a visit to his parents at the Brandstrup village school at Christmas in 1867, Tang Kristensen wrote down the songs of an elderly village woman. Over the next few years the 25-year-old man began to spend all his free time noting down songs and tales from the surroundings of the teacher's residence in Gellerup near Herning in central Jutland, at that time a typical heathland region. At first Tang Kristensen used his new preoccupation to assuage his grief over his young wife Frederikke's

---

2   For a more detailed overview of the history of the subject, see Holbek 1979: 57ff.

terrible death in childbirth the previous year, but from 1868 on he was captivated by the material and its meaning. In 1869, after a few modest attempts at publication, he contacted Svend Grundtvig at Copenhagen University, and their collaboration was to last until the death of the professor in 1883. Right from the outset it was clear that Tang Kristensen had no wish to be one of Grundtvig's many "recorders," even though several of them were qualified people. He wanted to publish for himself, and Grundtvig gave him a lot of help with his first books; probably because he saw in the young teacher an energetic and gifted, although occasionally stubborn collector with the ability to procure unknown and better sample material for research – and in quantities that exceeded those of all Grundtvig's other recorders. In all, Tang Kristensen managed to visit a good 3,300 informants in Denmark and personally recorded around 3,000 songs, 1,000 melodies, 2,500 tales and over 15,000 legends as well as dialect words and much other material from the culture of everyday life.[3]

With his first letter to Grundtvig Tang Kristensen had sent examples of ballads he had collected, which immediately aroused the enthusiasm of the normally critical scholar. Grundtvig thought that Tang Kristensen, with his "digging for treasure," had found "a goldmine" amidst the Jutland heath from which ballads had not hitherto been known, and he urged the teacher to continue "working the rich veins" that were available to him.[4]

These geological phrases were more than just metaphors: they were characteristic of the older scholarship's view of how new knowledge was created through collection. Grundtvig was a "Nordic philologist" and was himself working on his great edition of the ballads in a historical and comparative perspective. He was interested in old ballad forms and their manifestation in new variants, whether the sources were printed writings, older collections in libraries or "from the mouths of the folk." For the older philologists it was the ballads as texts that were the central issue, not so much the transmitters of the material, so Tang Kristensen's area of work could be compared to a mine from which he dug up his treasures. This was how Tang Kristensen learned to look at his activities, and he himself used these almost scientific terms. As a characteristic representative of his time and his environment, Grundtvig never abandoned the literary, not to say reified view of folk culture, but for Tang

---

3   However, ETK's collection is substantially larger, since it also includes a very large body of material that he received from other collectors, e.g. in his capacity as editor of the periodical *Skattegraveren* 1884-90.

4   Letter from S. Grundtvig to ETK, March 3, 1869. 1929/144I, Danish Folklore Archives (Dansk Folkemindesamling), The Royal Library, Copenhagen (subsequently DFS).

Kristensen a certain change took place slowly. This change had its background in the shift from more mechanical collection work towards true field research.

From Collecting to Fieldwork

At first Tang Kristensen recorded material in ever-widening circles among singers and storytellers around the teacher's house in Gellerup. Very characteristically, he spoke of his informants as oral sources: "In the winters immediately following ... I went to my sources as evening fell and when the weather more or less permitted, and rarely came home before one in the morning. I froze like a dog, but paid no heed to that. I waded through the flooded meadows and wandered over the trackless heaths, many times in peril of my life but always came safely back," as he later noted in some of his personal papers.[5]

The folklorists of Romanticism always saw collecting as a rescue mission to save something old that would otherwise be lost forever. Tang Kristensen wrote as early as 1870 to Grundtvig about this issue: "If it had been twenty years ago, then I feel there would have been a rich harvest, but now times are different for the venerable old plants, they are hated and downtrodden so thoroughly that it is odd that there is still so much left."[6] Gradually, as Tang Kristensen and Grundtvig considered the Gellerup area "emptied" of interesting material, larger collection areas had to be dealt with; this required proper expeditions, so that he could no longer stay at home.

From 1871 on, several grants from foundations enabled Tang Kristensen to employ an assistant teacher in the winter months, so he himself was free to go on longer collecting trips, often organized as tours lasting four to six weeks, with very short stays at home in between. In terms of the later Malinowski standards, these periods in the field were not long, but in fact they were as long as most American ethnographic work in the field up to the 1940s (Foster 1979: 4).

Not until 1888 did he receive a life annuity from the state for permanent folkloristic activity, and was able to give up teaching altogether. The first twenty years of collecting were the most well organized, and he experienced his breakthrough years as an expert in the 1870s.[7] In the beginning his expeditions explored the area between Herning, Ringkøbing, Holstebro, and Viborg, half of which was still heather-clad moorland. What was characteristic of this early, intensive fieldwork was

---

5   *Selvbiografier*, 1884a. 1929/142, DFS.

6   Letter from ETK to S. Grundtvig, March 25, 1870. 1929/1441, DFS.

7   In his old age ETK became more restless and was sometimes aware of this. From Værløs, for example, he wrote to Grete: "I am very well, but not getting much benefit from my tour, since I am getting through it too fast." Letter from ETK, July 26, 1889. 1929/153, DFS.

Figure 2. Tang Kristensen on the heath east of Holstebro in 1887. In the background, a woman tending sheep (woodcut from contemporary photograph, Danish Folklore Archives, Royal Library, Copenhagen).

that the field was not regarded – as in the time of international community studies in the twentieth century – as a defined locality, but as a larger landscape space in which the informants were localized, in the heathland areas often very far apart.

Thanks to Tang Kristensen's exemplary notebooks, which he called "diaries," his long letters to his second wife, Grete, and to Svend Grundtvig, we can see today how the work was organized. The most striking thing is how concrete the approach was, how few discussions of principle preceded it – and how fruitful the tours were. From the preceding years, it is true, he had gained good insights into interviewing and recording methodology, and any technical field problems were apparently solved along the way. The same unceremonious attitude to fieldwork preparations was apparently to be found in early American ethnography under Franz Boas (Mead 1959: 31ff).

Tang Kristensen liked to work his way forward by making good use of the informants' own contacts. Only a few people in an area were really good storytellers, but good ballad singers and narrators as a rule knew of one another's existence within a certain radius. For that reason he noted some of the informants' recommendations on the back of his notebooks or at the bottom of the pages (with the book turned 180 degrees) so he could visit them later. Tang Kristensen literally wandered out into this wonderland, and as a result of his new physical absence from domestic surroundings he could not avoid being enveloped by the intensity of differences that everyone experiences in such situations. Concentration enhanced observation and thinking, and through correspondence along the way, scholarly and practical problems could both be discussed to a certain extent. These circumstances made his work more qualified. Although his routes were rarely strictly laid down in advance, his collecting became more systematic and intensive than the norm among other field collectors of the period, who worked more at random and under less pressure. Neither Tang Kristensen nor Grundtvig commented on the subject, although Grundtvig made no secret of Tang Kristensen's exceptional abilities in the field. But in reality a new norm was established for how good ethnographic-folkloristic knowledge was to be created.

In the rest of Scandinavia, folklore was also being collected according to new guidelines. Both Jørgen Moe (1813-82) and P. Chr. Asbjørnsen (1812-85) in Norway did their early folktale collecting in the 1830s and 1840s during extended, planned journeys and were in close contact with the informants, but the rest of their working method and retellings resembled that of the other collectors in the first half of the century more than Tang Kristensen's disciplined recording and archiving. At the end of the nineteenth century Professor Moltke Moe (1859-1913) in Christiania, whom Tang Kristensen incidentally visited in 1895, also went on long and successful collecting trips, but his choice of material was more selective than Tang Kristensen's, and his later editing often more radical. In Sweden the Gotland collector P.A. Säve (1811-87) and the Halland doctor and folklorist August Bondeson (1854-1906) must be singled out for good informant data and primary documentation, but once more the differences are clear. This is particularly true of their more limited ambitions in geographical terms and the volume of their collections, and in Bondeson's case of his frequent literary use of the folklore, for example presenting it as examples in his own frame tales (Sandklef 1956: 70ff, Hodne 1979: 18ff, 40ff, Liestøl 1984: 79ff, Palmenfelt 1993). Despite clear parallels, Tang Kristensen was different from these collectors in that, at a very early juncture, he combined the different elements of fieldwork techniques and recording methodology in a way that only became common in the twentieth century.

Surviving "Out There"
Tang Kristensen thought winter was always best for serious fieldwork, because then he could meet people at home, and they had time to talk. Otherwise rural people

were restless: "When the weather becomes good and they can get out, they fidget and are always pottering about with something or other. Since their youth they have been used to constant activity, and they cannot settle down until they are forced to," he wrote in 1876 (Kristensen 1876a: VII). People were not honored to receive visits from strangers, and since the recording work required a certain calm, he had to visit them when they were not occupied outdoors. Tang Kristensen had no horse, and thought it inappropriate to arrive in a carriage at the homes of the poor people he wanted to talk to. For that reason most of his journeys throughout his life were undertaken on foot. If there were stagecoaches on stretches between the market towns he often took them, and later, when the rail network was developed, he used the railway, travelling third class where this was convenient. But the majority of the time he walked.

This meant that he could not take more equipment with him than he could carry in his shoulder bag and a holdall with a handle. With the other hand he supported himself on a stick and later an umbrella. Fairly quickly he procured long boots and a stiff rain cape, although this was bothersome in dry weather; he also wore an overcoat, and in the winter a woolen cap with a large shade, in the summer a broad-brimmed straw hat. Otherwise his headgear was a bowler. The holdall contained extra woolen socks, darning wool, a haversack for bread, boiled potatoes, and often bacon for a week or two. Later he took galoshes and morning or gym shoes, as well as some of his books for sale. In his shoulder bag Tang Kristensen carried diary notebooks, rough paper and stationery, pen and ink, a pocket book with a calendar, a knife and fork, a little tea and sugar, and tobacco.[8]

Tang Kristensen always started early in the morning, and he could walk over 30 kilometers in a day, even in rough weather. When he arrived at his destination he often went to the local village teacher or parish pastor on the chance of being introduced to local informants. He quickly learned that in unfamiliar areas it could be a good thing to have the teacher with him, since people were otherwise slightly suspicious about his business and were unwilling to open themselves up to complete strangers.[9] After making these first contacts, he maneuvered for himself. If at the same time he could obtain overnight accommodation that was fine, but otherwise it was important to make appointments with people for the evening and spend the last hours of light finding his bearings in the landscape, since he might not always

---

8   Information gathered from several sources, but the subject is well dealt with in *Selvbiografier*, 1884a-c. 1929/142, DFS.

9   Letter from ETK to S. Grundtvig, October 13, 1871. 1929/144I, DFS.

be able to find his way back to his lodgings in the dark.[10] By day he could ask for directions, but sometimes the local residents were rather surprised that he did not know the roads as well as themselves. Tang Kristensen was always keen to start the collecting work quickly, but also appreciated being invited to tea in the parish rectory – for the sake of the respect and relaxation it provided and to sound out the pastor's intellectual make-up.

Tang Kristensen was good at finding his way around and built up a vast local knowledge, especially of northern Jutland. All the same he could be taken by surprise, especially on the heath in drifting snow or during the night in rainy weather. Otherwise he walked in the moonlight guided by the stars and far-off gleams of light, before the quickset hedges blocked off the view. On the heath there were rarely proper roads, and often he had to go along the deer paths and other tracks that existed on the terrain. But in the winter he could not avoid having to wade out into bogs or lakes of dammed-up water, and the downpours of autumn posed difficulties. In November 1873 his first tour began with a lot of rain; the roads were muddy and his boots were wet: "I suffered much from sitting in those simple houses with cold feet … One evening I got stuck so much in the mud, almost to my knees, that I sprained one of my ankles." This kind of accident vexed the avid collector; he was just as irritated that it was impossible to record material in Christmas week, because people were busy with baking and other preparations for the festive days.[11]

In February of the next year Tang Kristensen spent the night in a house in Torsted Parish, where the water dripped through the roof into the bed, and in the morning he went looking for a new informant on the other side of the river Madum Å at Filsø. But when he came to the river the plank bridge had been washed away, and Tang Kristensen had to wade over in water up to his waist. Now he was wet for the second time, and he only had one pair of stockings to change into. It was rare for him to be as lucky as he was in an incident that occurred when he had become a well-known figure in 1891: the teacher in Holingholt, north of Herning, took off his own boots and stockings and carried Tang Kristensen over the river Storå on his back.[12]

One of the worst situations he encountered was arriving at a place in the evening after having traversed many miles in bad weather with a particular aim in mind, and people simply refusing to put him up. Then there was nothing to do but travel

---

10   Letter from ETK to S. Grundtvig, November 20, 1875. 1929/144II, DFS.

11   Letter from ETK to S. Grundtvig, November 26, 1873. 1929/144II, DFS. See also: Kristensen 1924-27: 2, 1924: 153 (subsequently *MO*, volume and year), letter from ETK to S. Grundtvig, November 20, 1875. 1929/144II, DFS.

12   Letters from ETK to home, February 8, 1874, June 11, 1891. 1929/153, DFS.

on or turn back, even if he was soaked. But often he succeeded in finding a mill, a hostelry, or usually private lodgings, where he always left a little money for people's trouble when he departed. Where possible, Tang Kristensen tried to procure a hot meal wherever he arrived, but on one afternoon in 1873 he had consumed the last of his food in the haversack. It had lasted over two weeks. The next day, at Sneftrup Huse, it was hard to get started on his work, for "the body had too little, I was truly starving, since the cottages out there were so wretched, and the people so poor that I could not bring myself to ask for food from them. Indeed, that day I really longed to be at home again," he wrote to his wife. He was ashamed of his own wealth compared with the conditions in which others lived, and he told Grete: "I [have] never before … journeyed in such deserted areas. It was heath and more heath almost everywhere, and only here and there around the few houses some cultivated spots."[13]

In general Tang Kristensen found that people were hospitable, especially in western Jutland, where he was often let in even when he knocked on the door in the middle of the night (Kristensen 1930: 39). All the same he was sometimes suspected of being a salesman, a Mormon minister, a plain-clothes policeman, or an ordinary vagrant. He encountered particular resistance among adherents of the new Pietist-Evangelical tendency, Indre Mission, which regarded folklore as the remains of ancient heathendom that were incompatible with a modern Christian life. Several times he was shown the door by the faithful, or found that they immediately refused to say anything to him.

In the latter half of the nineteenth century the younger people living in the countryside were turning away from their rural life towards a more modern, rational lifestyle. This meant that the older songs and stories were viewed with indulgence or even actually mocked as worthless "old rubbish."[14] Tang Kristensen therefore found that it often required a cautionary approach to get the old people to talk: "They are very suspicious, which is not strange, since anyone will be careful not to be made a fool of; but once one has convinced them of one's sincerity and love of their lore, the matter is then still not clear. The memory has to be aroused: the lore that has long since been buried or has lain dormant has to be brought forth, so the singer must, if I can put it this way, be placed in a different world from the present one," he wrote in 1871 in his first book.[15] Thus a purely mechanical collecting technique was not enough. On the whole Tang Kristensen found that if one wanted to engage in intensive fieldwork, familiarity with the informant was essential, and this first and

13 Letter from ETK to home, December 14, 1873. 1929/153, DFS. *MO.* 2, 1924: 156.
14 Letter from ETK to S. Grundtvig, March 16, 1872. 1929/144 I, DFS.
15 Letter from ETK to S. Grundtvig, March 16, 1872 1929/144 I, DFS. See also: Kristensen 1871: IX.

foremost required time and knowledge of the local culture. But then on the other hand the rewards were great: "Many times I have gained truly great insight into people's mental and emotional life after spending a day in a cottage, which could not otherwise have happened after a long period of superficial relations. One must as it were penetrate into the nooks and crannies of the soul," he noted in 1876 (Kristensen 1876b: VII).

However, there was also the practical and emotional cost of being away from home. Grete was not happy to be alone, and for most of her life pleaded with her husband to come home or at least to be at home more. But that would have clashed with the idea of the new fieldwork. He himself also thought that it was hard to deal with the school's farming work through correspondence with Grete, and it was particularly burdensome for him to be away from his small children. But there was rarely any doubt in his mind about what should be given the highest priority. In February 1874 Tang Kristensen wrote to Grundtvig from the region around Nissum Fjord: "The worst thing is that I overtax myself and get too little sleep, for I can by no means get used to sleeping in the many strange beds. I get very chilly from leaving at night ... from the places where I can get a bed for the night, and that happens almost every evening towards midnight. I am so heavy-hearted, and my mind is perplexed."[16] It is clear that contradictory feelings and experiences crowded in, but in his eyes it could hardly have been otherwise.

When his clothes became too ragged or dirty, Tang Kristensen wrote home for something new, like a jersey, clean stockings, or mended underwear. Later, when he began to sell his books, he also asked the family to send extra copies when he had sold out. After a few days he then fetched the things at a sub-post office, a station, or a private address, that is if the family had managed to send them off in time. His home acted in general as a storehouse for many practical necessities during his expeditions, although he was not afraid to darn stockings or sole boots himself on his tours. In February 1874 he had to change his route because there was no "bite" from the people in the area, and he asked Grete to send him five *rigsdaler*, a shirt, and the mittens he had forgotten as well as a little "linen, there is not enough, especially collars."[17]

In parallel with his supply line home via the letters to Grete, he kept up a scholarly connection with Grundtvig. In January 1875 Tang Kristensen had some doubts about what he should do with a Swedish woman's songs in the midst of a traditional Danish peasant tradition between Holstebro and Viborg. Grundtvig

---

16   Letters from ETK to S. Grundtvig, November 5, 1873, February 14, 1874. 1929/144 II, DFS.
17   Letter from ETK to home, February 4, 1874. 1929/153, DFS.

replied quite in keeping with the pan-Nordic thinking that was especially prevalent in Denmark and Sweden:

> In my view you must do your best to reproduce them [the ballads]. They are of interest in several respects, on the one hand in themselves as Nordic ballad texts and then also for Jutland folk songs, as will perhaps appear in the next generation in her neighborhood. There is still an exchange of folk poetry among the Nordic countries, and it is instructive to apprehend an individual bearer and inspect her stock.[18]

Tang Kristensen got a reply to his question and at the same time learned a little *Kulturkreislehre* about regional connections. This mixture of specific and general information was characteristic of the training that Tang Kristensen more or less consciously received.

On the basis of such examples the observer could easily get the impression that Tang Kristensen recorded all the folklore he encountered along his route, but this was not the case. Even within such a classic genre as the ballad, not everything a good singer knew found a place on the collector's paper. People sang both old and more recent ballads, but it was the non-urban influenced and especially the older ballads that were of folkloristic interest.[19] For example in 1874 Svend Grundtvig considered that Tang Kristensen's record of the ballad of "The Bishop's Daughter," despite its highly lubricous refrain, represented a development of an older ballad that was not previously known from Denmark, although it existed in the Faroe Islands. He therefore advised his Jutlandic friend that " ... nothing that exhibits ancient form must be scorned or neglected, even though it may be repellent in its present shape."[20]

Visiting the Cottages: The Materiality of the Situation
Besides the folklore, Tang Kristensen's visits frequently involved investigating the everyday local diet and sometimes also the accommodation. Next to the teachers' homes, where he could have a base for days, he preferred a bedroom or a bed in the large main room of a good farm. Often, however, he had to be content with less. He found his ballads and stories among farm workers, smallholders, servants, craftsmen,

---

18   Letters from ETK to S. Grundtvig, January 29, 1875, and from S. Grundtvig to ETK, March 10, 1875. 1929/144 II, DFS.

19   ETK himself spoke of the matter, cf. letter from ETK to S. Grundtvig, January 15, 1873. 1929/144II, DFS.

20   Letter from S. Grundtvig to ETK, February 3, 1874. 1929/144II, DFS.

teachers, and ministers, but the material that he and Grundtvig considered most exotic was to be found in "the cottages." This was Tang Kristensen's general term for everything from smallholder cottages to homes half-made of turf walls and rafters. Most of the houses of this kind have long since collapsed; they housed people who lived in a way that disappeared at the beginning of the twentieth century. It is therefore of special interest to follow his work in this very different reality, which has only survived through a few accounts. In this area Tang Kristensen demonstrated a decided feel for ethnographic visualization and documentation of sensory impressions.

In February 1874 he did not dare brave what he saw as the wild heath from a solitary house on the way to Tvis, and "there was only one bed in the whole house … but the woman of the house knew what had to be done. I had to lie with the man in their conjugal bed, and then she would lie under the stove and leave her bed to me. That was kindly done, and I could not have appreciated it more. She gathered together some sacks and some old clothes and out of them managed to make a bed on the floor, and then she lay down there in her clothes with a few sacks and other rags over her," he wrote later. Tang Kristensen also noted that he was in no doubt that the wife was cold on the earth floor, but he was also aware that she was used to a hard life. After breakfast they continued with the preceding day's collecting work together (Kristensen 1924, 2: 174).

A month later Tang Kristensen walked from the home of the cottager Jens Talund in Hallundbæk out over Feldborg Heath to the smallholder "Grey-Erik" (Jensen) and Maren Jensdatter in their "ramshackle hut," a dwelling that can be reconstructed as having had a total area of 33 square meters.[21] Maren sang, and Erik narrated. But when night came Tang Kristensen decided that it was impossible to find his way from there to inhabited regions. In the living room there was a bed space with some old clothes, and there he lay, fully dressed with his waterproof cape over him. But he did not get much sleep, for a cold draught came in through the crevices in the windows and the rafters in the ceiling. The next winter he visited the family again, and, suffering from toothache and rheumatism, he spent the night seated at the dining table with his head resting on its surface (Kristensen 1924, 2: 182; cf. 200).

On the whole, life among poor people was rather different from what he was used to, but he quickly learned to live with the conditions, although in the winter the condensation would drip from the ceiling and make his writing paper wet, and the

---

21 Branddir., Holstebro, Sogneprot. 1863-1868, Haderup parish, Erik Jensen, July 2, 1863. LAV, The Provincial Archives for Northern Jutland, Viborg.

earth floors were muddy both with the moisture from the walls and with the urine of children and small animals.

The food among the ordinary heath-dwellers also had its peculiarities, and Tang Kristensen made the acquaintance of the cabbage dish *kas*, which he described in his memoirs: "They slaughtered a small pig before Christmas, and perhaps a couple of lean moorland sheep, but that was indeed all the meat they butchered, and when they ate potatoes for dinner, which I saw often, all they ate with them was a little floury dip and a very small morsel of bacon. Of so-called cooked food they ate hardly anything but cabbage, but often not the least bit of pork or other meat was boiled in it, only a few grains of barley; and with such a dish, which was called *kas*, I was not rarely regaled – for example at the home of Jesper Skrædder in Sammelsted village, which was a cluster of very scattered cottages. The wife there wanted to give me a little to eat at dinnertime, and what she gave me was *kas* … " (Kristensen 1924, 2: 158). So they slurped their *kas* together.

In a letter to Grundtvig shortly before Christmas 1873, Tang Kristensen told him about his visit to the Skrædder family. The wife, Mette, offered him something to eat, and she had heard that fine folk used a cloth on the table. So she fetched one of her husband's ragged shirts, which she spread out and laid the food on. Mette wanted to do the stranger proud, so she also served the expensive rye bread: "Then she cut bread, but her son, a snotty little lad three years old with black hands and face, grabbed hold of one piece after another. She snatched them back from him and said: 'Nay, leave off, little Peter, that's for the stranger man'," wrote Tang Kristensen. And he ate the bread too.[22]

This acceptance of the eating habits of the local population was far from common among field researchers of the time. For example, when the Scottish scholar W.A. Craigie and his young wife visited Tang Kristensen in 1897, they went on a small trip together so that Craigie could see the surroundings of the storytellers that he knew from Tang Kristensen's books: the excursion quickly came to an end. Although a woman in Skovby near Herning offered her guests the best she had, Craigie's wife in particular felt that the field conditions and the food were quite unacceptable, and she and her husband quickly returned to Copenhagen (Kristensen 1927, 4: 143).

Through his work Tang Kristensen gained great insight into the living conditions of a variety of people. In general he was not impressed by the social hierarchy of the more prosperous farming community, and he learned to admire poor people's way of surviving. Only rarely did Tang Kristensen complain about the behavior of the poor. Yet he did point out that many people lived in what he called boundless filth,

---

22   Letter from ETK to S. Grundtvig, December 18, 1873. 1929/144 II, DFS. See also: *MO.* 2, 1924: 159.

and it vexed him to become infested with vermin. Even though he appreciated being invited to drink coffee, it could be hard to swallow because of the unhygienic habits of his hosts: "But the most vexatious of all for me was the fact that the brown sugar I had to take with it had mostly been in people's mouths before. As a rule it was so beslobbered with people's saliva that many times I took my pocket handkerchief and wiped it off first before I put it in my mouth. The thing is that people were used to the utmost frugality with sugar; there seem to have been many places where they got no more than a couple of pounds of sugar a year, and it was always brown sugar. Then when someone had had a lump of sugar in his or her mouth while slurping coffee, he or she would spit it out again and put it back neatly in the sugar bowl" (Kristensen 1924, 2: 158). Despite everything, this was too much for Tang Kristensen.

In the cottages in particular, the occupants expected to get a little money for their information: "quid pro quo is the peasants' motto." Tang Kristensen hated unnecessary expenditure, but according to the heath-dwellers, if a man like him was wasting his time collecting these things, they must be valuable, so certain compensation was reasonable.[23] In 1874-75 the payment ranged from three skillings for one narrator through nine skillings for Jens Talund to 24 skillings in a third place. A bed in a humble inn could cost 40 skillings, about the same as a summer day's wages for a man.[24] If he spent the night privately with ordinary people he usually left at least 20 skillings before he departed. These amounts meant a lot to people with little money, and taken together the payments for the informants were an appreciable expense in the accounts of the project. If on occasion he bought the dish of the day at a cheap hostelry, it would cost 16 skillings; a journey with the mail coach of 27 km, for example from Ikast to Silkeborg, amounted to 112 skillings. If he went to church on one of his trips, he would put four skillings in the poor box.[25]

Although Tang Kristensen could not help being a stranger, he tried to fall in with local social norms. Often, too, people knew how to take advantage of the unusual visit, and sometimes he helped them to formulate letters about legal problems and even acted as an arbiter between spouses.

Now and then Tang Kristensen came across very distinctive people, and although he appreciated good collecting conditions, he had to conform to the informants' ways of doing things if he wanted to get anything out of them. In one of the poorhouses on Ørre Heath in 1873 he visited Johanne Tygesdatter, who turned

---

23   Letter from ETK to S. Grundtvig, November 28, 1871. 1929/144 I; DFS. See also: MO. 2, 1924: 157.

24   *Lommebøger*, 1929/14, DFS. Lommebog 3, RA, The Public Record Office, Copenhagen. Stat. Dept., Arbejderbefolkningens vilkår 1872, Ringkøbing County, Hammerum District.

25   *Lommebøger*, 1929/14, DFS. Notes from pocket book 3; cf. also no. 1.

out to be able to sing hitherto unknown ballads with a medieval sound. In the letter to Grundtvig written from the heath Tang Kristensen characterized Johanne and another woman as follows:

> I have been with an old woman who besides being deaf trembled so violently that she could hardly speak. She stood right up against me to speak to me, and I could not myself avoid being so thoroughly shaken that I could scarcely write. Each moment I was given a push, and at the same time her head was in almost direct contact with mine. She started kneading dough, and all the while as she kneaded with her old bony hands, which were almost deformed, she came over and spoke some verses to me, while her husband grumbled because the dough was being neglected. Her name is Johanne Tygesdatter, and you can see [from the transcripts] that she has preserved very rare traditions ... I suspect that [Kirsten Jensdatter's ballad] 'Bjergmand red sin vejen frem' [The mountain wight rode on his way] and [Johanne Tygesdatter's] 'Her går dans o bjerge' [The dance goes over the mountains] are unknown to you; otherwise they would be in *Danmarks gamle folkeviser*, Part 2. Indeed, I have seen the refrain to one of them there.[26]

Besides the vivid account of the meeting in this letter, we can see the mixture of exotic experiences and scholarly discussion of the ballad material in the correspondence of the two men. Although Tang Kristensen spent much of his time in sparsely populated areas, he was in contact with current scholarly discourse through the post. Grundtvig wrote that the two ballads were finds of the first class, but he was glad that it was not he who had been close to the slightly odd Johanne in the house of 27 square meters, which in Tang Kristensen's words was both dirty and messy.[27] It was a well-known fact that Grundtvig himself did not care for the smell in the peasants' cottages, and felt uncomfortable with that kind of behavior.

There is no doubt that Tang Kristensen had an eye for the "differentness" he experienced, a quality that posterity would come to appreciate. When he published his material, sometimes he could not refrain from offering his observations on the presentation of the informants, but otherwise he kept the two types of ethnographic data separate. In this way he was himself a party to the reproduction of the

---

26 Letter from ETK to S. Grundtvig, December 18, 1873. 1929/144 II, DFS. S. Grundtvig's large source edition of the ballads (Grundtvig 1853ff) was concluded 123 years later.

27 Letter from S. Grundtvig to ETK, December 29, 1873. 1929/144 II, DFS.

contemporary framework for folkloristic knowledge, while at the same time his personal field impressions called for a different presentation.[28]

Quality and Working Technique

At the end of the nineteenth century "the quality of the material" came to mean more the accuracy of the records than the rarity of the material. This shift was related to the researchers' far greater overview of the various types of ballads and tales. In the comparison of type variants, precision was crucial when scholars were to determine whether one ballad version was more complete than another, or whether they belonged to independent variants. A later era placed more emphasis on diversity and meaning, but in earlier folklore studies such a precise reproduction of what had been sung and narrated was crucial if the material was to be included in the comparative textual studies that also included archival sources.

Like so many others, Tang Kristensen was subjected to criticism. Critics have claimed that he noted far too many variants; on the other hand, other folklorists have complained that from some storytellers he only noted the initial lines of a tale or legend if he had heard it many times before. Like all other researchers of the time, in his printed manuscripts Tang Kristensen also filled out obvious "gaps" in ballads or tales with material from parallel but more complete versions.[29] He has also been reproached with not commenting on the contexts of which the collected ballads and stories formed a part. But neither did anyone else at that time; it was only later that the requirement of *culture in context* was raised. As a very young man Tang Kristensen also experienced objections from the music professor A.P. Berggreen (1801-80), who asserted in 1869 that his ballad melodies could not possibly be correctly notated, since they conflicted with the theories that were then current in musical studies. However, many years later Tang Kristensen's early records of singing in the Herning area gained full recognition in musical circles (Kristensen 1871: X, 380, Laub 1904: 195).

The most instructive account of Tang Kristensen's textual recordings was given by the folklorist Bengt Holbek (1933-92), who compared the philologist Axel Olrik's and Tang Kristensen's parallel recordings of the same tale in 1888. In many areas the preserved notes show full agreement, but the two researchers' notation techniques were different, and Olrik (1864-1917) did not have Tang Kristensen's experience.

---

28 Cf. the still relevant discussion of how disciplines are often more preoccupied with reproducing their own cultural forms than with representing other cultural forms, in Clifford 1986.

29 Cf. Rey-Henningsen 1997. The criticism formulated here is however too imprecise to be adequate. The ballad material and its editing are best discussed in Koudal 1983-84: 101f.

Olrik's manuscript is also a fair copy of his original, but scrapped notes, where he inserted missing words and passages from memory. Tang Kristensen's version is the original recording, which he made by following the storyteller all the way through the story and writing down the account sentence by sentence (Holbek 1969, 1987: 73). This seems to be a general feature of Tang Kristensen's full length original notes of ballads and stories. Only when he was very familiar with a particular story did he choose to shorten it by simply noting the differences from the more common form in the informant's individual narrative. The more than twenty thousand pages of original notes from between 1871 and 1917, mainly made in notebooks whilst in direct contact with people and later bound in 32 thick volumes, are what he should be judged on. What he and others later adjusted in their transcripts and in published editions belongs to the editing process and is bound to be clouded by the requirements of linguistic accuracy and communication.

Although Tang Kristensen also received large quantities of material from others, it was in his own primary work that he stood out. Both in Denmark and in other European countries, earlier researchers had scrapped their primary notes; many had only written the stories down when they came home, and several of the most famous folklorists, like the Grimm brothers, made what they saw as "improvements" to the material before it was presented to the public. It is in his own notes and in the quantity of examples contained in those notes that Tang Kristensen stands strong.[30] In terms of the number of researchers in Scandinavia who focused on primary ballad or tale types and in some cases on the informants' repertoires, his zeal in recording the variants in the real world probably calls more for appreciation than criticism. His distinctiveness can be summed up in the following points:

1 Tang Kristensen lamented that he never learned real shorthand like the Swedish folklorist August Bondeson. On the other hand, at an early stage he developed a very precise abbreviation system that made it possible for him to write just as quickly as the storytellers spoke.[31] In addition he had consistent and easily readable handwriting.
2 As mentioned, he wrote the material down in the presence of the storytellers, and he always kept these original notes. Where this was difficult, for example in interviews with people who were threshing or with craftsmen at work, he went at intervals into an adjacent locality to take notes, if he did not write directly in the

---

30 H. Grüner-Nielsen also singled out the correctness of ETK's ballad recordings, and his large body of example material. Cf. *Vejle Amts Folkeblad*, January 25, 1923, on the occasion of ETK's 80th birthday.
31 Re. shorthand, see Kristensen 1884: VIII.

workshop, standing by a workbench (Kristensen 1924, 2: 152, 209, Kristensen 1943: 37). On the whole he was in closer contact with his informants than was usual. For example one winter night he accompanied "Grey-Erik" 35 kilometers from his house to Viborg, during which time they spoke together of many matters, and Tang Kristensen later regretted that he was unable to sit down at the side of the road along the way and make notes.[32]

3   Early on, Tang Kristensen learned to furnish his notes with names and provenances and adhered more or less to this rule, which makes it possible today to cross-reference them, and in some cases to link the otherwise rather anonymous informants with other contemporary material.

4   From 50 years of almost continuous collecting he left such comprehensive amounts of material that its representativity must be said to be more than satisfactory, even for most repertoire studies.

5   Tang Kristensen had been familiar since childhood with the rural environment, and had the West Jutland dialect as his mother tongue, which helped him both socially and in terms of understanding conversation with the rural population.

However, the question of accuracy was far more complex than immediately appears. Tang Kristensen started by wanting to publish the tales *exactly* as they were told, but was forced to moderate this requirement. No informants narrated as if they were reciting a written text. Good storytellers always had their own special way of narrating, which in his words gave the story "color" or "scent," and at the same time the orality of the tale was accompanied by facial expressions and body language which cannot be shown in print. The written version could reproduce the content of the story, but not its form. For that reason the literary presentation would always only be an approximation of the reality, and moderate editing was necessary if the text was to cohere and be reasonably grammatical. This was also part of the background for Tang Kristensen's and Svend Grundtvig's decision not to note the material in dialect, but in standard Danish, although as far as possible with the storytellers' sentence structure and choice of words. Sometimes Tang Kristensen came across words in the ballads that neither the informants nor he understood, and so how were others then to grasp what they meant? This was the reason why, from the summer of 1874, he began to study Old Norse and the Danish language in Copenhagen. In the folklore studies of the time knowledge of the older language was considered a good, practical necessity for modern fieldwork. Finally, Tang Kristensen discovered that the singers often made up fragments of song themselves when they could not

---

32   *MO*, 2, 1924: 201.

remember a verse, and their predecessors had perhaps also done the same. A story could also be told a little differently, depending on who the listeners were. It was consequently impossible to maintain the requirement of one precise authenticity in the living cultural tradition.[33]

Sometimes Tang Kristensen used the rather heavy-handed expression that the material had to be "tormented" or "pumped" out of some storytellers. For example he said later of his contacts with Johanne Tygesdatter that he followed her all over the house, but since she was good in other respects, " ... I managed to torment five ballads out of her." Since the old woman apparently knew ballads that no-one else had remembered before, Svend Grundtvig came back to the subject in 1874 using a similar turn of phrase: "You must squeeze as much out of that Johanne Tygesdatter as is to be had," he wrote to Tang Kristensen.[34] These expressions were not always meant negatively, but described how the material was viewed as something irreplaceable, that *had* to be wrenched out of informants while there was time, in order to be included in another context that was alien to them.

Once Tang Kristensen had followed a potential informant indoors, people often claimed that they did not know much about the oral tradition, or that they had forgotten the old ballads. Then he himself had to begin to sing or give examples of tales to get the occupants going. This made Grundtvig source-critically watchful, but Tang Kristensen assured him: "I must ask your Lordship to be *fully assured* that I never put words into their mouths. I have always been afraid of precisely that, since I have been able to see the danger that lies in it."[35] The technique was necessary, but it had to be used with caution.

Following on from this, it must be mentioned that Tang Kristensen was among the few folklorists who at an early stage also wrote down the ruder everyday words like "fart", "shit", and "arse" that appeared in the songs and stories of the peasantry. Sometimes he also published these expressions, which offended some of the cultivated public. He argued that the words were elements in the rural people's everyday language, and people meant no harm with many of the expressions, just as children knew them from ordinary conversation. He omitted only blasphemous swearwords and the most sexual phrases.

---

33   Cf. Kristensen 1884: IX. See also the reflections on accuracy via checking with a phonograph in *MO*, 4, 1927: 346.

34   *MO*, 2, 1924: 160, letter from S. Grundtvig to ETK, February 14, 1874. 1929/144 II, DFS.

35   Letters from S. Grundtvig to ETK, December 8, 1870, from ETK to S. Grundtvig, December 15, 1870. 1929/144 I, DFS. ETK's emphasis.

Advice on Fieldwork – and his own Habits

In the 1880s Tang Kristensen spoke in several places about the methods and techniques of fieldwork. Several of the guidelines he gave were part of the general exchange of information in this new area of work, and his advice was based on the experience he had garnered over the preceding fifteen years. In 1883 and 1891 he published some rules for ethnographic work in postscripts to a couple of his books, which can be summarized thus:

- It is necessary to be accustomed to walking and to have a quick pen.
- It is important not to put on airs with the informants.
- One must be patient in relations with others and must be able to accept being led astray time and again.
- It is important not to press the informant to tell a tale on command, and one gets most out of people if one is willing to be with them for several days.
- It is necessary to win people's confidence and learn about their interests, thinking, and lifestyle.
- It is essential to be able to live without demanding the conveniences of life and to eat what one is offered, as well as to accept people's simple gifts if they want to give you something.
- It is of importance to be familiar with various dialects and to have some knowledge of music.
- One must not be afraid of getting scabies (in contact with people and by lying in very dirty beds).[36]

Knowledge of people's language was especially important to Tang Kristensen. In 1921, when the younger folklorist H. Grüner-Nielsen (1881-1953) was preparing for a journey to the Faroe Islands, the almost 80-year-old Tang Kristensen urged him to learn the language properly, for otherwise he would never have a good yield. Otherwise his expedition would be like those of the Copenhageners touring Jutland, where communication was hindered by both inadequate knowledge of Jutlandic and a different way of behaving. This repelled the local population, claimed Tang Kristensen.[37]

In West Jutland, he thought, it was best to be familiar with the special regional mindset if the fieldworker was to have any hope of getting people to open up to a

---

36 Kristensen 1883: 388f; Kristensen 1891: 320ff. Several of the points can also be found in ETK's letters to S. Grundtvig after 1871, 1929/144 I, DFS.

37 Letter from ETK to H. Grüner-Nielsen, April 14, 1921. 1929/152 I, DFS.

man outside the peasant class. He summed up some of the advice as follows: "It is an invariable rule that one must gain the confidence of those one wants to collect from, and it just would not do to come driving out to them or to demand the comforts of life from them. You had to be able to eat and enjoy eating the same food as they ate; and if you had no other lodgings available for the night, you had to be content to lie in their wretchedly equipped beds."[38] Although Tang Kristensen himself claimed that he was introverted by nature, he was good at winning most people's confidence when he was out in the field. But he did not try to make himself more common than he was. And he never acquired a taste for any "populist" style of dress of the kind he saw at Scandinavian teachers' meetings, where some participants appeared in "rustic" costumes. As was proper for polite people, he used the formal "De" (you) with the rural people – who were otherwise always "Du" (the familiar 'you') – and most of them answered him in the same way, unless they said "I" (plural or archaic "you"), "he," "the parish clerk," or a hasty "Du." In 1870 figures of authority with an old-fashioned consciousness of rank could still say "Du" to the peasants, which the rural population considered patronizing.

There was a difference between simply talking to people about casual subjects and deliberately recording folklore in the more rigorous forms of the tale and especially the proverb and ballad. This old "lore" required a special technique to be reproduced properly. Svend Grundtvig's suggestion that a tale could be written down from memory or a first draft noted on the spot before being checked and supplemented a week later on a new visit turned out not to be optimal. Tang Kristensen considered that ballads and good tales more or less had to be taken "on the fly" in the sense that they should preferably be committed to paper the first time they were sung or narrated. The first version was always the supplest, and if you asked the informant to repeat himself or herself, perhaps not the content, but always *the form* was poorer than the first time (cf. Ellekilde 1913: 62, Kristensen 1884: IX). In that respect his recording technique paid dividends. However, the concentrated process of transcribing was also a strain.

In many ways Tang Kristensen was conservative, and it took a lot before he would deviate from acquired views. He tried to get artists to travel with him so they could draw informants and folkloristically interesting localities, but when this failed, he overcame his resistance to the new technology of photography. In 1887, a photographer took photos of the outdoor environment for him for the first time. The great breakthrough came in 1895, when he and Olrik implemented a proper project to have older informants photographed by a professional photographer who

---

38 *Selvbiografier*: 1884a: 7f. 1929/142, DFS.

Figure 3. Photo from the 1895 expedition of one of Tang Kristensen's good informants, the almost blind Maren Jensdatter, born in 1812. Maren was married to the former smallholder, now on poor relief, "Grey-Erik," who is mentioned in the text above (photograph from Danish Folklore Archives, Royal Library, Copenhagen).

could tolerate the walking tours and the weight of the equipment. Olrik presented the successful – and in Copenhagen exotic – results at a meeting of the Royal Society of Northern Antiquaries.

Later Tang Kristensen bought his own camera. In 1907 he participated in reproducing the sound of melodies from old ballad-singers on wax cylinders for the first time. He did this with Grüner-Nielsen, who operated the phonograph recorder. Despite his skepticism the recordings were a success, many years before the tape recorder became common.

Occasionally fieldworkers experience that some informants claim co-ownership of the material or demand special payment for participating. As a rule Tang Kristensen took care of this with his small payments which compensated his informants for some of the disturbance he caused. Once when he was out on tour, one of his good informants appeared back at his home. The cottager Mikkel Skrædder from Torning had learned that Tang Kristensen had printed his stories in a small book the previous year, and now he wanted 50 copies of the book.[39] Tang Kristensen always tried to show understanding, especially for the poorer informants, but at the same time he

39   *MO*, 3, 1925: 363.

Figure 4. The only original picture of Tang Kristensen in conversation with an informant, although the situation is atypical. In 1915 the 72-year-old folklorist re-interviewed a descendant of the so called "potato Germans" (wearing a cap), whom Tang Kristensen had first visited in his home in 1874. The place is now the schoolteacher's garden in Over-Feldborg, planted with pine and spruce as shelter from the wind (Private collection).

was extremely sparing with free copies. After all, the books put money in his pocket. But Tang Kristensen was also sorry for Mikkel, who had at first been shown the door, and apparently he got his 50 copies, which he sold.

In 1884, the year after the death of Svend Grundtvig, the 40-year-old Tang Kristensen tried to take stock, and wrote an extended description of himself and his visits to people:

> I never feel better than when I am walking out in the middle of a great heath, have some snacks in my bag so I do not need to be afraid of starving for the first few hours, and I know where I will be laying my head at night. For I have constantly had to deal with these two worries – food and bed – on my journeys. And once I have overcome the shyness with people that has been in me since childhood and come into a poor cottage where I get into conversation with the occupants, then I am really in my element, since I know how to pump people and penetrate to the corners of their souls as a result of my knowledge of people and my rather sharp eye, then I can soon get people where I want them; I have almost always been allowed to study them in depth and have pressed in where no priest

or pastor has been able to go. Such an investigation is pure joy to me. In such cases I am no longer shy or taciturn, and in the great majority of meetings with true possessors of folklore I have been rewarded particularly well.

But it must not be forgotten either that in this activity I have been lucky. When I went out this way at random, it would have been strange if no good morsel had flown into my mouth. When I came to a place where there was otherwise nothing to be had, because I had been disappointed by others' statements, it has happened that just then a stranger has arrived who had exactly what I was looking for.[40]

Here the reader is introduced to the luck that no fieldworker can do without, as well as to Tang Kristensen's various worries. His pleasure in an almost inquisitorial interview technique is also clear. Many of these pieces of advice on how the good collector should behave are still useful today, and were progressive for his time. At the same time this understanding fieldworker also had less sympathetic qualities, which posterity has had a tendency to play down. But it is important to bring out these qualities too to paint a reasonably realistic picture of the famous man. Often great national collectors have been acclaimed for their superhuman sacrifices and their common touch with the "noble folk" whose traditions they wanted to save, but just as often we know very little about how they acted on an everyday basis. In Tang Kristensen's case it is possible to see a little behind the facade.

The older and more knowledgeable Tang Kristensen grew, the more impatient he was to get to the heart of the matter when he visited people. In time he thought he had heard most of it before, and in practice he was much keener on getting the information down on paper than when he advised others. Colleagues who observed some of his (later) recording situations have said that they could have the character of interrogations, and that Tang Kristensen ruthlessly stopped people when in his view they were wandering from the point (Grüner-Nielsen 1929). When he used words like "pumping" or "tormenting" the informants there was probably some truth in the expressions, and among friends he could admit that now and then it was necessary actually to scold. In 1874 Tang Kristensen wrote to Grundtvig that he had been in Hodsager to torment the pensioner Niels Mikkelsen, a man whom he knew from the preceding year: "Yes, he literally has to be tormented, and I had to make myself angry and scold him somewhat before I got him to open his mouth."[41] He himself was not blind to his robust manner, but excused himself by saying that he did it all in a good cause.

40 *Selvbiografier* 1884b. 1929/142, DFS.
41 Letter from ETK to S. Grundtvig, February 28, 1875. 1929/144 II, DFS.

A historically interested resident of Ulfborg, who visited with Tang Kristensen in 1873 a local potter who came from a family of "wise folk" (i.e. healers), described how the impatient folklorist rummaged through the potter's drawers when he did not find the man at home. He found part of a "Cyprianus" (a manuscript book of spells), which he took with him as the potter's wife watched.[42] Although Tang Kristensen knew the man from before, in his enthusiasm he breached all norms of acceptable behavior, and later when the potter came home, he was offended both because of the way his wife had been treated and because of the missing book.

Tang Kristensen's zeal to save the valuable old material from perishing could take on almost grotesque forms. A few acquaintances have told the story of how he sought out dying people to investigate whether they could tell him something hitherto unknown. In the family of the writer Thyregod, whom Tang Kristensen knew, the story was told of how he could directly order bedridden, apparently dying people not to die before they had told him what they knew.[43]

It would not be appropriate simply to weigh the accommodating and the less friendly side of Tang Kristensen's field practice against each other. He possessed several complex qualities in his character. Although his results speak their own language of efficiency and solidity, there is no reason to conceal the fact that he could also behave in a way that later times would condemn – if they knew about them. Just as the icon of anthropological fieldwork, Bronislaw Malinowski, could confide in his unofficial field diary that he sometimes hated the Trobriand Islanders, Tang Kristensen could also write in letters about his distaste for visiting "the dregs of the people" and could both gain their confidence and keep them "at arm's length."[44]

Like other fieldworkers who worked intensively, Tang Kristensen was not spared a guilty conscience about his wife and children, and his letters to Grete in particular bear the marks of almost constant marital conflicts in that respect. Sometimes the family problems were aggravated by the professional strain and the physical hardships of the field, so that he even felt like giving it up completely. In such a situation he wrote in January 1877 from the south of Hammerum Herred to Grundtvig: "Never before have I felt such a longing for home as now, and at dark moments I hardly think I can stand being away from it so long. It is a certain sense of duty that keeps me from travelling home at once (...). And it is clear that there has to be an end of it now. My wife has also got the promise from me that I will no longer leave her this way; she

---

42 Letter from Kr. Lassen Vestergaard to Gunnar Knudsen, September 16, 1919. 1906/134, DFS.

43 Thyregod 1938 [article 2/4]. The same or a similar story is said to have come from the painter V. Jastrau, who was one of the artists who went on a tour with ETK, see 1929/159Ø, DFS.

44 Malinowski 1967: 264, letter from ETK to S. Grundtvig, November 6, 1871. 1929/144I, DFS.

walks in constant fear for me, and that is not pleasant."[45] But when he came home and experienced the troubles of family life more directly, the tours began again, although in moderation; after a few years he was in full swing again, exploring new regions.

Conclusion: Moving Beyond Previous Insights

The folklorists both of the time and of posterity have been keen to characterize Tang Kristensen primarily as a collector of folklore.[46] He himself crept into the role, for he knew that the concept of collecting was at the time defined quantitatively, although it was not considered as prestigious as the title of scholar or researcher. And as a collector he was sure of his place in the first rank. Although his large quantity of types and variants is awe-inspiring and often useful, it is not the volume of his material that is most interesting in itself.

It was the character and quality of Tang Kristensen's material which made him stand out at the end of the nineteenth century, and this difference must mainly be attributed to his different way of working in the field. To the amazement of Grundtvig, Tang Kristensen found far more, denser, and rarer material than the other collectors of his day. This had its explanation in Tang Kristensen's talent and persistence, but also in the fact that he organized his fieldwork professionally, staying among the informants for long periods. Through this type of work there came a quiet shift in the view of folkloristic knowledge and how this knowledge was created.

Grundtvig and Tang Kristensen both regarded the research process as beginning with the cultural researcher's extraction of the old ballad or story from "the source" – or "the mine." This hiding-place could be the archive, the library, or *the folk* in the form of the various informants. In such a perspective, knowledge was something concrete that was stored away or hidden until it was uncovered, and often the collector almost literally had to dig the new information up. Very characteristically, the two folklorists could also speak of being out "hunting" and hope for a "good catch."

During Tang Kristensen's more persistent work in the field, often with the same informants, the information was exchanged at a slower rate than in cases where the recording process was managed as a more mechanical transfer of oral articulation to writing. Although Tang Kristensen was often restless and always thought himself that he was busy, this slowness was probably crucial for his own further involvement in the knowledge process. Through conversation and social relations with the informant, and by giving his own samples of parallel material, he helped to uncover

---

45 Letter from ETK to S. Grundtvig, January 29, 1877. 1929/144 III, DFS.

46 This was true of almost everyone as long as the classic academic consciousness remained intact. As a characteristic example, see the discussion in Boberg 1953: 179.

new layers of the informant's memory and thus to bring forth the knowledge that later appeared on paper. Without misunderstanding the word, one can say that in his closer relations with the informants he himself helped to "create" the new information, which he happily told Grundtvig he had found. Theoretically, this way of constituting knowledge is not unproblematic. However, modern fieldwork has shown that it is necessary, because the informant is a human being and cannot simply be compared to a speaking machine or a mine from which one has to wrench one's treasures. The term "collect" is in fact not adequate either, since the word lacks the reciprocity and dynamism that is important if one is to understand a little of what is really happening.

Tang Kristensen sensed at an early stage that if he really wanted to get to the heart of the people, one-way communication was insufficient. There had to be a kind of dialogue between questioner and informant, and in a letter from 1874 he noted that in almost every place he entered he was asked about his own civil and domestic life – which more refined people would have considered impolite. But Tang Kristensen knew that it was not meant as such, and he talked about himself when the situation demanded it – at least when he was younger.[47]

True, Tang Kristensen continued to use the scholarly language he had learned in his early youth, and referred without great reflection to his good registrations as "finds." But the letters from his fieldwork in the 1870s show that he did not only find the material. He often tracked it down through the informants' own network, and frequently, through correspondence with Grundtvig, had scholarly strategies behind his search for new "sources." The folkloristic information turned out to be nowhere near as limited as he and Svend Grundtvig had first assumed, and through persistent work in the field, new material could apparently keep appearing. When Tang Kristensen at first spoke of having "emptied" a particular area of relevant material, this often turned out to be a rather unfortunate expression. As early as the 1870s he revisited many of his good singers and storytellers, often at an interval of a year, and still took new material home with him in his bag.

Svend Grundtvig trained Tang Kristensen in a Romantic perspective on the folk culture material and in empirical methodology, and both of these were characteristic of the scholarly folklore studies of the time. We know only a little about how most of Grundtvig's middlemen collected their material, but there can be little doubt that Tang Kristensen, thanks to his complete reproduction of his initial transcriptions, improved the empirical authenticity of the work, even though his exclusion of the non-folkloristic ethnography followed the usual norms.

---

47   Letter from ETK to S. Grundtvig, May 20, 1874. 1929/144 II, DFS.

However, Tang Kristensen's 24,000 pages of handwritten original notes make up only a part of the information with which he came home from the expeditions, and as a rule wrote down over the next few months. The ballads and the tales were at first the basis of the comprehensive text editions he published. But he also took home a quantity of knowledge about people's thinking and lifestyles, which was used in the introductions to the source editions and in the wealth of articles and small books that he wrote alongside these. Most of these impressions were not at first committed to paper in the same way as the classic folkloristic text genres, for by prevailing standards they did not represent cultural knowledge at the same level as the clearly identifiable ballads and tales. Tang Kristensen never commented on this difference between the knowledge so accurately written down in his diaries and the unregistered impressions that he nevertheless used in his descriptions – for example in the little book *Heden* (The Heath), which was originally written in 1880 as a contribution to M. Galschiøt's large work *Danmark* 1-2 (Kristensen 1930), in which he very much wrote on the basis of his extra, unwritten knowledge – and with great success.

Every anthropologist today is aware that he or she knows more than can be read in their interviews and other written notes, and so it was for Tang Kristensen. Unlike Grundtvig, Tang Kristensen slowly became interested in a broader account of rural life and the surroundings, domestic life, and general view of life of the informants. But as in pre-functionalist anthropology, he was unsure of how one approached such descriptive analyses. In their early fieldwork under foreign skies around 1890, several anthropologists looked for a kind of refuge in the measurements of physical anthropology and in folklore, because these subject areas were more manageable than life as it was lived.[48]

Tang Kristensen did not break with Romanticism's concept of the lore in the ballads and the tales as concentrated forms of (national) culture, but in practice he opened an avenue for a more spacious understanding of culture. This broader view was, in his eyes, necessary if one wanted to paint a true picture of people's lives. This could be seen clearly in his articles and partly in the work *Det jyske almueliv* (Jutland Peasant Life; Kristensen 1891-1905). Here culture was expressed as a documentation of regional differences in human life and relations in a fuller, more descriptive sense than in the genres of ballads and tales. But he was not satisfied with the result. For him, the legendary material in particular pointed to *belief* (that is the ancient folk

---

48   Stocking 1992: 21. On his first field tours in the mid-1880s to Baffin Land, F. Boas focused on photographing, geographical and geological registrations, cranial measurements, and musical transcriptions, as well as noting local language and collecting objects for New York's American Museum of National History, cf. Jacknis 1996: 189.

beliefs), and this should not be separated from, but linked with the description of *life* as it was lived – but how did one do this in a written account (Kristensen 1892-1901, 1980: 493)? Although Tang Kristensen never gave an account of broader folk culture research in monographic form, he did transcend some of the earlier folklorists' narrower view of culture by conducting comparative analyses of the text genres.

Tang Kristensen himself did not delve too deeply into the highly varied paths he trod, and it was purely practical circumstances that led him into new paths for collecting. On the basis of his rather rationalistic view of everyday life he measured only the differences in the volume of recorded material. Although he represented a new departure in ethnographical fieldwork, the concept of the modern is hardly the right one to use in connection with Tang Kristensen. Clearly he stood with one foot in each camp, the old and the unfamiliar new. Professionally, he thought in terms of a philological tradition, but Tang Kristensen's intimate contact with the informants and their milieu – which only a few folklorists had at the time – meant that he sometimes came to different results from those that were already known, and on this basis he could also formulate new ideas.

In the postscript to *Sagn og overtro fra Jylland* (Legends and superstitions from Jutland), for example, he wrote that "The bearers of folklore are not even aware of what they know [of material]; much of what I want to bring forth lies down in the lumber-room among many other things, and when one rummages among these things it is very possible that the very best lies at the bottom" (Kristensen 1883: 388). Tang Kristensen expressed himself against the background of a conventional view that the existing folklore was the remains of one original, magnificent body of national poetry. Through his practical work, though, he had also learned that the contemporary culture was far from constituting any well-ordered totality. Nor did it need to be something of which the folk were themselves conscious, and the question was whether the traditional material had ever made up one totality. Without over-interpreting the formulation, his thinking suggests that he was no stranger to the idea that for the individual, cultural ideas could be both chaotic and unconscious. This would not surprise the researchers of today, but was a rather disturbing realization at the end of the nineteenth century.

No-one in his time took notice of the remark, and the view of rural culture as a harmonious whole lived on for many years. The most interesting thing was perhaps that none of Tang Kristensen's fieldwork norms, and in particular his full-time work in the field, had major consequences. Everyone acknowledged his activities, but the Danish scholarly milieu does not appear to have been able to absorb his new insights into its practice. For that reason his field efforts were only described as exceptionally comprehensive collecting.

In the generation after Tang Kristensen, Svend Grundtvig's way of collecting through middlemen continued, later such that the informants themselves wrote down their material and sent it to the scholars. Folkloristic armchair ethnography thus came to endure until the middle of the twentieth century, not only in Denmark, but also in many other places. Not until the last part of the twentieth century did scientific fieldwork gain a foothold, with inspiration mainly from the American and British tradition of research in the Third World.

References

Boberg, I. 1953. *Folkemindeforskningens historie*. Copenhagen: Einar Munksgaards Forlag.

Clifford, J. 1986. "Partial truths." In: J. Clifford & G. E. Marcus (eds.) *Writing Culture. The Poetics and Politics of Ethnography*. Berkeley: University of California Press.

Ellekilde, H. 1913. "Lærerne og folkemindearbejdet." In: C. Poulsen & W. Th. Benthin (eds.) *Lærerne og samfundet*. Copenhagen.

Foster, G. M. 1979. "Introduction." In: E. Colson, G. M. Foster, T. Scudder & R. V. Kemper (eds.) *Long-Term Field Research in Social Anthropology*. New York: Academic Press.

Grundtvig, S., A. Olrik, H. Grüner-Nielsen, K.-I. Hildeman, I. Piø, E. Dal, E. Abrahamsen, H. Thuren, T. Knudsen & N. Schiørring. 1853-1976. *Danmarks gamle folkeviser*. Vols. 1-12. Copenhagen.

Grüner-Nielsen, H. 1929. "Evald Tang Kristensen 1843-1929." *Berlingske Tidende*, April 10.

Hodne, B. 1979. *Eventyret og tradisjonsbærerne. Eventyrfortællere i en Telemarksbygd*. Oslo: Universitetsforlaget.

Hodne, Ø. 1979. *Jørgen Moe og folkeeventyrene. En studie i nasjonalromantisk folkloristikk*. Oslo: Studentsamskipnadens Servicesentral.

Holbek, B. 1969. "To optegnere og én meddeler." *Folkeminder*, 14.

Holbek, B. 1979. "Folkemindevidenskab." In: P. J. Jensen (ed.) *Københavns Universitet 1479-1979*. Vol. 11. Copenhagen.

Holbek, B. 1987. *Interpretations of Fairy Tales. Danish Folklore in a European Perspective*. Helsinki: Akateeminen kirjakauppa.

Jacknis, I. 1996. "The Ethnographic Object of Ethnology in the Early Career of Franz Boas." In: G. W. Stocking (ed.) *Volksgeist as Method and Ethic. Essays on Boasian Ethnography and the German Anthropological Tradition*. Madison, WI: University of Wisconsin Press.

Kofod, E. M. 1983. *Evald Tang Kristensens syn på folkeminderne*. Copenhagen: Dansk Folkemidesamling.

Koudal, J. H. 1983-84. "Evald Tang Kristensen som folkeviseindsamler og udgiver – belyst ud fra Jens Mikkelsen og Niels Albretsen i Kølvrå." *Musik & forskning*, 9.

Kristensen, E. T. 1871. *Jydske folkeminder 1: Jydske folkeviser og toner*. Copenhagen: Gyldendal.

Kristensen, E. T. 1876a. *Jyske folkeminder 2: Gamle jyske folkeviser*. Copenhagen: Gyldendal.

Kristensen, E. T. 1876b. *Jyske folkeminder 3: Jyske folkesagn*. Copenhagen: Gyldendal.

Kristensen, E. T. 1883. *Jyske folkeminder 6: Sagn og overtro fra Jylland*. Copenhagen: Gyldendal.

Kristensen, E. T. 1884. *Jyske folkeminder 7: Æventyr fra Jylland*. Copenhagen: Gyldendal.

Kristensen, E. T. 1891. *Jyske folkeminder 11: Gamle viser i folkemunde*. Copenhagen: Gyldendal.

Kristensen, E. T. 1891-1905. *Gamle folks fortællinger om det jyske almueliv*. Vol. 1-6. Aarhus: Forfatterens forlag.

Kristensen, E. T. 1892-1901. *Danske sagn*. Vol. 1-7. Aarhus & Silkeborg.

Kristensen, E. T. 1924-1927. *Minder og oplevelser*. Vol. 1-4. Viborg.

Kristensen, E. T. 1930. *Heden saaledes som den var ca. 1880*. Copenhagen: Woel.

Kristensen, E. T. 1981 [1927]. *Gamle kildevæld. Nogle billeder af visesangere og eventyrfortællere*. Copenhagen: Nyt Nordisk Forlag.

Kristensen, J. E. T. 1943. *Evald Tang Kristensen 1843-1943*. Copenhagen: Munksgaard.

Laub, T. 1904. "Vore folkevise-melodier og deres fornyelse." *Danske Studier*.

Liestøl, K. 1984. *P.Chr. Asbjørnsen. Mannen og livsverket*. Oslo: Johan Grundt Tanum.

Malinowski, B. 1967. *A Diary in the Strict Sense of the Term*. London: Routledge.

Mead, M. 1959. "Apprenticeship under Boas." In: W. Goldschmidt (ed.) *The Anthropology of Franz Boas*. Menasha, WI: The American Anthropological Association.

Olwig, K. F. 2002. "Det etnografiske feltarbejde: antropologiens arbejdsmark eller faglig slagmark?" *Norsk antropologisk tidsskrift*, 133.

Palmenfelt, U. 1993. *Per Arvid Säves möten med människor och sägner*. Stockholm: Carlssons.

Rey-Henningsen, M. 1997. "Evald Tang Kristensens eventyr." In: E. M. Kofod & E. K. Mathiesen (eds.) *Traditioner er mange ting. Festskrift til Iørn Piø*. Copenhagen: Foreningen Danmarks Folkeminder.

Sandklef, A. 1956. *August Bondeson. Folklivsforskaren – forfattaren*. Lund: Gleerup.

Stocking, G. W. 1992. *The Ethnographer's Magic and Other Essays in the History of Anthropology*. Madison, WI: University of Wisconsin Press.

Thyregod, O. 1938. "En jyde i liv og gerning." *Politiken*, April 2.

Jeremy Vetter

Field Life in the American West
Surveys, Networks, Stations, and Quarries

What is it like to do science in the field? Since "field life" has never been a unitary thing, no single account of field practice will suffice. Yet the diversity of field practices has not been simply an undifferentiated mass: close historical analysis reveals patterns of work that lie between the particular case study and generalization to all the field sciences. To develop an analytical language that can characterize the variety of work practices found in the field sciences at this middle level, I will refer in this essay to several "modes of knowledge production." Rather than using this term in the orthodox Marxist sense of a series of successive and mutually exclusive historical stages, I will use it instead to describe the different, co-existing ways that experiences of nature have been converted into knowledge through work. This essay provides a descriptive, historical taxonomy of field practice by sampling all the scientific fieldwork that occurred in a particular period and region: the U.S. Central West during the railroad era of the late nineteenth and early twentieth centuries. By "Central West," I mean the region opened up by the building of the eastern (Union Pacific) half of the transcontinental railroad in the late 1860s, encompassing the present-day states of Wyoming, Colorado, Utah, South Dakota, Nebraska, and Kansas.

The quintessential mode of knowledge production in the modern world has been the laboratory (Latour & Woolgar 1981), which has proven enormously productive and powerful in shaping the contours of the knowledge system in the same way that large-scale factory production has reshaped the material economy (Dierig 2001, Todes 2002). Like lab scientists, field scientists have aimed to make knowledge production more systematic, rigorous, and epistemically strong. In addition to expanding and continuing the long and flourishing tradition of natural history practice, field scientists found new ways to engage with the natural world that responded to the epistemic challenge presented by the laboratory. Modes of knowledge production in the field have differed from their laboratory counterparts in seeking a middle ground between the appeal of universality and the practical usefulness of relating to particular regional or local environments and cultures.

The four modes of field production that emerged most strongly in the U.S. Central West between the late 1860s and 1920 were what I call the survey, the lay network, the station, and the quarry. These modes, gradually recognized as possessing characteristics that identified them as "fieldwork" in distinction to the laboratory during its own nineteenth-century rise, evolved out of earlier modes of natural history collecting and exploratory expeditions. Surveys and lay networks have been extensive modes of knowledge production, covering relatively large geographical areas, though selected localities within them may of course be studied more intensively. A survey deployed one or more teams of trained and supervised scientific travellers to follow a route through the field. The lay network, by contrast, linked geographically dispersed observers or collectors who already lived in the places to be studied, and who often made a living in some way other than through a scientific research career. The other two modes, stations and quarries, have been more locally intensive. The station was a more or less permanent outpost in the field at which researchers conducted ongoing research work in its surrounding environment, while the quarry involved the extractive mining of scientific materials that are further processed for knowledge production elsewhere.

Surveys

The deep historical roots of surveying lie in property boundary survey practices, which stretch back centuries. As American settlers moved west, land surveying proceeded in several distinct zones (Hubbard 2009). For example, in much of the Central West, townships were mapped out based on a series of standard parallels north and south of the boundary between Nebraska and Kansas – at precisely forty degrees north latitude – which served as the base line, and a series of guide meridians east and west of the Sixth Principal Meridian, located 108 miles west of the intersection of the baseline with the Missouri River. Most of the area east of the Sixth Principal Meridian was completed by 1861. After a slowdown due to the Civil War, land mapping once again moved west, covering most of these states' western halves by the mid-1870s and the remainder by 1895 (Grim 1985: 130). The General Land Office also extended its rectangular surveying fieldwork throughout the rest of the Central West during the late nineteenth century, although some of the most inaccessible areas were not completed until the switch from contract work to direct surveying by government employees in 1910 (Cazier 1985: 100-105). With the shift to the direct system, the General Land Office established ten districts, as well as a national Field Surveying Office in Denver.

Land surveying provided the model for a much broader "survey" mode of knowledge production in the field. In the western U.S., the land surveys of the

General Land Office were one important pre-existing model for survey science, along with the U.S. Army, which undertook numerous topographical surveys in the mid-nineteenth century. By the 1870s, broader government-sponsored scientific surveys were proliferating in the same region. Four relatively large ones, led by Ferdinand V. Hayden, Clarence King, John Wesley Powell, and George M. Wheeler, competed for federal patronage, and sometimes territory as well (Turner 1987, Bartlett 1962). Although Wheeler and a few smaller operations continued the U.S. Army tradition of topographical mapping – and King was a civilian under nominal military authority – field scientists gradually emancipated themselves from military control during this period. The direct practical foundation of all the so-called Great Surveys, civilian or military, was the set of underlying topographical survey techniques developed by the U.S. Army. Topographical surveying itself produced a tremendous volume of low-level factual details whose aggregation in map form provided the basis for depicting other kinds of scientific data collected in the field. These topographical measurements are preserved in archives as seemingly endless pages of primary triangulations, traverses, calculations, and survey notes.

Gradually, elements of geological mapping and natural history collecting were bolstered and increased in visibility. Even the Wheeler Survey tried to keep up with changing standards by adding more civilian scientists to its rosters. This scientific work, in addition to topographical mapping, resulted in significant results that could be circulated in the wider scientific community. At the same time, the multi-purpose organization of the Great Surveys could create tensions on the ground when the practical needs of topographers, geologists, and collectors conflicted. Topography needed to be done basically everywhere, but the particular local environments most conducive to geology and natural history collecting were often more specific. Still, the Great Surveys made room for quite specialized kinds of research, even including such auxiliary work as Alpheus Spring Packard's explorations of Clinton's Cave in Utah during the Hayden Survey work of 1875-76 (Bocking 1988: 432). In 1879, the remaining Great Surveys were consolidated into a single U.S. Geological Survey (USGS).

After the founding of the USGS, its leaders tried (unsuccessfully) to bring federal land surveying under its own topographic branch. However, the General Land Office defended its control over land surveying by emphasizing the distinction between scientific surveys and cadastral surveys, and its historians have agreed. "Topographic surveys are made for informative purposes," one argues, while "cadastral surveys are based upon, interwoven with, and regulated by the law." Moreover, the growth of new knowledge in science was quite different from the stasis of property boundaries: "Scientific surveys made for informative purposes can be changed as more information is made available, as conditions change, or as better equipment becomes available. By way of contrast, cadastral surveys mark the lines and corners of the public lands according to specific law, and *for all time"* (Cazier 1976: 160). While such a contrast between scientific (including topographic) surveying and land surveying for legal purposes is certainly valid, there were also plenty of historical examples that combined the two. The survey of the 49th parallel boundary between the U.S. and Canada in the 1870s from Lake of the Woods to the Rocky Mountains, for example, was fundamentally designed to map a fixed line legally defined by treaty, yet it also included substantial topographic mapping and even natural history collecting (Rees 2007). Nevertheless, such a precedent did not allow the USGS to prevail in its desire to absorb all federal land surveying.

Accordingly, USGS geological and topographical field teams were the main vehicles of the agency's survey science. But other kinds were also under USGS authority, including forestry surveying (for a brief period in the late 1890s); hydrography (surface waters); and hydrology (underground waters). Gradually, the USGS topographers completed the mapping of the Central West, also reporting the data ("spirit leveling") in tables for each state. Even though the USGS represented a form of national authority over the region, most of the work done by USGS field parties was either popular or at least uncontroversial in most parts of the West. The most conspicuous exception was irrigation survey work, whose unpopularity due to withdrawals of land from legal settlement led to Congressional investigations in the early 1890s.

Many USGS field parties reported to the Rocky Mountain Division during its existence as a regional center for the production of knowledge. The Division flourished in the Denver area in the 1880s, establishing a semi-permanent resident corps of research scientists. Under the direction of Samuel F. Emmons, they conducted field work across the mountains of Colorado and neighboring states, including a highly-regarded report on the mining region around Leadville (Emmons 1880-81). This field research was accomplished in close collaboration with the mining industry, through such co-operative organizations as the Colorado Scientific Society. However, within a few years of Powell taking over the directorship of the USGS from King in 1881, he moved to consolidate the branch offices in Washington, D.C. – though he still included the Colorado Division as one of twelve geologic divisions between 1884 and 1893. Despite the Rocky Mountain Division's short-lived status, by accumulating some scientific authority in the West itself, Emmons and his colleagues developed a decentralized version of federal survey science.

Along with the USGS, which later also sponsored the Yellowstone Division of Arnold Hague, the other offspring of the consolidation of Great Surveys in 1879 was the Bureau of American Ethnology (BAE). Although the word "survey" was not in the BAE's title, it nevertheless engaged in considerable survey work, including mapping. As one leading historian of the BAE has argued, the "survey impulse" especially predominated in BAE activities until the publication of the national map of Indian languages, produced in 1891 (Hinsley 1981: 151). This linguistic mapping project was a major obsession of its first director, Powell, who during much of this period also directed the USGS. "The immediate purpose in instituting these researches and in organizing the Bureau in 1879," argued Powell in an early 1890s report, "was the discovery of the relations among the native American tribes, to the end that amicable groups might be gathered on reservations," which he believed could best be done through the study of Indian languages (Powell 1892-93: xxvii).

Another important federal patron of survey science in the Central West was the U.S. Department of Agriculture (USDA), whose research activities exploded in the 1890s and into the early twentieth century. At the USDA, the most important branches for survey work during this period were C. Hart Merriam's Biological Survey and Milton Whitney's Bureau of Soils. Just as Powell sought to construct a national map of Indian languages through the survey work of the BAE, so too did a mapping project constitute the center of research at the U.S. Biological Survey. For Merriam, this task was the preparation of a national life-zone map, first published in preliminary form in an 1890 report on his fieldwork in the Southwest, then in revised form in 1894 (Merriam 1890, 1894a, 1894b).[1] Within a few years, it was supplemented by a crop-zone map, and the two were published together in 1898. "The Biological Survey aims to define and map the natural agricultural belts of the United States," wrote Merriam, "to ascertain what products of the soil can and what can not be grown successfully in each, to guide the farmer in the intelligent introduction of foreign crops, and to point out his friends and his enemies among the native birds and mammals, thereby helping him to utilize the beneficial and ward off the harmful kinds" (Merriam 1898: 9).

To supplement his own life-zone map, Merriam also included maps of where cereal crops were already grown. This list and set of crop maps were prepared by the director of the Indiana Agricultural Experiment Station at Purdue University, C. S. Plumb. In this case, because the maps to a great extent reflected actual farming practice in the 1890s, most of them were focused on the eastern half of the U.S. (Plumb 1898). Yet despite this public-minded focus on practical utility in the reports of the U.S. Biological Survey, the actual fieldwork on the ground was dominated

---

1 A black and white version appears in Merriam 1894c. On Merriam's fieldwork in the San Francisco Mountains of Arizona to develop the life-zone concept, see Maienschein 1994.

by a zoological agenda focusing on the geographical distribution of various animal species. Fieldwork was accomplished by very small survey groups of two or three men, with the most extensive work in the West done by Vernon Bailey in the late 1880s and 1890s.

Likewise, in 1898, the U.S. Soil Survey initiated its fieldwork (Weber 1928, Helms 2002). The next year, what would become known as the National Co-operative Soil Survey began operation, directing co-operative work between the USDA and individual states. Intensive soil surveys, usually of individual counties but sometimes based on other geographical units such as river valleys, expanded in number during the 1900s and 1910s as experienced field surveyors proliferated. In the Great Plains region, Whitney initiated a more extensive form of soil survey covering entire halves of states such as South Dakota, Nebraska, and Kansas. Soil samples followed a trajectory from the field to the chemical and physical laboratories of the USDA Division of Soils. When, beginning in 1912, funds for chemical investigation of soils were appropriated by Congress, the chemistry division analyzed soils and published bulletins describing "soils of geographic areas or the soils on certain parent materials" (Flach & Holzhey 2002: 67). Also, between 1913 and 1917, Bureau scientists undertook summer reconnaissance survey work in many national forests, ostensibly to "locate agricultural land," but also involving the production of soil maps (Roth 2002: 192). This dramatic expansion in survey work by the Bureau of Soils made it one of the most active federal agencies for producing environmental knowledge of the region in the early twentieth century.

In addition to the Biological Survey and Bureau of Soils, other parts of the USDA also funded field survey work. During the 1890s, for example, the Division of Botany hired many field agents to conduct summer surveys, including several students of Charles Bessey at the University of Nebraska. Longer lasting was the U.S. Commission on Fish and Fisheries, established by Spencer Baird in 1869 in response to concerns about the decline of the Atlantic Ocean fishing industry. While the Fish Commission did very little of its work in the Central West, in the spring of 1890 it sent the young ecologist Stephen Forbes on a summer field survey trip to Yellowstone along with field associate Edwin Linton, a professor at Washington & Jefferson College in Pennsylvania, as well as "a guide, a teamster, two packers, and a cook" (Croker 2001: 133-134). Such an extended labor force was indeed common to many examples of survey science in the American West.

By the early twentieth century, surveys were also increasingly funded by states in the West, asserting a knowledge-making domain over and promoting economic development in their own geographical territories. State geological surveys developed out of earlier state expert positions that had generally lacked funds for extensive fieldwork. The first state surveys in the Central West were undertaken in Kansas in the mid-1860s by B. F. Mudge and G. C. Swallow (Buchanan 1981, Page 1984, 1996). These were not fully comprehensive surveys, however. The early Kansas surveys focused their field survey work almost entirely on the eastern half of the state – "on account of Indian troubles," according to Swallow – and were concerned above all with the location of coal resources, which were such a vital energy source for the settlement of the Plains. "In a State, where, to a great extent, prairie covers the surface of the country," Mudge observed, "the question of fuel becomes of the first importance" (Mudge 1866: 17, Swallow 1866: 29). On the other hand, most territorial and early state governments in the Central West contributed no funding for geological surveys until later in the century; it was only in the 1890s and 1900s that the other states began to fund large-scale survey work (see Table 1).

State	Directors
Kansas (1864-65, 1889-)	B. F. Mudge (1864) G. C. Swallow (1865) Erasmus Haworth (1895-1915) W. H. Twenhofel (1915-16) Raymond C. Moore (1915-37)
Wyoming [Office of the State Geologist] (1890-)	Henry C. Beeler (1901-08) Edwin Hall (1909-10) Claude E. Jamison (1911-13) Loyal W. Trumbull (1913-19) Glenn B. Morgan (1919-23)
South Dakota (1893-)	J. E. Todd (1893-1903)* Ellwood C. Perisho (E dist, 1903-07; all 1907-14)* C. C. O'Harra (W dist, 1903-07) Freeman Ward (1915-26)
Nebraska (1899-)	Erwin H. Barbour (1899-1918) George E. Condra (1919-21)
Colorado (1907-)	Russell D. George (1907-27)
Utah [University Geological & Resource Survey] (1919)	Frederick J. Pack (1919)

Table 1. State Geological Surveys in U.S. Central Plains and Rocky Mountains to 1920. Sources: Socolow, A. A. (ed.) 1988. The State Geological Surveys: A History. Grand Forks: Association of American State Geologists; Vanorny, P. M. 1970. A History of the South Dakota Geological Survey. Vermillion: University of South Dakota; Bryans, W. 1986. A History of the Geological Survey of Wyoming. Laramine: The Geological Survey of Wyoming. *Different term transition year given in Socolow 1986.

While geology dominated the state survey scene, other kinds of survey fieldwork sometimes found support, although usually from university and agricultural experiment station budgets and staff, rather than directly from government appropriations. Francis W. Cragin at Washburn College in Topeka, Kansas formed a biological survey of that state in the mid-1880s. It was hailed as the first biological survey anywhere in the country, possibly the world, at least in name. Officially launched in 1884 after a few months of preliminary operation, it was intended, in Cragin's words, to be "an informal biological survey of Kansas" that would "probably require five, and perhaps ten, years for the accomplishment of its objects." The survey would include botany and zoology, but not geology. Cragin invited the co-operation of "all intelligent and public-spirited citizens of the State" through collection and donation of specimens. While local collectors from across Kansas volunteered, many early participants were out of state natural history collectors from the Midwest and East Coast (Cragin 1884: 2-3). By 1890, Cragin was pleading with the college trustees to reimburse him for publishing the *Bulletin of the Washburn College Laboratory of Natural History*, in which the survey's results appeared.[2] As he pointed out to them, only one (very small) issue of the *Bulletin* out of the last five had been fully funded by the College, leaving Cragin to pay the rest of the costs himself.[3]

Other colleges in the region followed with biological surveys of their own.[4] In Nebraska, significant plant survey work took place under Charles Bessey and his students in the University of Nebraska's Botanical Seminar ("Sem Bot"), founded in 1886. The Sem Bot launched the Nebraska Botanical Survey in August 1892, after some general planning earlier that summer ("The Botanical Seminar" 1906; "Preliminary" 1892).[5] Likewise, faculty and students at Colorado Agricultural College undertook collecting trips around their state during the 1890s, leading to

2   F. W. Cragin to Board of Trustees, August 2, 1890, Washburn University Archives (subsequently WUA), Box 2.7.

3   Receipt for payment, April 3, 1890; F. W. Cragin to Board of Trustees, June 24, 1891; both in WUA, Box 2.7. Upon Cragin's departure the following year, the survey ceased to exist. A new "Kansas Biological Survey" was only established a few decades later. Based at the University of Kansas, it does not appear to have sent out summer field parties until 1910-12 (Bunker 1913, Gunthorp 1913, Iseley 1913, Hungerford 1919).

4   The shift to biological surveys was a nationwide trend, but the best supported early efforts were in the Midwest, especially in Illinois. For the argument in favour of focusing on biological surveys as a way to move beyond "purely practical" work at the agricultural experiment stations, based on a talk delivered at the meeting of the Association of American Agricultural Colleges and Experiment Stations, see Weed 1891. See also Kohler 2002.

5   See also "Minutes and Proceedings, 1886-1900", Sem Bot (Botanical Seminar) Club Records, #12/7/5, Box 1, University Archives, Love Library, University of Nebraska, Lincoln. For further details, see Tobey 1981: 14-21.

the production of a *Flora of Colorado* by Per Axl Rydberg of the New York Botanical Garden in 1906 (Rydberg 1906). University of Colorado Museum curator Junius Henderson, in collaboration with zoologist T. D. A. Cockerell and botanist Francis Ramaley, led summer expeditions to various parts of Colorado in the early 1900s (e.g. Dodds et al. 1907). This work developed into a "Colorado Biological Survey" by the 1910s, with fieldwork undertaken by Cockerell, Ramaley, and their students.[6] This work eventually culminated in synopses by the two researchers in the late 1920s (Cockerell 1927, Ramaley 1927). In Wyoming, botanical survey work centred around Aven Nelson and his Rocky Mountain Herbarium, for which he collected not just around the state but throughout the entire region (Williams 1984).[7] Neither Utah nor South Dakota seems to have had much in the way of state-wide plant survey work at all – though, in South Dakota, the geological survey expanded in 1903 under new director Ellwood Perisho to cover plants, animals, and artifacts (Vanorney 1970: 17).

Not all survey work was done by government agencies and state universities, however. By the end of the nineteenth century, some urban museums were funding more field expeditions. While the primary aim of such efforts was to collect specimens, in some cases museums attempted such comprehensive geographical coverage of particular regions that their work took on the complexion and terminology of survey science. This was especially true for work on contemporary human beings, which was not typically pursued with any vigor by state survey agencies. The survey era in the anthropology has often been obscured in accounts of the transition from armchair theorizing to field ethnography. However, even such an ethnographic luminary as Franz Boas did fieldwork in the Northwest Coast region in the survey mode (Cole & Long 1999).

One elaborate example of survey anthropology was the Plains Indian research of the American Museum of Natural History in New York, organized by Clark Wissler in 1907. Connected to this museum effort at fieldwork on contemporary Plains Indians was an archaeological project called the Huntington Southwest Survey, organized by Wissler in 1909 (Snead 2001). As a mode of knowledge production, the survey is a preeminently regionalist project, leading to systematic study of

---

6   Despite the formal name and its connotation of more systematic coverage, most research was done around either Boulder itself or the University's biological station in the Rocky Mountains at Tolland (Vetter 2010).

7   In addition to Nelson's own field travels, students sometimes collected under Nelson's direction across the Rocky Mountain region and were guided in much the same way as survey staff members at other institutions, as revealed in his letters written to them in the field, such as Aven Nelson to Mr. Edward Payson, July 26, 1913 and Aven Nelson to Edwin [Payson], May 8, 1914; both in University of Wyoming, Botany Records, #545001, Box 29, Folder 13, American Heritage Center, University of Wyoming, Laramie (subsequently AHC-UW).

geographically delimited areas. Thus, Wissler's work in the Plains, the Southwest, and other regions led eventually to his own survey introduction to American Indian anthropology with classification by culture areas, including Plains, Plateau, and Southwest (Wissler 1917, 1920). Other museums also engaged in some survey anthropology, although not as systematically. George Dorsey, who was curator of anthropology at Chicago's Field Museum from 1899 to 1915 (after serving for a few years as assistant and acting curator), engaged in fieldwork and collecting among a variety of Indian groups in the American West, including the Arapaho and Pawnee.[8] Insofar as these museum researchers were responding to contemporary standards for systematizing their collecting of artifacts and observations over particular tribal groups and geographical areas, they were participating in a variant of survey science – and indeed, in some cases they explicitly claimed to be doing so.

The pervasiveness of heightened standards for surveying practice was evident in the Central West by the early twentieth century. Even regional museums were becoming involved. At the Colorado Museum of Natural History in Denver in the mid-1910s, director J. D. Figgins (1913: 20) sent biological survey teams into the field with instructions that closely followed the metropolitan practices of collecting large series and focusing on common species as much as rare ones. Moreover, his devotion to higher standards precipitated frequent conflicts with his collectors in the field. This case is interesting in part because it shows a museum engaged in regional survey work on subject matter – birds and mammals – that was also the focus of the national Bureau of Biological Survey. As a museum in a growing provincial metropolis trying to increase its standing to the upper ranks of urban museums, the Colorado Museum chose to focus initially not on collecting specimens from around the world, but on implementing systematic and rigorous survey methods within its own state boundaries.

Lay Networks

Unlike surveys, which sent out trained scientific observers and collectors to cover a geographical territory, lay networks relied on people already living in the area of interest, or to those who travelled to new places for other purposes. The issuing of instructions to travellers for making observations dates back at least to the early Royal Society under Robert Boyle and the earlier project of Samuel Hartlib that inspired it (Carey 1997). By the nineteenth century, some disciplines such as geography and anthropology had begun to rely more heavily on data reported from

---

8   F. J. V. Skiff, "History of the Museum, Department of Anthropology" (1916), Library, Field Museum of Natural History, Chicago (susequently FMNH).

networks of observers in the field, whether travellers or long-term residents. This tradition was well developed in the British Empire, where questionnaires circulated from scientific associations were an attempt to systematize the information that could be collected from individuals scattered around the world. In 1874, the British Association for the Advancement of Science issued its anthropological *Notes and Queries*, which had antecedents by the Ethnological Society of London in 1843, and the British Association itself in 1851 (Urry 1972). *Notes and Queries* went through several more editions into the mid-twentieth century, including two in the 1890s. The Royal Geographical Society likewise produced its *Hints to Travellers*, first issued in 1854 by Captain Robert Fitzroy and Henry Raper with a focus on land surveying in the field. By 1906, the *Hints* was already in its ninth edition (Collier & Inkpen 2003, Driver 2001).

In the Central West by the late nineteenth century, relying on local people or travellers to make observations and collections was often a matter of necessity given the lack of funding to send out trained personnel. This was especially true at the state level, where the collection of observational data through networks substituted for more expensive surveys by state employees. The late nineteenth century was a period in which state government science was usually the province of designated "experts," such as the state chemist, state entomologist, and state geologist, working alone without significant support for staff or fieldwork. Many of these state experts were not even paid adequate salaries, instead receiving only small sums for preparing brief reports every year or two. Under such circumstances, state experts had to rely on inquiries to local people scattered through their state domains, either by circular or individual solicitation. J. Alden Smith, who was Colorado State Geologist in the early 1880s, depended not just on USGS scientists stationed in Denver as part of the Rocky Mountain Division, but also on a whole network of observers who reported the minerals in different places throughout the state's mountainous section. Smith gave the names of these observers at the beginning of the list, noting that a "single initial after the notice of [each] mineral indicates the name of the gentleman who first observed it" (Smith 1883: 137).

Federal agencies also used networks to produce knowledge from the field. The USGS, for example, under lead geographer Henry Gannett, collected its altitude information by relying on a wide variety of "authorities," including railroad companies, military and civilian expeditions, and others (Gannett 1884). In a more focused approach, the knowledge on each different mineral industry compiled by the USGS Division of Mining and Mineral Resources was mostly gathered by undertaking "a complete canvass of the subject among all the producers in the United States," supplemented by field agents only when absolutely necessary. Government employees sustained this network by maintaining a list of mineral producers, which was updated by asking each of them for the names of nearby producers. Only after three failed requests for information would the USGS go to the effort and expense of sending out a field agent to get the information (*United States Geological Survey* 1904: 48-49). For example, the reports on coal assembled by Charles Ashburner for 1885 included compositional analyses at various mining sites, as well as aggregate production and employment statistics. The extent and quality of the data depended on both environmental resource endowments and the co-operation of mining producers to provide the information, so that while Colorado had ample data from many different sites, indicating fifty productive coal mines in the state employing approximately 2200 miners, the section on the Dakota Territory was very brief and reported very little data (Ashburner 1885).

Departments of the USDA such as the Bureau of Biological Survey also used circulars to collect national information. In his 1893 report on spermophiles, for example, field surveyor Vernon Bailey relied not just on his own excursions into the field but also on a whole network of corresponding local observers. These responses were elicited from circulars distributed throughout the country, including an initial, more general circular in 1886 followed by a more specific one in 1888-89 on spermophiles and pocket gophers. Based on the replies to the second circular, Bailey mapped the ranges of these two animal species "with greater accuracy and detail than otherwise could have been accomplished. Each locality from which specimens were examined was marked on a large map of the United States devoted to the species." The geographical data from these maps, in combination with facts observed on the ground through survey work, were then converted into distribution maps. The large size of the network made it virtually impossible for Bailey to give individual credit to those who participated. "It would be impracticable, even were it desirable," he wrote, "to mention the large number of correspondents who have contributed data." Local collaborators had submitted "specimens, other valuable notes on the range or habits of the several species," and replies to specific questions. Only a few correspondents merited mention as having contributed observations on especially rare varieties, such as the striped spermophile (Bailey 1893: 11-14, 34-38).

Another avid user of networks to produce knowledge within the USDA was the Weather Bureau, which became part of the Department in 1890 after several decades under military authority. "Weather observing networks, probably more than any other kind, [were] continuously dependent on a regular geographical distribution of collaborators," because weather phenomena were often large in scale rather than localized. The weather observing network was a key example of a lay network based on the telegraph and centralized control (Vetter 2011: 261).

Like their counterparts at the national level, state agencies also collected information through networks. One major category of state-centered network was for agricultural experiments. Like weather forecasters, agricultural researchers wanted to test varieties in particular locations spread throughout a geographical area. Due to funding limitations, only a relatively small number of branch station sites could be operated directly by state researchers, but scientists could vastly extend the area covered by relying on interested local farmers. It was not usually difficult to find farmers willing to co-operate by testing varieties. In the experience of Samuel Salmon in Kansas, for example, "the number of cooperating farmers was usually limited only by the capacity of the experiment station to handle them. They might number 100 or more in a single year" (Salmon n.d.: 32). Results obtained from the variety tests at local farms would then be reported as data to the organizers of the network. Officials at the USDA actively encouraged such arrangements as an alternative method of testing varieties, instead of spending scarce agricultural research funds on branch stations or model farms.

Government agencies were the primary institutions to produce knowledge through systematic lay networks of observation and reporting. It is true that researchers at museums and universities also relied on local observers and collectors. At the American Museum in New York, for example, anthropological fieldwork under Clark Wissler sometimes relied on relationships with local observers who were not career scientists, such as James R. Walker, who was physician to the Pine Ridge agency in South Dakota (Walker 1980). In such cases, however, particular individual relationships of mutual trust functioned in much the same way that state and federal authority did in the case of government networks. Observational networks for museums and other similar entities also tended to be less comprehensive in their geographical scope: *ad hoc* and particular rather than systematic and national or state-wide.

Stations

Unlike surveys and lay networks, with their geographical breadth, field stations were designed to produce knowledge more intensively, in and around a particular environmental location. The most common and best funded type of field station in the Central West during the formative period before 1920 was the branch agricultural experiment station. These stations were associated with individual states, though much of their funding was provided at the national level through federal government measures, especially the Hatch Act (1887) and the Adams Act (1906). But state-operated agricultural experiment stations did not spring up full blown without precedent; they had precursors. Many of these incipient field stations were sponsored by either the U.S. Army or railroad companies, which were both crucial institutional forces in the settler expansion into the West.

U.S. army posts often included experimental farms, although these could be severely challenged by environmental conditions in the Central West (Adams 1989: 42-43). These farms and kitchen gardens grew out of attempts to provision distant outposts locally; their "experimental" nature was not explicitly designed as such, but was the practical effect of trying to grow food in new western environments (Stewart 1982). However, despite some success with military farms in eastern Nebraska and Kansas along the Missouri River, similar U.S. Army attempts in locations in the West, such as Fort Kearny in Nebraska and Fort Laramie in Wyoming, did not do as well. "Despite their attempts to provide some food for travelers on the Oregon Trail," one historian writes, "the respective Nebraska and Wyoming posts lacked the manpower for agriculture, and they did not possess the water supply and fertile soil that Fort Leavenworth [in eastern Kansas] had" (Tate 1999: 158). Even when large scale farms proved infeasible, post gardens were common. Due to the importance of nutritious food in keeping soldiers healthy, gardens at military forts were typically operated by the post surgeons. And the building of railroads and stage lines slowly eliminated the need for experimental farms and gardens. Wyoming's Fort Fetterman, for example, after "several successive years of bad weather ... became increasingly reliant on crops purchased from civilians at Greeley and Fort Collins, Colorado," while Fort Dodge in western Kansas ended its "several years of marginal rains and intense summer heat" by relying on the Atchison, Topeka, and Santa Fe Railroad after it reached the fort in late 1872 (Tate 1999: 162).

Yet even after the arrival of railroad lines, some posts continued to experiment, adapting to the arid environment by making advances in well drilling and irrigation in 1870s and 1880s. In some cases, nearby surface water was simply channeled into post gardens. At western Nebraska's Sidney Barracks, for example, Lodgepole Creek "provided year-round sustenance" for 3.5 acres of farm land. The experimental work, especially attempts to grow crops on alkali soil and pioneering attempts at irrigation, provided a model for agricultural improvement groups in the eastern part of the state as they began to cast their eyes to the more challenging West (Tate 1999: 163). When adequate surface water was not available forts sometimes experimented with deep well drilling, such as the irrigation system based on windmill and pump at Fort Wallace in western Kansas in the 1870s.

Even as the railroad companies extended their tracks into the West, thereby facilitating long-distance provisioning of U.S. Army forts in place of local experimental farms, they also inaugurated their own experimental farming projects. These functioned to a large extent as model farms for public display in order to attract settlers to the region. The problem of displaying results to the public in the history of science has a long pedigree, especially the experiments conducted in laboratory sciences. "Demonstrations have the power to convince," contends Harry Collins, "because of the smoothness of performance, distancing the audience from the untidy craft of the scientist – caging Nature's caprices in thick walls of faultless display" (Collins 1988: 728). Likewise, in their model farms, railroads attempted to convince the public of the efficacy of settler agriculture, although the field environment was no doubt harder to control than the laboratory performance. As shall become clear below, the agricultural stations that came later did not wholly retreat from this emphasis on model or demonstration farms.

Settler farmers' individual efforts to learn what kinds of plants grow well in the Great Plains environment have sometimes been described as "experiments." Elam Bartholomew, for instance, tried small plots of "Egyptian or Rice corn" in 1879, along with some experimental orchards consisting of ten different apple tree varieties (Bartholomew 1998: 37). Sometimes, men of science reported the results of what settler pioneers discovered about their new land. In the late 1870s, farmers in the drier part of Kansas were producing new knowledge of the land, as reported at the Kansas Academy of Science by H. R. Hilton: "Our Western farmers have been experimenting in the past few years, and the stimulating effect of the recent drouth [sic] has not been without fruit. New discoveries are being made each season of some new plant more especially adapted to the climate and soil." In particular, a man named Mr. Bennyworth at Larned in the central part of the state spent $25,000 demonstrating the possibility of sugar cane manufacture, while a farmer named Mr. W. H. Gill experimented with the deep plowing of eighty acres of wheat in the same county. Still others were building an irrigation ditch near Garden City (Hilton 1879-80: 44). The latter project attracted additional notice at the Academy by another presenter named F. G. Adams, who reported how it had "excited the interest of the people" in that part of the state (Adams 1879-80: 79).

Early farming experiments by army posts, railroads, and individual boosters were important for the public image of the West, but few results were ever turned into knowledge that could circulate in the scientific community. Still, they were part of the continuity of agricultural experiment station development in the region. When states established official stations after the passage of the federal Hatch Act of 1887, they usually portrayed their efforts as fundamentally new. As nodes in a network linked to the international scientific community through their personnel and reports, they certainly were significantly different from earlier experimental efforts. Yet they also shared some similarities. Most important, early agricultural experiment stations, especially the branch stations located in particular environmental regions,

Figure 1. USDA Office of Dryland Agriculture field stations in the U.S. Central West region in the early twentieth century. Cropped from a map appearing in E. C. Chilcott, J. S. Cole, and W. W. Burr, Spring Wheat in the Great Plains Area: Relation of Cultural Methods to Production, Bulletin of the U.S. Department of Agriculture, no. 214 (1915), p. 2.

were often operated as model farms at least as much as centers for the production of new knowledge. Across the Central Plains, branch agricultural experiment stations operated in tension between the demands of rigor for scientific publication, reinforced by the supervisory efforts of the USDA Office of Experiment Stations, and the need to demonstrate practical techniques to the settlers who were pouring into the region in the early twentieth century.

The USDA itself established some stations in the early twentieth century, many of them in collaboration with the states. Especially prominent on the Central Plains were the stations affiliated with the USDA Office of Dry-Land Agriculture (see Figure 1), which were intended to produce knowledge for the region's semi-arid environmental conditions (Quisenberry 1977). Dispersed across the Plains, these stations "were located with a view to having each of them on soil representative of extensive areas," and they were intended to "represent nearly all of the important soil types to be found in the strictly dry-farming portions of the Great Plains" (Chilcott, Cole, & Burr 1915: 5). Such an extensive, co-ordinated effort would enable claims to be made at a larger regional level than was usually possible through station-based research.

Agricultural stations might have been the most common example of this mode of field production in the Central West, but they were by no means alone. A long-standing agrarian interpretation of American expansion should not blind us to the reality of the industrial making of the West (Andrews 2008, Isenberg 2005). Far from being the sole exemplar of development and the economic need for knowledge, agriculture was just one of many important economic activities. Other resources were important as well – and not just in the industrial sector, but also ancillary extractive work with minerals and forests.

Some states also applied the station model to urban industrial problems by establishing engineering stations. After these stations first emerged in the industrial belt of the Midwest, engineering experiment stations were also founded further west, at agricultural colleges in Kansas (1912), Colorado (1917), and Utah (1918) (Seely 1993: 350). Despite the failure of a national movement during World War I to provide federal funding for an engineering experiment station in each state after the model of the Hatch Act (Kevles 1971), states did in many cases provide patronage for such stations. In their early years, these research sites were much smaller than the agricultural stations and did not necessarily involve much additional expense for the states. The station in Logan, Utah, for example, was staffed by faculty members already at the college and very little original research was conducted there until the late 1950s.[9]

Yet, even more than urban manufacturing, the great economic activity of the West was mining, including in the Rocky Mountains and Black Hills. The Colorado School of Mines established two stations for mining experiments outside of its Golden campus: the Edgar Mine at Idaho Springs, which served as a "working laboratory for the purposes of instruction, experimentation, and tests" and a geological and petroleum camp at Wild Horse Park, fifteen miles northeast of Pueblo (Hoyt 1949: 5). The U.S. Bureau of Mines, founded in 1910 primarily to address safety issues in Eastern coal mines, also undertook station research in the West, opening several experiment stations there, including its first one on the University of Utah campus in Salt Lake City (1914) and a Rare and Precious Metals Station a few years later on the Colorado School of Mines campus (LeCain 2004).

---

9 Richard Merrell, "Engineering Experiment Station" (1961), Utah State University Historical Materials, #10.2, Box 18, pp. 1-2, Tanner Reading Room, Merrill Library, Utah State University, Logan.

In addition to industrial and mining stations, several other kinds of places functioned as sites for knowledge production under the station mode of field science. Like the agricultural stations, these research sites involved attempts to manage living organisms. Most similar to the station farm, though usually located in a more urban setting, was the botanical garden. One prominent example in the Central West was the botanical garden at the University of Denver, established by Ira E. Cutler of the biology department. After the founding of a school of pharmacy, with encouragement from the new school's dean and the Denver Pharmaceutical Association, Cutler plotted in the fall of 1916 to establish a permanent site for the botanical garden on the campus. Like many of the agricultural stations, the botanical garden was organized and managed with the assistance of the USDA.[10]

For Cutler, the entire University of Denver campus became in effect a field station for research on living organisms in the regional environment. Cutler's experimental work – though on animals rather than plants – began even earlier, when he treated the University campus itself much like a rudimentary urban agricultural experiment station. Around 1902, he began work on pheasants, in particular "a pair of Chinese Ring-neck pheasants purchased from an old German who lived in the city." Cutler aimed to "so far domesticate the birds that they might be allowed their liberty around a home without fear of their turning feral," and he systematically recorded the results of each season's attempt to raise pheasants under domestication.[11] He also propagated rose plants around the campus from a single plant given to the Chancellor in about 1900. And during the First World War, Cutler performed trials on Indian corn varieties for the U.S. government on a "corn experimental plot" he maintained nearby. But above all, his work was identified with the botanical garden itself, which would later be described as "the largest botanical and pharmaceutical garden in the west" ("Dr. Cutler" 1973: 118-119).

While research on forests took place in the Rocky Mountains – as islands of arboreal vegetation in a vast sea of grassy plains – there were also tree-growing experiment stations in what would seem to be the most unlikely places. Across the Central Plains, attempts to transform the environment through tree planting were legendary. By the late nineteenth century, these efforts had evolved into systematic efforts to create forests on the vast, treeless expanses of western Nebraska and Kansas. In 1887, Kansas established two forestry experiment stations in Trego

---

10 "History of Establishment of Botnical [sic] and Pharmaceutical Garden" (n.d.), Ira E. Cutler Collection, Box 1, Folder 2, Special Collections, Penrose Library, University of Denver, Colorado (subsequently SC-PL-UD).

11 "Experimental Work with Pheasants" (n.d.), Ira E. Cutler Collection, Box 1, Folder 2, SC-PL-UD.

and Ford counties, co-operatively administered by the state forestry commissioner and Kansas State Agricultural College in Manhattan (Drake 2003: 24). Efforts in Nebraska were at first more *ad hoc*. In 1891 botanist Charles Bessey convinced his entomologist colleague Lawrence Bruner to let USDA forester Bernard Fernow provide funds and materials for tree planting experiments on property he owned in Holt County in the Sandhills. During the next few years, Bessey continued to promote the idea of planting trees in the Sandhills to the state agricultural and horticultural associations (Overfield 1993: 140-141).

This preliminary work eventually led to what was undoubtedly the most notable effort at Plains reforestation: the planting of the Nebraska National Forest near Halsey in the middle of the sandhills region beginning in 1904. Linked to the establishment of a forestry school at the state university in Lincoln, the Nebraska National Forest became a major site for forestry research, especially the science of growing forests from scratch (Gardner 2009, Pool 1953, Scott 1903). In 1905, Kansas followed with its own plans for a 30,000-acre Kansas National Forest near Garden City (Drake 2003: 30). Meanwhile, in the Rocky Mountains, where forests already existed in abundance, forestry stations and forestry schools sprouted up together, such as the one established in 1906 at Colorado College in Colorado Springs. It included a forest reserve located up in the mountains beyond Ute Pass, twenty-five miles from campus. "Camp Colorado" operated as the school's forestry station (Reid 1979: 63).

But not all forest stations were operated by regional institutions. By the early twentieth century, the U.S. Forest Service was becoming a major force not only in managing forest resources, but also in establishing field stations in the West to produce new knowledge. In 1908 the Forest Service undertook a major decentralization of its work, creating regions (initially called districts) with offices and trained researchers living and working there. Region 4, for example, included Utah and western Wyoming and began "to adopt scientific and professional techniques of range management." Two units, the Caribou National Forest (1910) in western Wyoming and nearby Idaho and the Great Basin Experiment Station (1912) in central Utah, became "models for the implementation of research validated methods" (Alexander 1987: 410).

A co-operative project with the Weather Bureau begun in 1910 extended Forest Service work into the realm of "forest-flood relations" at a site called Wagon Wheel Gap in Colorado (Steen 1976: 131). Finally, in 1912 the U.S. Forest Service chief made the agency's research work more formalized by establishing a Central Investigative Committee, with members responsible for investigations on silviculture, grazing, and forest products. This committee structure was replicated at the district level and then more tightly centralized in 1915 with the creation of a more autonomous Branch of Research to replace these committees (Steen 1976: 137). Related to this Forest Service work was the USDA Division of Agrostology's effort to create range management experiment stations, with the first site for experimental work set up in Texas in 1898 (Box & Taylor 1990).

Up near the forests, in the high valleys of the Rocky Mountains, a quite different kind of field station emerged. In the early twentieth century, a few leading state universities established seasonal biological field stations, especially in the Rocky Mountains, for summer research, teaching, and recreation (Vetter 2010). One of these was the University of Colorado Mountain Laboratory at Tolland, established in 1909 by Francis Ramaley of that institution. Another was the Alpine Laboratory established by Frederic Clements, who started out at the University of Nebraska doing summer work around Pikes Peak from 1899 before moving to the University of Minnesota, where he transformed the station into a "Graduate School of Ecology." At the Alpine Laboratory, Clements led advanced study in sophisticated ecological methods, assisted by Edith Clements, R. J. Pool, and H. L. Shantz ("Mountain and Alpine Laboratories" 1913). Both stations gave researchers access to a wide variety of environments, from foothills to alpine peaks, for repeated study over a period of years.

What further distinguished the summer biological field stations from other kinds of stations was their seasonal occupation. Situated high in the Rockies, they experienced heavy snowfall and cold, windy conditions (not to mention the many alpine plants and other species that are essentially dormant in the winter), which limited use of this environment to the summer season. This pattern harmonized quite well with the collegiate calendar, which gave both students and teachers the time to relocate to the mountains during the summer. On the other hand, being away from the station during most of the year meant leaving buildings unoccupied and equipment untended, both of which were subject to damage by harsh weather or invading fauna. Upon arriving at the Alpine Laboratory in June 1918, for instance, Pool found that the "rats had complete control" of the upper cabin.[12] Despite such small-scale depredations, however, researchers found that operating seasonal biological stations offered great opportunities for intensive research in appealing locations with diverse environmental resources.

With forestry and biological stations, we have almost come full circle back to the agricultural station. In the geography of stations across the regional landscape,

---

12   J. E. Weaver to Dr. Clements, June 10, 1918, Edith S. and Frederic E. Clements Papers, #1678, Box 1, Folder 2, AHC-UW.

agricultural stations were at middle level between peripheral forests and mountain parks on the one hand, and urban gardens and engineering stations on the other. While much of the scientific research undertaken in the railroad era in the Central West was done at agricultural experiment stations, we should not forget that other kinds of stations were spread across the landscape of the West, producing knowledge from all the diverse environmental zones of the region.

Quarries

If stations were intensive field sites where researchers produced knowledge in and around them, quarries were also intensive but resulted in the extraction of collected materials for further processing and scientific study elsewhere. The prototypical example was the fossil quarry, which was a rich source for one of the most distinguished American field sciences of this period. The earliest fossil quarry sites emerged out of the context of the collector networks of E.D. Cope and O.C. Marsh. As these Eastern fossil warriors cultivated and extended the reach of their surrogates in the field, a few of them came across caches of fossil material so rich that the sites started to change from places where one would merely collect a few surface specimens to enduring and semi-permanent locales with much more intensive fossil digging. Three such sites were located along the eastern base of the Rocky Mountains – also, significantly, the location of the densest early settlement and railway development. These sites included Como Bluff, along the transcontinental Union Pacific railroad in southeastern Wyoming; Morrison, in the foothills hogback ridges just west of the emerging regional metropolis of Denver; and Cañon City, further south near Pueblo and in what would become the industrial hearth of Colorado. As the rich promise of these sites, most of which were found in the late 1870s, became evident, Cope and Marsh scrambled to find reliable surrogates to exploit and defend them.

Other important fossil quarry sites in the Central West were both east and west of the Continental Divide that follows the spine of the Rocky Mountains. West of the Rockies was the Green River Basin, which included a variety of quarries in the vicinity of Fort Bridger. East of the Rockies were not only the foothills and hogback fossil beds of Colorado and Wyoming, as already noted, but also the celebrated Badlands area of South Dakota and its extensions into nearby parts of Nebraska and Wyoming. Despite the long history of fossil collecting there, the Badlands still seemed like a promising field site. Indeed, one observer in the early 1890s wondered: "Will the Bad Lands ever be exhausted of fossils? The treasures of this region ... adorn the museums of the east, and every year collectors are at work. Of course the specimens which are weathered out can soon be picked up, but there are quantities left. In fact, the beds may be said to be inexhaustible. Each spring a new crop may be expected" (Kingsley 1891: 968). Such an optimistic outlook on the sustainability of fossil collecting in the Badlands did not even take into account the possibility of deeper excavation by field parties with the resources to undertake fully fledged quarrying operations, as institutions such as the American Museum and Carnegie Museum would do by the turn of the century. Just about everywhere else, however, and even in most areas of the badlands by the early twentieth century, fossils were perceived as a finite resource. Intense competition for the most valuable finds was the norm.

More arenas for competition over fossil resources were located further east on the Plains, in the extensive fossil beds of northwestern Kansas. These places included both the Smoky Hill Cretaceous chalk in the Gove County area and the Sternberg quarry – so named because it was discovered by the intrepid collector in 1882 – near the town of Long Island, which contained Pliocene mammals and land turtles (Lanham 1973: 198). Unlike the foothills region where the Rockies and Great Plains meet or the Dakota Badlands region, with their rich and accessible outcroppings, the Sternberg/Long Island quarry was in a more typical grass-covered Plains area. Before finding the site – "the greatest deposit of fossils that I have ever discovered" – Sternberg remembered he "had not met with much success, as the rocks here disintegrate so easily and hold moisture so readily that the whole country is covered with grass." But after going further east than he had planned and pitching his tent where "the three requisites, grass, wood, and water, were at hand," he was surprised to find the fabulous quarry (Sternberg 1909: 126-128).

One of the few early fossil quarries that played no role in the Cope-Marsh fossil war was the Florissant site west of Colorado Springs. Although not far from Cañon City and Morrison, it contained mainly fossil insects, trees, and leaves, rather than mammals or dinosaurs. First reported by A. C. Peale in the 1874 annual report of the Hayden Survey, the Florissant site was an ancient lake basin located a few miles west of Pikes Peak in a high mountain valley. Its most famous scientific visitor was Samuel H. Scudder of Harvard, who collected an enormous number of fossil insect specimens. University of Colorado biologist T. D. A. Cockerell, who later worked at the site himself, proclaimed "the general excellence of [Scudder's] work, while its amount is simply astonishing" (Cockerell 1911). In 1882 other nearby fossil beds were discovered in a different Colorado mountain park, near Fairplay, by Arthur Lakes, who taught at the Colorado School of Mines. Lakes sent a few specimens to Scudder, prompting him to make "a more thorough exploration of the locality" the next summer (Scudder 1890: 433). Because the objects extracted from the Florissant site and other high Rocky Mountain quarries were rather small and much easier to collect than dinosaurs or large mammals, its status as a significant, long-term quarry rests more on the sheer quantity of material, which encouraged lengthy stays there and repeated visits to the site by many field parties over many decades.

The next great wave of quarry sites emerged in the 1890s with the onset of vigorous competition between Eastern museums to secure fossil material for their public displays. It was this later wave, in fact, that constituted the main trend in the opening of quarries for multiple years rather than just small amounts of fossil materials being mined over a period of weeks. Without the resources and large trained research teams of urban museums, such an era of quarry science would have been inconceivable. Important new quarries were located in Wyoming, Nebraska, and South Dakota. Later, significant work would also begin in the Uinta Mountains of northeastern Utah, undertaken primarily by the Carnegie Museum. Besides dinosaurs and mammals, some sites were noteworthy because of fossil fishes. One such quarry was on the edge of the Green River region: Fossil Butte in southwestern Wyoming, worked for several decades by commercial collector Lee Craig.

Four leading American museums for paleontology around this time were the American Museum in New York, the Carnegie Museum in Pittsburgh, the Field Museum in Chicago, and the U.S. National Museum (Smithsonian) in Washington, D.C. The American Museum was first on the ground, as paleontologist Henry Fairfield Osborn sent Jacob L. Wortman (a former Cope collector) and O. A. Peterson to collect fossils at mammal quarries in Wyoming, Utah, South Dakota, and New Mexico in the early 1890s. Then, in 1897 the American Museum's Barnum Brown re-opened the dinosaur quarries at Como Bluff in southeastern Wyoming (Rainger 1991: 68-69). Not far from Como Bluff, competing museum teams opened new quarries in the Freezeout Hills. The American Museum operated the Bone Cabin Quarry (so called because of a rough cabin made partly out of the abundant fossil materials lying all over on the ground) from 1898 to 1905, as well as the nearby Nine Mile Quarry and Six Mile Gulch (Rainger 1991: 95). Shortly thereafter and in the same general area, the Carnegie Museum opened up their own headquarters, which they called Camp Carnegie.

Attempts to find rich fossil quarries in South Dakota and Nebraska – an area extending south and west from the Dakota Badlands, which themselves had been the object of collecting expeditions since the mid-nineteenth century by such field scientists as Hayden and Meek – culminated in the opening of new sites. These included one quarry at the Agate Springs Ranch in Nebraska in 1904 opened by the Carnegie Museum, and another one along Porcupine Creek on the Pine Ridge Indian Reservation in South Dakota opened by the American Museum a few years later. Although these quarries would become massive sinks for museum research funds, they also promised compensatory payoffs in valuable material. Opening such quarries enabled the Eastern museums to compete not just for the richest dinosaur fossils (Brinkman 2010) but also for the fantastic Miocene mammals at sites such as the Agate quarry in Nebraska (Vetter 2008), which were equally desirable from a research point of view and only slightly less desirable than dinosaurs for public displays.

The Field Museum in Chicago was a relative latecomer to this game and had to content itself with picking over the remains and less desirable quarry sites. In 1898, Elmer S. Riggs led a fossil expedition to the Black Hills and Badlands of South Dakota, but these sites had long been exploited by other collectors (Brinkman 2000). The next year, Riggs showed up in Medicine Bow, Wyoming looking for fossils, this time following more directly on the heels of the American and Carnegie museums.[13] Riggs also undertook significant work in the area around Grand Junction, Colorado (Brinkman 2005), though the Field Museum remained well behind the American Museum and Carnegie Museum. The U.S. National Museum was even more constrained by its lack of dedicated field funds, depending as it did on specimens contributed by other government agencies such as the U.S. Geological Survey and its predecessors.[14] No other museums really had the stature or funds to effectively compete with the big players.

Paleontology was not the only discipline that depended on quarry science during this period. While archaeological research sites were rarely called "quarries," in practice they shared many of the same issues and problems of their paleontological counterparts. Researchers at quarries negotiated for access with local people, relied on them for reconnaissance and hospitality, and often hired many of them for heavy excavation work on-site. As finite, extractive natural resources, both archaeological specimens and fossils experienced the same kind of vigorous competition among institutions. Eastern and western interests collided and fought over access on explicitly regional grounds.[15] There were some distinctive features of archaeological quarries, including a focus on restoration of ruins in addition to extraction, the greater significance of student labor, since artifact quarries often functioned as

13 E. S. Riggs to Dr. Farrington, September 5, 1899, Directors Group, General Correspondence, Box 18, E. S. Riggs Folder, FMNH.

14 The first U.S. National Museum fossil quarry expeditions to the West, other than Alaska, did not occur until the early 1920s, with work at Nebraska's Agate Springs quarry led by J. W. Gidley and Utah's Dinosaur National Monument quarry led by Charles W. Gilmore (Gilmore 1941: 327-328).

15 A fuller account of Southwestern archaeology – for which Mesa Verde represents only the tip of the iceberg – is beyond the scope of this chapter. Unlike the fossil sites of the Central West, archaeological quarries were historically part of a Southwest region that encompassed much of Arizona and New Mexico in addition to southern Colorado and Utah (Snead 2001, Elliott 1995, Fowler 2000).

archaeological field schools, and more acute problems with controlling access to research sites to prevent amateur relic hunting. These issues were not completely irrelevant to paleontology, but they were far more important in archaeology. The restoration of ruins, for example, could be applied to the *in situ* display of a fossil quarry wall at Dinosaur National Monument in northeastern Utah, but this came much later and has never been pervasive in paleontology. In archaeology, by contrast, the restoration of architectural structures at sites such as the many cliff dwellings of Mesa Verde was a central imperative in the early twentieth century that dominated the field practice of archaeologists in the region.

Thus, just as the competition among expanding urban museums to secure vertebrate paleontology materials at the turn of the century stimulated work at fossil quarry sites, so too did those same museums' archaeology departments undertake major new research projects at artifact quarry sites. The American Museum was very active, the Carnegie Museum less so. And in archaeology the Field Museum of Chicago lagged behind almost as much as it did in vertebrate paleontology. Whether or not this was due to "the exclusion of Chicago's museum people from the East Coast networks of power," as one historian has suggested, it meant that at the turn of the century only federal government work really rivaled the American Museum's efforts (McVicker 1999: 38). Establishment of national park or monument status for many of the key quarry sites was crucial. Mesa Verde, for example, a spectacular assemblage of cliff houses, plateau-top structures, and abundant artifacts in the southwest corner of Colorado, was proclaimed a national park in 1904, hastening the control of its resources by government excavators under the Bureau of American Ethnology (Bretemitz 1983, Harrell 1987, Johnston 1979).

Intensive field ethnography also merits inclusion as an example of quarry science. Partly this is because the museum context of collecting ethnographic artifacts was so much at the center of museum fieldwork at the turn of the century.[16] But even the ethnographic fieldwork that concentrated on recording ceremonies and stories, in addition to the artifacts associated with them, was essentially conducted in quarry mode. The cultures of American Indian groups were viewed as scarce resources on their way to rapid extinction, thus heightening the need to document and preserve them for science.

British parallels are helpful in understanding intensive field ethnography as an example of the quarry mode of field science. One early advocate of field ethnography, W. H. R. Rivers, argued in 1913 for a shift from "survey work" to what he called

---

16 For an insightful account of this artifact-based ethnography as practiced in the more popular Northwest Coast region of the U.S. and Canada, see Cole 1985.

"intensive work." While surveys were able to cover more ground, "observing and comparing the customs of different tribes and places," in his view they also inevitably relied on "superficial information." By contrast, intensive work achieved "intensity and thoroughness" by accepting "limitation in extent" as the trade-off. "A typical piece of intensive work," he contended, "is one in which the worker lives for a year or more among a community of perhaps four or five hundred people and studies every detail of their life and culture." He conceded that "preliminary surveys" might help researchers select the best places for intensive work – and that surveys could help fill the gaps between such studies – but intensive work was clearly the best kind, in his view (Rivers 1913: 6-7, 13-14, see also Kuklick 2011).

In the United States, intensive ethnographic field work was also gaining a foothold in the early twentieth century, although the main exemplar of this approach (Boas on the Northwest Coast) was not in the Central West. Still, the work of James Mooney for the Bureau of American Ethnology in South Dakota and Oklahoma involved the beginnings of field ethnography in the quarry mode. For Mooney as for other ethnographers who wanted to study the American Indians of the Plains region scientifically, this usually meant either visiting the Sioux confined to reservations in South Dakota or the Cheyenne and Arapaho tribes in Oklahoma. (Members of these tribes had formerly inhabited the High Plains of Colorado, Kansas, and Nebraska.) Mooney also expended considerable effort collecting ethnographic objects for display at exhibitions and museums to illustrate the ways of life of Plains Indians. In this way, his efforts were a kind of quarry science not just in their intensive, almost extractive character, but also in their focus (for patronage if nothing else) on circulating physical objects out of the region (Moses 1984).

One final example of quarry field work in the Central West was the collecting of material for habitat dioramas and life groups in museums. The survey and network collecting activities of institutions such as the Colorado Museum in Denver have already been noted. Yet the correspondence of the museum's director J. D. Figgins in the 1910s indicates that much of the museum's fieldwork was oriented around collecting material for display, not research. In some cases, this meant returning to the same site many times in order to mine it for usable materials. The Barr Lakes area, an important bird nesting location just northeast of Denver, was one such site that became both a site for mining biological materials for display and the name of that type of display in the museum's exhibit halls.[17] Given its proximity and the

---

17 Some collecting for the "Barr Lakes Group" was not done at the actual Barr Lakes at all, but instead at another "biological quarry" named San Luis Lakes, near Hooper, Colorado, where the natural features were similar.

convenience of access for Denver-based scientists, the frequent use of Barr Lakes itself was understandable. Even in the dead of winter (though "fair, and warm for this time of the year"), in late January 1913, museum assistant Ray Hersey was able to get out to Barr Lakes for eleven days of collecting from this biological quarry, during which he "took 58 mammals, of which [he] made 46 into skins."[18] When museum zoologists returned again and again to the same place, knowing that they could reliably find the materials they wanted there, what on the surface might appear to be a sort of survey work was starting to turn into quarry mode.

Conclusion

I make no claim to have created an exhaustive inventory of the modes of knowledge production that have emerged in the history of science. In addition to labs, lay networks, surveys, stations, and quarries, there may very well be other important modes that have flourished in particular historical and social contexts. I do think it is likely that these four field modes have been important in many scientific frontier zones that were opened during the modern period of the late nineteenth and early twentieth centuries, especially where similar rail-based transportation and telegraph-based communication infrastructures existed. Finally, like the various "ways of knowing" outlined by John Pickstone, the modes of production allow us to appreciate commonalities across disciplinary boundary lines in the history of science (Pickstone 2000). Unlike Pickstone's ways of knowing, however, the modes of production have the advantage of generating a taxonomy based on the work organization of the practices themselves, at the ground level. By examining the historical emergence of these modes of knowledge production in the field across a wide variety of scientific disciplines, I hope to have suggested the larger possibilities of this approach within the history of science.

---

18   L. R. Hersey to Mr. Figgins, February 14, 1913, Jesse Dade Figgins Correspondence, Box 3, Denver Museum of Nature and Science (Colorado Museum of Natural History), Denver, Colorado.

References

The Botanical Seminar. 1906. *University journal (Lincoln, Nebr.)*, 3, (December): 34.

"Dr. Cutler and the Denver University Rose." 1973. *The green thumb (Denver Botanical Gardens)*, 30(4): 118-121.

"Mountain and Alpine Laboratories, Colorado." 1913. *Journal of Ecology*, 1, 157.

"Preliminary: The Plan and Scope of the Survey." 1892. *Botanical Survey of Nebraska*, 1.

The United States Geological Survey: Its origins, development, organization, and operations 1904. U.S. Geological Survey Bulletin 227.

Adams, F. G. 1879-80. "Irrigation in Kansas." *Transactions of the Kansas Academy of Science*, 7: 77-83.

Adams, G. M. 1989. *The Post near Cheyenne: A History of Fort D. A. Russell, 1869-1930*. Boulder: Pruett.

Alexander, T. G. 1987. "From rule of thumb to scientific range management: The case of the Intermountain Region of the Forest Service." *Western Historical Quarterly*, 18: 409-428.

Andrews, T. G. 2008. *Killing for Coal: America's Deadliest Labor War*. Cambridge, Mass.: Harvard University Press.

Ashburner, C. A. 1885. "Coal." *Mineral Resources of the United States*, 1885: 10-73.

Bailey, V. 1893. *The prairie ground squirrels or spermophiles of the Mississippi Valley*. U.S. Biological Survey Bulletin 4.

Bartholomew, D. M. (ed.) 1998. *Pioneer Naturalist on the Plains: The diary of Elam Bartholomew, 1871 to 1934*. Manhattan, Kan.: Sunflower University Press.

Bartlett, R. A. 1962. *Great Surveys of the American West*. Norman: University of Oklahoma Press.

Bocking, S. 1988. "Alpheus Spring Packard and cave fauna in the evolution debate." *Journal of the History of Biology*, 21: 425-456.

Box T. W. & I. R. Taylor. 1990. "Grasses on the Plains: The first range management experimental stations." *Journal of the West*, 29(4): 75-81.

Breternitz, D. A. 1983. "Mesa Verde National Park: A history of its archaeology. *Essays in Colorado History*, 1983(2): 221-34.

Brinkman, P. D. 2000. "Establishing vertebrate paleontology at Chicago's Field Columbian Museum, 1893-1898." *Archives of Natural History*, 27: 81-114.

Brinkman, P. D. 2005. "Henry Fairfield Osborn and Jurassic dinosaur reconnaissance in the San Juan Basin, along the Colorado-Utah border, 1893-1900." *Earth Sciences History*, 24: 159-174.

Brinkman, P. D. 2010. *The second Jurassic Dinosaur Rush: Museums and Paleontology in America at the Turn of the Twentieth Century*. Chicago: University of Chicago Press.

Buchanan, R. 1981. "Science and the 'disciples of progress': Creation of the first Kansas Geological Survey, 1864." *Kansas History*, 4: 146-153.

Bunker, C. D. 1913. "The birds of Kansas." *Kansas University Science Bulletin*, 7: 137-158.

Carey, D. 1997. "Compiling nature's history: Travellers and travel narratives in the early Royal Society." *Annals of Science*, 54: 269-92.

Cazier, L. 1976. *Surveys and Surveyors of the Public Domain, 1785-1975*. Washington: Government Printing Office.

Chilcott, E. C., J. S. Cole & W. W. Burr. 1915. *Crop production in the Great Plains Area: Relation of Cultural Methods to Yields*. Departmental Bulletin of the U.S. Department of Agriculture, no. 268.

Cockerell, T. D. A. 1911. "Samuel Hubbard Scudder." *Science*, 34: 338-342.

Cockerell, T. D. A. 1927. *Zoology of Colorado*. Boulder: University of Colorado.

Cole, D. 1985. *Captured Heritage: The Scramble for Northwest Coast Artifacts*. Seattle: University of Washington Press.

Cole D. & A. Long 1999. "The Boasian anthropological survey tradition: The role of Franz Boas in North American anthropological surveys." In: E.D. Carter (ed.) *Surveying the Record: North American Scientific Exploration to 1930*. Philadelphia: American Philosophical Society, 225-249.

Collier P. & R. Inkpen 2003. "The Royal Geographical Society and the development of surveying, 1870-1914." *Journal of Historical Geography*, 29: 93-108.

Collins, H. M. 1988. "Public experiments and displays of virtuosity: The core-set revisited." *Social Studies of Science*, 18: 725-48.

Cragin, F. W. 1884. "Prospectus of the Biological Survey of Kansas." *Bulletin of the Washburn Laboratory of Natural History*, 1.

Croker, R. A. 2001. *Stephen Forbes and the Rise of American Ecology*. Washington: Smithsonian Institution Press.

Dierig, S. 2003. "Engines for experiment: laboratory revolution and industrial labor in the nineteeth-century city." *Osiris*, 18: 116-34.

Dodds G. S., J. Henderson, T. D. A. Cockerell., H. Markman & F. Ramaley 1907. "Scientific expedition to northeastern Colorado." *University of Colorado Studies*, 4: 145-165.

Drake, B. A. 2003. "Waving 'a bough of challenge': Forestry on the Kansas grasslands, 1868-1915." *Great Plains Quarterly*, 23: 19-34.

Driver, F. 2001. "Hints to travellers: Observation in the field." In: *Geography Militant: Cultures of Exploration and Empire*. Oxford: Blackwell, 49-67.

Elliott, M. 1995. *Great Excavations: Tales of Early Southwestern Archaeology, 1888-1939*. Santa Fe: School of American Research Press.

Emmons, S. F. 1880-81. "Geology and mining industry of Leadville, Lake County, Colorado." *U.S. Geological Survey. 2nd Annual Report*, 201-290.

Figgins, J. D. 1913. "Report of the director." *Annual Report of the Colorado Museum of Natural History*.

Flach K. W. & C. S. Holzhey 2002. „History of the soil survey laboratories." In: D. Helms, A. B. W. Effland & P. J. Durana (eds.) *Profiles in the History of the U.S. Soil Survey*. Ames: Iowa State Press, 65-100.

Fowler, D. D. 2000. *A Laboratory for Anthropology: Science and Romanticism in the American Southwest, 1846-1930*. Albuquerque: University of New Mexico Press.

Gannett, H. 1884. *A dictionary of Altitudes in the United States*. U.S. Geological Survey Bulletin, 5.

Gardner, R. 2009. "Constructing a technological forest: nature, culture, and tree-planting in the Nebraska Sand Hills." *Environmental History*, 14: 275-297.

Gilmore, C. W. 1941. "A history of the Division of Vertebrate Paleontology in the United States National Museum." *Proceedings of the U.S. National Museum*, 90: 305-377.

Grim, R. E. 1985. "Mapping Kansas and Nebraska: The role of the General Land Office." *Great Plains Quarterly*, 5: 177-197.

Gunthorp, H. 1913. "Annotated list of the Diplopoda and Chilopoda, with a key to the Myriapoda of Kansas." *Kansas University Science Bulletin*, 7: 161-182.

Harrell, D. 1987. "'We contacted Smithsonian': The Wetherills at Mesa Verde." *New Mexico Historical Review*, 62: 229-48.

Hilton, H. R. 1879-80. "The rainfall in its relation to Kansas farming." *Transactions of the Kansas Academy of Science*, 7: 37-46.

Hinsley, C. M. 1981. *Savages and Scientists: The Smithsonian Institution and the Development of American Anthropology, 1846-1910*. Washington, D.C.: Smithsonian Institution Press.

Helms, D. 2002. "Early leaders of the soil survey." In: D. Helms, A. B. W. Effland & P. J. Durana (eds.) *Profiles in the History of the U.S. Soil Survey*. Ames: Iowa State Press, 19-64.

Hoyt, M. E. 1949. *A Short History of the Colorado School of Mines*. Golden: Colorado School of Mines Library.

Hubbard, B. 2009. *American boundaries: The Nation, the States, the Rectangular Survey*. Chicago: University of Chicago Press.

Hungerford, H. B. 1919. "The biology and ecology of aquatic and semiaquatic Hemiptera." *Kansas University Science Bulletin*, 11: 3-328.

Iseley, D. 1913. "The biology of some Kansas Eumenidae." *Kansas University science bulletin*, 8: 235-309.

Isenberg, A. C. 2005. *Mining California: An Ecological History*. New York: Hill and Wang.

Johnston, P. C. 1979. "Gustaf Nordenskiold and the treasure of the Mesa Verde." *American West*, 16(4): 34-43.

Kevles, D. J. 1971. "Federal legislation for engineering experiment stations: The episode of World War I." *Technology and culture*, 12: 182-189.

Kingsley, J.S. 1891. "The Hat Creek bad lands." *American naturalist*, 25: 963-971.

Kohler, R.E. 2002. *All Creatures: Naturalists, Collectors, and Biodiversity, 1850-1950*. Princeton: Princeton University Press.

Kuklick, H. (2011). "Personal equations: Reflections on the history of fieldwork, with special reference to sociocultural anthropology." *Isis*, 102: 1-33.

Lanham, U. 1973. *The Bone Hunters*. New York: Columbia University Press.

Latour B. & S. Woolgar 1981. *Laboratory Life: The Social Construction of Scientific Facts*.

LeCain, T. J. 2004. "From eastern coal to western copper: The Bureau of Mines and industrial mining in the West." *Journal of the West*: 43, 20-28.

Maienschein, J. 1994. "Pattern and process in early studies of Arizona's San Francisco Peaks." *BioScience*, 44: 479-85.

McVicker, D. 1999. "Buying a curator: establishing anthropology at the Field Columbian Museum." In: A. B. Kehoe & M. B. Emmerichs (eds.) *Assembling the Past: Studies in the Professionalization of Archaeology*. Albuquerque: University of New Mexico Press, 37-52.

Merriam, C. H. 1890. *Results of a Biological Survey of the San Francisco Mountain Region and Desert of the Little Colorado, Arizona*. U.S. Biological Survey, North American Fauna No. 3.

Merriam, C. H. 1894a. "Report of the ornithologist and mammalogist." *Report of the Secretary of Agriculture, 1893*. Washington: Government Printing Office, 227-332.

Merriam, C. H. 1894b. "Law of temperature control of the geographic distribution of terrestrial animals and plants." *National Geographic*, 6: 229-38.

Merriam, C. H. 1894c. "The geographic distribution of animals and plants in North America." *Yearbook of the Department of Agriculture*, 1894, 207-214.

Merriam, C. H. 1898. *Life Zones and Crop Zones of the United States*. U.S. Biological Survey Bulletin 10.

Moses, L. G. 1984. *The Indian Man: A Biography of James Mooney.* Urbana: University of Illinois Press.

Mudge, B. F. 1866. *First Annual Report on the Geology of Kansas.* Lawrence: Speer.

Ostrom J. H. & J. S. McIntosh 1999. *Marsh's Dinosaurs: The Collections from Como Bluff.* 2nd ed. New Haven: Yale University Press.

Overfield, R. A. 1993. *Science with Practice: Charles E. Bessey and the Maturing of American Botany.* Ames: Iowa State University Press.

Page, L. F. 1984. "Benjamin F. Mudge: The first Kansas geologist." *Earth Sciences History,* 3: 103-111.

Page, L. F. 1996. "George Clinton Swallow, the *other* Kansas state geologist." *Transactions of the Kansas Academy of Science,* 99: 134-145.

Plumb, C. S. 1898. *The Geographic Distribution of Cereals in North America.* U.S. Biological Survey Bulletin 11.

Pool, R .J. 1953. "Fifty years on the Nebraska National Forest." *Nebraska History,* 34: 138-79.

Powell, J. W. 1892-93. "Report of the director." *Annual Report of the Bureau of American Ethnology,* 14.

Quisenberry, K. 1977. "The dry land stations: their mission and their men." *Agricultural History,* 51: 218-228.

Rainger, R. 1991. *An Agenda for Antiquity: Henry Fairfield Osborn and Vertebrate Paleontology at the American Museum of Natural History, 1890-1935.* Tuscaloosa: University of Alabama Press.

Ramaley, F. 1927. *Colorado Plant Life.* Boulder: University of Colorado.

Rees, T. 2007. *Arc of the Medicine Line: Mapping the World's Longest Undefended Border Across the Western Plains.* Lincoln: University of Nebraska Press.

Reid, J. J. 1979. *Colorado College: The First Century, 1874-1974.* Colorado Springs: Colorado College.

Rivers, W. H. R. 1913. "Report on anthropological research outside America." In: *Reports upon the Present Condition and Future Needs of the Science of Anthropology.* Washington, D.C.: Carnegie Institute of Washington.

Roth, D. 2002. "Soil survey and the U.S. Forest Service." In: D. Helms, A. B. W. Effland & P. J. Durana (eds.) *Profiles in the History of the U.S. Soil Survey.* Ames: Iowa State Press, 191-214.

Rydberg, P. A. 1906. *Flora of Colorado.* Colorado Agricultural Experiment Station Bulletin 100.

Salmon, S. C. n.d. *The odyssey of an agricultural scientist: The Memoirs of Samuel C. Salmon.* Hyattsville, Md.: Samuel C. Salmon.

Scott, C. A. 1903. "Foresting the Nebraska Sand-Hills." *Forestry and Irrigation*, 9: 454-457.

Scudder, S. H. 1890. "The insects of the Triassic beds at Fairplay, Colorado." In: *The Pretertiary Insects of North America, including Critical Remarks on and Description of, some European Forms*. New York: Macmillan, 433-448.

Seely, B. 1993. "Research, engineering, and science in American engineering colleges: 1900-1960." *Technology and Culture*, 34: 344-386.

Smith, J. A. 1883. "Catalogue of the principal minerals of Colorado with annotations on the local peculiarities of several species." In: *Report on the Development of the Mining, Metallurgical, Agricultural, Pastoral, and other Resources of Colorado for the Years 1881 and 1882*. Denver: Tribune, 137-158.

Snead, J. A. 2001. *Ruins and Rivals: The Making of Southwest Archaeology*. Tucson: University of Arizona Press.

Steen, H. K. 1976. *The U.S. Forest Service: A History*. Seattle: University of Washington Press.

Sternberg, C. H. 1909. *The Life of a Fossil Hunter*. New York: Holt.

Stewart, M. J. 1982. "To plow, to sow, to reap, to mow: The US Army agriculture program." *Nebraska History*, 63: 194-215.

Swallow, G. C. 1866. *Preliminary Report of the Geological Survey of Kansas*. Lawrence: Speer.

Tate, M. L. 1999. "Uncle Sam's farmers: Soldiers as frontier agriculturalists and meteorologists." In: *The Frontier Army in the Settlement of the West*. Norman: University of Oklahoma Press, 150-173.

Tobey, R. C. 1981. *Saving the Prairies: The Life Cycle of the Founding School of American Plant Ecology, 1895-1955*. Berkeley: University of California Press.

Turner, S. D. 1987. "The survey in nineteenth-century American geology: The evolution of a form of patronage." *Minerva*, 25: 282-330.

Urry, J. 1972. "*Notes and queries on anthropology* and the development of field methods in British anthropology, 1870-1920." *Proceedings of the Royal Anthropological Institute*, 1972: 45-57.

Vanorny, P. M. 1970. *A History of the South Dakota Geological Survey*. Vermillion: Science Center, University of South Dakota.

Vetter, J. 2008. "Cowboys, scientists, and fossils: The field site and local collaboration in the American West." *Isis*, 99: 273-303.

Vetter, J. 2010. "Rocky Mountain high science: Teaching, research, and nature at field stations." In: Vetter, J. (ed.) *Knowing Global Environments: New perspectives on the History of the field sciences*. New Brunswick, N.J.: Rutgers University Press, 108-134.

Vetter, J. 2011. "Lay observers, telegraph lines, and Kansas weather: The field network as a mode of knowledge production." *Science in Context*, 24: 259-280.

Walker, J. R. 1980. *Lakota Belief and Ritual*. R. J. DeMallie & E. A. Jahner (eds.) Lincoln: University of Nebraska Press.

Weber, G. 1928. *The Bureau of Chemistry and Soils: Its History, Activities, and Organization* Baltimore: Johns Hopkins Press.

Weed, C.M. 1891. "The biological work of American experiment stations." *American Naturalist*, 25: 230-236.

Williams, R. L. 1984. *Aven Nelson of Wyoming*. Boulder: Colorado Associated University Press.

Wissler, C. 1917. *The American Indian*. New York.

Wissler, C. 1920. *North American Indians of the Plains*, 2nd ed. New York.

Neha Gupta

Before Creation
Competing excavators, Imperial Interests and the Making of the Indus Civilization in 1920s India

A recent news focus on Pakistani and Indian archaeology appearing in the pages of *Science* suggests a largely isolated development of the discipline in the two countries (Lawler 2008). This very significant interest in Indus studies (Basu 2008) presents an opportunity to discuss the development of this field in the twentieth century and to examine the factors that have shaped the practice of contemporary archaeology in Pakistan and India.

Lawler's examination, based on a combination of visits to Indus field-sites currently under investigation and interviews with influential archaeologists in local and foreign institutions, includes the argument that the creation in 1947 of Pakistan and Independent India left the study of the Indus "largely an academic backwater for nearly a half-century" (Lawler 2008, p. 1277). Stagnation in Indus studies, Lawler maintains, was distended by the relative isolation of Cold War Pakistan and India from Western academia. In so doing, Lawler presents an overly simplistic understanding of the relationship between archaeology and society.

Instead, I argue here that it was precisely Imperial interests in the twentieth century that shaped the study of the Indus civilization, and that relationship in turn, influenced the practice of Indian archaeology. Available data shows greater complexity in Indus studies than is assumed by Lawler. A compilation by the noted archaeologist Gregory Possehl of the University of Pennsylvania of Indus field studies (excavation and survey) in operation between 1902 and 1996 shows show that Indus studies did not stagnate after the creation of Pakistan and administratively independent India (Possehl, 2002). The spatial and temporal dispersion of these archaeological investigations – particularly in the fifty years since Independence – paints a complex dynamic. Possehl's research shows both investigations before and after 1947, and points out that archaeologists actively conducted archaeological field studies for Indus sites. In addition, new investigations began east from the Indus River, in territories that ceded to Independent India. Archaeological research tended to focus on northern territories. There is no clear break in archaeological field studies in 1947, rather strong evidence of change as well as continuity in archaeological practice throughout much of the twentieth century.

In this chapter, I trace the meandering fate of Indus studies as it snaked through popular thought and scientific knowledge between 1922 and 1931. During Indian archaeology's formative years, while excavators and scholars from closely related disciplines debated the origin of the ancient society and its organization, the popular image was that of a recovered indigenous golden age. This popular rendering resonated so deeply that it shaped the practice of Indian archaeology until the mid-1990s.

With a sensational headline for *The Times of London* in December 1924, "Important finds in India – possible link with Sumerians," John Marshall, Director General of Archaeology, unveiled Mohenjodaro in District Larkana, Sind, and its older northern sibling, Harappa in District Montgomery, Punjab, as an "Indo-Sumerian civilization" (*The Times* 1924). Yet in 1928, Marshall publicly replaced his startling appellation "Indo-Sumerian civilization" with "Indus civilization." Why did Marshall modify the name in 1928? And how did Marshall's Indus compare with his earlier Indo-Sumerian civilization?

Archaeology in Imperial India

The ancient society was first announced in 1924 after the discovery, by a team of ASI archaeologists working in the Lower Indus, of seals bearing an unknown script and brickworks of an unknown age. The announcement by John Marshall came at the heels of internal and external geopolitical reorganization in Imperial India. During the opening decades of the twentieth century, territories west of the Indus River were the stage for intense Imperial interest and competition over sovereignty. The Crown raised the ante during World War I with increased political, economic, and military persuasion in Mesopotamia, relying heavily on Indian ground troops to secure oil-rich territories (Jeffery 1984). Simmering tension in the Province of Punjab reached boiling point when, in the spring of 1919, Imperial forces in Lahore and Amritsar attacked civilians protesting the infringement of their liberties. Hundreds gathered in Jallianwala Bagh, Amritsar to protest the Rowlatt Act, and were fired upon by Imperial troops. The act, passed in March 1919, extended the restrictive wartime Defence of India Regulations Act. Implemented for World War I, the provision made anti-Imperial activities punishable without trial or appeal, including public rallies.

It was in this tumultuous geopolitical climate that Marshall and his officers set out to undertake archaeological field studies along the Indus in Sind and in Punjab. At that time, the Archaeology Department had concentrated its efforts on conserving and restoring monuments, leaving exploration for the Public Works

Department and Provincial agencies (Marshall 1930a). Field studies conducted by the Department were often at historic sites previously described in explorer accounts and epigraphy, which Europeans had been recording since the early eighteenth century. Understandings of those sites were cast in relation to ancient Greece. Marshall, for example, had conducted archaeological fieldwork at Taxila, where a tope or *stupa*, amongst the earliest stone buildings of the Buddhist period, had been recorded. Taxila had significance as the place where Alexander had camped on his Imperial campaign. Marshall had worked in the sensitive North West Frontier since his knighthood in 1914. With a keen advocate in the influential Viceroy George Nathaniel Cuzon, Marshall frequently garnered resources for the Department.[1]

In Marshall's first year as Director General, the Department hired additional surveyors, assistants, and superintendents, including Marc Aurel Stein, for archaeological exploration in the recently created Northwest Frontier Province. Additionally, the Department introduced recruitment and training scholarships. The close relationship between archaeology and language studies at that time is ascertained by the first Government awards offered – two in 1903 – one for Sanskrit and one for Persian and Arabic. It was under those auspices that Daya Ram Sahni joined the Department as an excavation officer. Another excavation officer, Rakhal Das Banerji, was appointed when in 1910 Sahni was promoted to assistant superintendent (Lahiri 2006: 195). It was Banerji's exploratory work at Mohenjodaro that led to the first press report of a "pre-Buddhist" period (*The Times*, 1923).

The excavations at Mohenjodaro and at Harappa between 1923 and 1927 constituted the largest undertaking by the Archaeology Department in Imperial India. During the winter months, excavators from the Northern and Western Circles converged at the two sites, enlisting up to 1200 laborers to dig, manage, and record archaeological finds. Each excavator was responsible for his discrete section at the excavation site and submitted his report for publication in the *Annual*. Marshall introduced the Departmental publication when he took office in 1902. He oversaw both the administrative and popular dissemination of Department's groundbreaking work in the 1920s: at the time, Marshall was the public face of archaeological field studies.

Marshall's approach to archaeological work in Imperial India was echoed in Curzon's 1904 *Ancient Monuments Preservation Act*. The Act authorized the Imperial Government *alone* for the discovery, classification, conservation, and the maintenance

---

[1] The two men worked together when Curzon appointed Marshall to the ASI in 1902. Curzon's tenure as Viceroy ended in 1905, yet he remained an active supporter of the Department in Parliament until his death in 1925 (Curzon 1911, Marshall 1939).

of Indian heritage. Curzon had reasoned that possessions the Crown had secured were vulnerable to exploitation by other Europeans. Thus, preventative measures were central to Imperial interests (Curzon 1907). Locally, the act was leverage for Department officers serving as surveyors in Indian States and as superintendents in Provincial Governments. The protectionist approach delineated terms on which other investigators and museums, foreign or otherwise, worked in Imperial India.

Until Mohenjodaro's identification by Banerji, the long-standing questions on Indian antiquities ranged from the origin of civilization, the impact on the Indian peninsula from the migration of Aryans from the north and west, to prehistory or the time before the Greeks came. The accepted story of India told of an Aryan migration not very long ago into the Indian peninsula. Riding through the north and the west, the northern Indo-Aryan pushed the Dravidian inhabitants south. Thus, the most archaic institutions were found static and unchanged among the southern Dravidians. Stone tools and burials that had been recorded since the early nineteenth century in the southern parts of the peninsula were assumed to demarcate the geographical extent of these people. The northern Indo-Aryans changed and developed due to subsequent migrations, and thus, history began in the north. The earliest recorded antiquities were coins, Asoka stone pillars, Buddhist brick *stupas*, and rock-cut temples. Early explorers had collected coins from *stupas*, which aided the study of numismatics. In turn, those scholars delineated a relative chronology for migrations from the time of Alexander onwards. Other monuments classified as "Jain," "Hindu," "Mohammedan," and "British" reflected successive empires to the most recent. That was *known* history. But a void still existed in the history of the region: what came before the Buddhist ruins? Where were the imposing ruins like at Luxor and Thebes? What else had the Aryans built?

"Long-forgotten" Mohenjodaro

The years between the end of World War I and the beginning of World War II were marked by growing dissatisfaction amongst Western-educated Indian elites. Priorities within the Empire changed in the face of deepening economic crises felt both in Britain and in its possessions (*The Times* 1922a, b). The pre-World War I climate, celebrated in lavish Empire Day festivities in all its Imperial holdings, was a thing of the past. Where Indian elites had once enthusiastically partaken in the economic opportunities of the Empire, the Crown's foreign interests now made such opportunities increasingly inaccessible to Indians (*The Times* 1922c). A lack of voice in governance was cause for rising local elites to question both the authority and the legitimacy of the Government of India; this in turn placed emphasis on ethnicity as the fuel for nationalist movements in Imperial India. It was in this visceral climate that Mohenjodaro surfaced.

Relative Dating at Mohenjodaro
On June 29, 1923, John Marshall wrote to the Editor of *The Times*. Responding to a press report on recent archaeological work in Sind, Marshall corrected the Bombay Correspondent on the nature and character of "seals with hieroglyphics" collected at Mohenjodaro. Over the winter of 1922, Banerji, then superintendent of the Western Circle, conducted preliminary investigations on three of the five mounds at Mohenjodaro (*The Times* 1923). His objective was to verify the most remote inhabitants of the western province, in order to delineate the geographical extent of those inhabitants along the lower Indus. To test his hypothesis, Banerji examined the presence and absence of antiquities.

The relative chronology at Mohenjodaro was the core of Banerji's analysis. It was key to his understanding of the site and its significance. While there he examined the highest Buddhist *stupa* and observed a thick layer of ash below its base. The stratigraphy suggested to Banerji that the platform was built on the ruins of an earlier structure, possibly another temple. Further examination of the superimposed layers showed that beneath the ash deposit laid a brick building. Its identification confirmed to Banerji that a culture existed in Sind that predated the Buddhist culture.

From the mounds, Banerji collected coins and categorized them into four successive periods, the most recent being the *stupa* (Banerji, 1923). The coins confirmed to Banerji the accepted culture history of the province. Other antiquities including stone tools, clay and shell bangles, copper and bronze beads, painted pottery, and funerary urns being widespread, the mound was recorded as Site 1. It was there that Banerji collected three seals similar to those excavated by Sahni at Harappa two years earlier. Two other mounds were examined against Site 1 or the *stupa* mound. Banerji classified pottery from the mounds by their physical characteristics – red slip, white slip, and glazed. The *stupa* mound alone yielded the superior glazed type.

Noting the absence of Mohammedan coins from the mounds, Banerji surmised that by the eighth century A.D., Mohenjodaro had been abandoned due to the Indus changing course eastward toward its present location. Banerji also identified burials on the mounds. He classified these remains as "Aryan post-cremation period" cist, coffin and jar burials (Banerji 1984: 114). Unlike the *stupa*, relative dating of burials was not based on stratigraphic principles. Rather, Banerji described each as the historical record of contemporary peoples. For example, jar burials were thought to represent Munda customs (Banerji 1984: 115-116). Urns with ashes at the site indicated predominant Hindu "post-cremation" customs. Banerji further noted diversity in the shapes and sizes of these urns; diagnostic shapes and sizes assumed continuity through time. This meant that prehistoric times were not very different from the present.

Banerji's emphasis on chronology and ethnicity is suggestive of prevailing thought on prehistory at that time. Early excavators assumed that the monuments and artifacts they were studying were records of past accomplishments and that the golden age would be recovered by their efforts. By identifying types or discrete groups in archaeological material, Banerji conflated language, race, and religion. He aimed to show that Mohenjodaro was older than any known Buddhist site at that time. Banerji explained idiosyncrasies and change through a combination of degeneration, catastrophic events, and migration. These explanations effectively denied local relationships and innovation. In turn the close spatial relationship between artifact types was de-emphasized, as was the changing relationship between the monument and the settlement. Rather, excavators often sought out artifacts that made a connected history available, such as coins and seals with inscriptions.

Mohenjodaro and Harappa as the "Prehistoric Civilization"

For Marshall, Mohenjodaro bore features that no other excavated site in Imperial India had revealed. In his popular writings, antiquities, particularly seals, evidenced "a widespread culture which must have flourished for many centuries in the plains of the Indus" (Marshall 1924). Thus, culture diffused over territories. At Mohenjodaro was a "flourishing" brick city with roads, a shrine, and a royal palace, lying deep below an ancient site, all dating to a time before the Greeks came. Before Mohenjodaro's identification, prehistory in the colonial provinces had revealed scattered stone tools, "cyclopean walls," and cromlech burials. Excavators were taken by seals inscribed with a script unlike anything seen in India. Brick tombs with grave goods rather than the expected cremation urns exemplified differences in culture. Excavators marvelled at similarities in painted pottery, ornaments, and decorative arts with those found at sites in the colonial provinces of Punjab and Baluchistan (Marshall 1926a).

Public Appeal

The relationship between archaeology and the public was central to Marshall's aims for the Department. Wary of fanciful writing by the public on the Department and its works, Marshall opted to present ongoing work as a regularly published government report rather than through popular publications (Marshall 1904). For most of his tenure and up until 1923, Marshall did not publish the Department's activities in popular journals. Thus, the sensational reports that Marshall published on Mohenjodaro and Harappa were a significant change in direction for the Department.

Marshall's approach coincided with new constraints on the Department. The Crown had recently centralized key Indian Departments directly under the Government of India. Until that time, the bulk of the Archaeology Department's annual budget came from Provincial Governments. This meant that each provincial superintendent prepared a progress report on his work of the previous year. Those reports were submitted to Marshall, who then summarized all provincial work for the *Annual*.

With the recent Montagu-Chelmsford Reforms, the Government of India alone would support archaeological work. The Reforms, passed in 1919, were intended to ease via diarchy the transition from Crown colony to Independent statehood. In effect, they prepared the colony to be a springboard for Imperial expansion. While this meant the transfer of some government offices including Education to Provincial councils, the Crown kept a tight reign on key government offices, including Defence and Foreign Affairs (Curtis 1920). Despite its close relationship with Provincial Governments, the Archaeology Department was taken over by the Imperial Government. Consequently, from this time onwards the Department prepared one annual comprehensive report for India (Spooner 1924).

In the first year of this new administrative structure, six new appointments including a superintendent for the Indian Museum accompanied the changes at the Archaeology Department. This was followed by a reduction in the Department's conservation budget. Facing monetary constraints, Marshall sought ways to sustain the Department. Within months of his *Illustrated London News* article, the Education Department allocated 80,000 Rs/ £ 5,333 for the continuation of archaeological field studies in Sind and Punjab (*The Times* 1925).

Archaeology for the Public
The excitement generated by Marshall and the Department's work shaped the practice of archaeology in Imperial India. The vista that Mohenjodaro opened had presented new opportunities and challenges. Marshall's public exhibition of the civilization encouraged both Indian interest and foreign excavators eager to harvest the "limitless field" (Marshall 1926a). Up until that time, Provincial Governments had discretion over how the Preservation Act was interpreted and locally implemented. In practice, the Act emphasised the conservation of standing monuments; thus, so long as the monument was safe from damage, local excavators were free to dig and decide what to do with recovered materials (Marshall 1930a). This flexible arrangement allowed

Superintendents to purchase antiquities that were of interest to the Department from local excavators.

Facing increasing pressure from eager foreign excavators, Marshall proposed two amendments to Section 20 of the Preservation Act – one that extended the protection of "ancient monuments" to include virtually any place that might yield buried antiquities and a second that vested legislative authority on excavations and on the ownership of antiquities in the Governor General in Council (Marshall 1930a).[2] Much to the disadvantage of local communities and irk of Provincial authorities, the amendments brought known sites and as yet unknown places of interest into the direct regulation of the Government of India. Antiquities were at the core of Marshall's modifications and from that point on the Government of India weighed antiquities by their "national importance."

In Marshall's public announcements, similarities between engraved steatite seals collected from Mohenjodaro and those from Harappa were the key to the identity and organisation of the civilization. Marshall emphasised the distance between Mohenjodaro on the Lower Indus and Harappa on Upper Indus, asserting that the 400 miles between the sites added to the civilization's significance. He believed that the ancient civilization extended not only along the Indus but also further west of the river towards the politically sensitive Kingdom of Iraq. Marshall hypothesized that the culture at the two sites was part of a "much wider sphere" that included India, Persia, southern Iraq, and Central Asia and likely extended as far west as the Mediterranean (*The Times* 1924, Marshall 1926b: 49). This implied that there was an underlying unity across the territories.

Marshall argued that the indigenous culture had deep roots in the Indus: he put forth the possibility that in the Indus valley lay the root of Babylonian, Assyrian, and Western Asian culture (*The Times* 1924). It was precisely to establish the extent of the Indo-Sumerian civilization that Marshall sent Department Officer Harold Hargreaves westward to the Kalat State and Baluchistan for an archaeological survey (Hargreaves 1927).

Indus = Indo-Sumerian?

Was Mesopotamia or India the cradle of West Asian culture? That was the question posed when Marshall unveiled Mohenjodaro in *The Times* in 1924. The sensational announcement had its place. C. Leonard Woolley had been excavating at Ur, a site south of Baghdad, in the new Kingdom of Iraq. In the wake of World War I and the

---

2   The two are amendments to Section 20 of the 1904 Ancient Monuments Preservation Act. The second amendment came into effect in 1932.

break-up of the Ottoman Empire, the Provinces of Baghdad and Basra had fallen into the League of Nations mandate. The Crown, having Imperial interests there, strongly supported unification of the two provinces.[3] The Kingdom was formed in 1922, when the Crown picked Faisal ibn Husynin as monarch (Winstone 1990).

Archaeology for Whom?
Just as the *Illustrated London News* brought its readers Howard Carter's Tutankhamen, Woolley gave them the Tower of Babel in the Iraqi ziggurat (Woolley 1924). Woolley often published his findings in popular journals over the twelve years of his excavations at Ur. Financed by the British Museum in London and the University Museum in Philadelphia, excavators were trusted to act in the best interests of the Museum and to acquire unique pieces from the field for display. Future funding, if not annual allowances, depended on the project delivering these goods. This was the scene of intense rivalry between field archaeologists prior to the professionalization of the discipline.

Further Investigations
The Persia-Baluchistan and Afghanistan-North West Frontier frontline were of special concern to the Crown. At the turn of the twentieth century, the British Crown launched a reorganization of its possessions with the aim to secure them. The Government of India signed an agreement with the Afghan Amir, and established an international frontline between Afghanistan and the northwestern most Crown-administered territories. The Crown believed that this would protect its Indian possessions from foreign invasion. At that time, the Crown considered the northwest frontier its Achilles heel in maintaining the empire. Policy makers believed that there lay key passages through the otherwise impassable "natural barriers". Because this was so, acquiring specific places and routes was imperative to repel invasions.

It was precisely these geo-political concerns that led to the creation in 1901, of the North West Frontier Province on the sensitive border (see Figure 1). At the same time, in securing these territories, the Government of India brought relative peace and stability to neighboring Provinces and Native States, including the Province of Punjab.

During and after World War I, the Crown and the Government of India weighed the cost of maintaining that international border. But the tenuous Afghan-Frontier frontline would remain until India's independence in 1947. This in turn led to the suggestion for cost-effective alternatives to Indian troops for missions

---

3   Oil-rich Mosul, also under mandate, ceded to the Kingdom in 1926.

abroad, in particular greater investment in a Royal Air Force. The close relationship between archaeology and the military is suggested by the first airborne surveys of archaeological sites: the RAF conducted the survey along the Indus in 1925 (Marshall 1927: 1931).

During his survey in March 1925, Hargreaves explored mounds in Baluchistan and Kalat State. With help from the resident Political Agent in that State, Hargreaves examined mounds for evidence of the Indo-Sumerian culture – seals, glazed pottery, and brickworks. At Nal, previously excavated in 1904 by Marshall, Hargreaves observed large mounds and stone embankments. Hargreaves dug several excavation trenches at Sohr Damb. There he found stone walls and burials with funerary objects, including copper tools. Hargreaves categorized human internments as complete burials or "fractional burials," in which funerary vases were found with disarticulated bones (Hargreaves, 1928, 64-70). Of the six trenches he excavated, three had well defined funerary deposits, including infant graves. Hargreaves collected over 200 specimens of "Nal" pottery from the site. He argued that the ancient remains belonged to an advanced stone-using people unlike the nomadic inhabitants he had encountered there before. Hargreaves concluded that the Kalat sites were Copper Age and supposed that they were different from Mohenjodaro and Harappa (Hargreaves 1928: 72).

Scholarly Discussion
Meanwhile, academic dialogue raged in popular journals on the origin and nature of the Indo-Sumerian civilization. Within days of Marshall's *Illustrated London News* piece, Cyril Gadd and Sidney Smith, both from the British Museum and both working with Woolley at Ur, reacted to the newly discovered Indo-Sumerian civilization. That followed a publication by eminent Assyriologist and Professor at Oxford University Archibald Sayce, who remarked on similarities in the script on seals from Babylonian, Elamite, and Indus sites (Sayce 1924). At that time, the prevailing view was that foreigners had made the spectacular monuments that Woolley was excavating in Ur. It was believed that Sumerians were distinct from Iraqi locals in their language, institutions, and race. Akin to the Aryan that migrated into India not long ago, the ancient Sumerian was thought to be "intrusive" and that he had originated from outside.

The authors pointed to a close kinship between Ur and Mohenjodaro. Gadd and Smith argued that at third dynasty Ur and at Mohenjodaro, "pictographs of bulls" on seals were identical, as were brickworks (Gadd & Smith 1924). They further compared Sumerian signs with those on Indus seals and found sixteen to be identical or related. They concluded that Indus inhabitants had borrowed Sumerian script

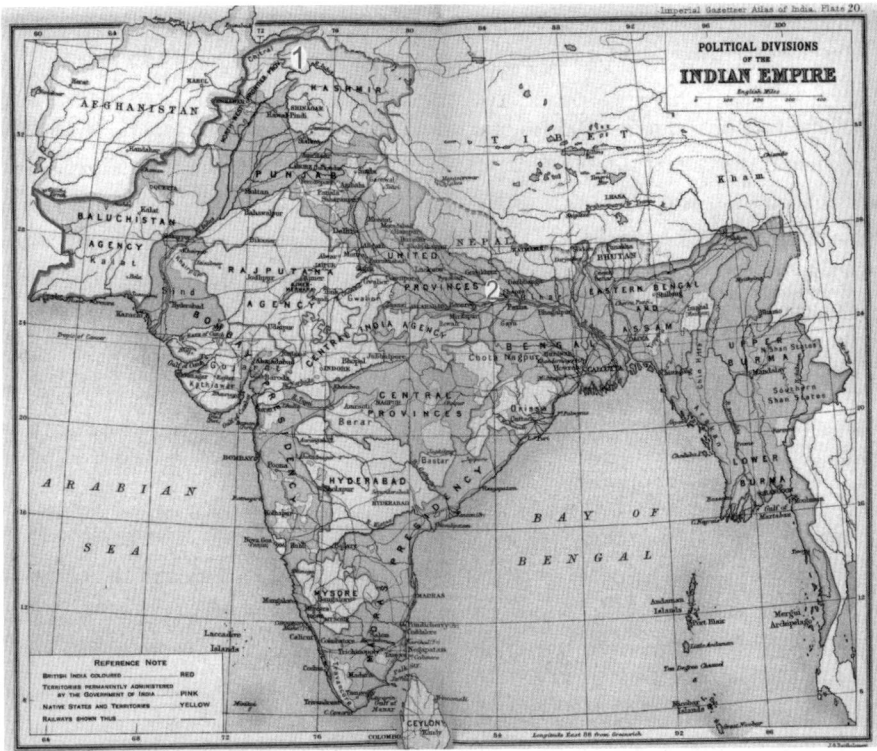

Figure 1. Imperial Gazetteer Atlas, Plate 20, map illustrating geo-political territories in 1909. The frontline with Afghanistan is highlighted in the top corner to the left to illustrate the position of the North West Frontier Province. The Indus River [1] and Ganges River [2] are highlighted.

and art between 3000 and 2800 B.C. Gadd and Smith assumed that the Sumerian culture had remained largely unchanged since the third millennium B.C. and that the widespread Aryans were influential in the development of ancient Ur. Thus, neither Ur nor Mohenjodaro were local developments. Both authors subsequently contributed to Marshall's edited volumes on the Indus script (Gadd & Smith 1931).

The Switch

Economic woes continued despite a greater understanding of the cost of maintaining widespread Imperial possessions. This was accompanied by growing disillusionment about progress and the capacity for change. By the close of the 1920s, protests led by Indian elites roared for *swaraj* or self-rule. Nationalist leaders no longer pursued the Government of India for a voice in decision-making – they sought representative government. A key issue for nationalists was separate versus joint electorates for the Indian voting public.

Upon its introduction into public space, the ancient society was received as a Brahmanic antecedent to Buddhist and later empires (Rai 1929). Foreign excavators and scholars saw the Indus sites as an offshoot of a larger Sumerian-based empire. Influential scholars believed the pictographic bull and the script were identical to those on Sumerian seals. Pottery from Mohenjodaro was similar to that from Kish, a site in Iraq (Marshall 1926b: 48). Mohenjodaro was assigned a relative date based on its diagnostic seals, which were subsequently found at four other sites in Iraq and in Persia (Mackay 1925). The ancient societies were therefore accepted to be contemporary and somehow associated. It was in light of these discoveries that the "Indus civilization" was born.

Explaining Similarities, Exploring Differences
Marshall and the ASI's works were intimately tied to the Empire and the Crown's changing relationship with its possessions. In his 1928 publication in *The Times*, Marshall discarded the term "Indo-Sumerian" and publicly replaced it with "Indus" civilization (Marshall 1928). Until then, Marshall had argued that the culture of the Indus had a close relationship with the Sumerians. Their accord was critical in shaping the early study of the civilization. Marshall had explained resemblances in antiquities through a common, single source of origin. On that assumption, he claimed the existence of an Indus-based empire bearing a cultural wingspan to the Mediterranean. Emphasizing similarities in seals, he relied on archaeology to construct a connected history of the "long-forgotten" Indo-Sumerian civilization at Mohenjodaro and Harappa.

By the end of the excavations in 1927, Marshall had established the existence of the Indus civilization. He christened it in public the following year (Marshall 1928, 1930b: 53). Marshall explained that his excavations showed a long presence of civilization in India. He argued for the separate civilization on three counts: first, that Mohenjodaro and Harappa were large cities with multiple strata; second, the similarity in material culture at the widely spread sites and; third, their distance from the sea. All three, combined with the surface discovery of seals, proved that a culture akin to (if not more advanced than) the Sumerian and Egyptian civilizations had developed along the Indus. That meant the Indus culture was widely spread and that it was largely homogenous. Idiosyncrasies in material culture were a result of traditions specific to locality.

Marshall once argued that resemblances in the Indus and Sumerian antiquities supported the theory of a single civilization. However, now he proposed that there were two separate civilizations. Similarities in antiquities at Mohenjodaro and Ur were now evidence that the two civilizations had "intimate commercial and other intercourse" (Marshall 1928: 13). It was with these assumptions that Marshall and his officers emphasized a distinctive Indus culture in the field.

In the Field

Anticipating several field seasons at Mohenjodaro, Marshall commissioned a permanent laboratory, offices, storage, and workrooms, a museum, and summer housing for officers and assistants. This investment followed a construction project by the Bombay Provincial Government to build a road from Dokri, the closest railway station, to the excavation site (Marshall 1928: 73). By the beginning of fieldwork in 1925, the local Public Works Department had completed most of the infrastructure for ASI officers.

Keenly aware of public interest in the Indus sites, Marshall pulled superintendents from nearby Circles to join the Mohenjodaro excavations. He had assigned his officers, Kasinath Narain Dikshit and Madho Sarup Vats, an area each for excavation. Marshall worked on the *stupa* mound with an excavation officer. Up to 1200 laborers were recruited through Sindhi contractors for the excavation. The following year, new site plans were made for the excavators. Marshall arranged for Surveyor A. Francis from the Survey of India (Cantonment) to conduct a topographical survey of the entire site. The datum for recording was set at 181 feet above mean sea level at a point in the northeast of the site (Marshall 1928: 74). The resulting 100 feet to 1 inch scale map was published with the 1931 Mohenjodaro report.

That season, Ernest John Henry Mackay, previously field director for the Field Museum-Oxford University Archeological Expedition to Mesopotamia-financed excavations at Kish in Iraq, joined Marshall's team. Mackay went on to excavate at Mohenjodaro for six seasons and well after Marshall left for Special Duty in 1928. In his first season, Mackay took over work east of the *stupa* in Dikshit's excavation area.

Paying greater attention to intra-site organization, Marshall introduced a hierarchical classification system to differentiate between areas, sections, blocks, and rooms. Rooms, being the smallest unit of measurement, were represented with small Arabic numerals in site plans and records (Marshall 1930b). Lanes and streets were sequentially numbered. Wells and drains were indicated as such on plans. During the excavation, ASI surveyors plotted plans of buildings on a scale of 10 feet to 1 inch. These plans were published in black and white for the *Annual*. But Marshall believed that the structures likely reflected two or more superimposed stages.

Thus, to represent the temporal relationship between one layer and another, he had proposed that the plans be made in several colors to represent successive layers of buildings. This would enable the reader to readily distinguish differences from one period to the next. The color coded plans were published with his final report on the excavations.

A Direct-Historical Approach
Mackay was the first to establish culture-chronological recording at Mohenjodaro (Mackay 1930). The archaeological method had not been used before in Imperial India. At that time, it was believed that prehistoric times were largely homogenous and represented only a short duration: change was sudden and catastrophic. Culture-chronology, however, assumed that idiosyncrasies at a site indicated temporal differences (Trigger 1989). Similar culture traits were thought to belong to one and the same chronological period. Thus, traits shared by many cultures were explained as older, while idiosyncratic ones were recent and particular to one culture. This assumed that there was continuity with only slight modification. Common origin remained the explanation for similar cultures; this in turn emphasized a common ethnicity.

In his analysis, Mackay identified artifact types and examined them for the presence and absence of particular traits. The trait list was used to define phases and stages. He classified Indus occupation at Mohenjodaro in three successive stages – Early, Intermediate, and Late, which were in turn sub-divided into phases. Mackay recorded the height below datum of all artifacts and he used these measurements to characterize successive phases of occupation. Central to Mackay's understanding of Mohenjodaro was a direct historical approach. He believed that culture differed only slightly between ancient Mohenjodaro and present day inhabitants in Sind (Mackay 1931). Mackay therefore emphasized regularity and continuity in the material culture at the site. He developed these ideas in his 1938 publication, *Further excavations at Mohenjo-daro*.

The (as yet) Un-Deciphered Script
By Marshall's last season at Mohenjodaro, over 13 acres had been opened up (Marshall 1930b). It was during this season that the team excavated a large, rectangular tank or "bath" close to the *stupa*. The *stupa* was not excavated. It was generally accepted amongst the excavators that underneath the monument lay an ancient temple and that brickworks in its immediate proximity likely served a religious function too (Marshall 1930b). Excavators aimed to distinguish Indus brickworks from Sumerian. But why did Sumer matter?

Much of Woolley's works at Ur, including descriptions of houses, were based on cuneiform texts that scholars had deciphered since the late nineteenth century. Commonly referred to as "omen texts," cuneiform tablets aided the study of languages. Biblical scholars believed the tablets were records of past accomplishments (Smith 1884). Influential scholars believed these tablets would yield clues on the as yet un-deciphered Indus script, and "long-forgotten" civilization. It was within this context that the Indus excavators examined the presence at the site of bitumen or crude oil. Excavators at Ur had identified its use as a waterproofing for brickworks, and it was thought to represent a uniquely Sumerian trait. The trait's appearance in the archaeological record in distant locations was explained by migration. Idiosyncrasies at the Indus sites included bathrooms, brick flooring, and an "elaborate system" of drains that would carry sewage into street tanks (Marshall 1928). Ur, as Woolley described it, lacked these features (Woolley 1927). The Indus excavators investigated Mohenjodaro on terms that were closely related to excavations at Ur, but excavators emphasized the uniqueness of their respective sites. As a corollary, investigation of the changing relationship between the ancient city and its hinterland was neglected.

Mohenjodaro Report
Marshall elaborated his ideas on the Indus civilization in his edited volumes. The seven excavators at Mohenjodaro described their respective excavation areas. To the volumes' thirty-two chapters, Mackay contributed fourteen, primarily on the subjects of the monumental remains and artifact types, such as beads, pottery, figurines, seals, and tools. In addition to Smith and Gadd's chapter, Stephen Langdon, Professor of Assyriology at Oxford University, wrote on the Indus script and its relationship with the Sumerian cuneiform. Colonel R. B. Seymour Sewell and B. S. Guha, both of the Zoological Survey of India, discussed human and animal remains.

Marshall examined the extent and the identity of the civilization. He argued that terracotta figurines and phallic objects found at Indus sites were representative of pre-Aryan cults. He believed that there was a long continuity of those religious traditions at Mohenjodaro and that some of these features were still preserved in non-Aryan "jungle-tribes" of southern India, Western Asia, and Iraq as well as around the Aegean (Marshall 1931: 76-78). At that time, Marshall's ideas were heterodoxy to the popular view that Vedic Aryans developed Hinduism and the Indus cities.

Marshall doubted Langdon's assertions that the Indus script was antecedent to Brahmi, the language found on *stupas* and prevalent in early historic India. He questioned the basis for an Aryan-developed Indus civilization. Rather, Marshall credited the development of the civilization to pre-Aryan inhabitants, likely the Dravidians, and identified their religion as the "lineal progenitor of Hinduism"

(Marshall 1931: 76-78). He explained that uniformity in the Indus civilization was a consequence of its long development at Mohenjodaro and Harappa.

Based on Guha's skeletal analysis of races at Mohenjodaro, Marshall reasoned that Punjab and Sind were diverse and thus society was likely to have been heterogeneous in ancient times (Marshall 1931: 109). He rejected suggestions that skeletal remains would correspond to a Dravidian racial type if this were the case, arguing instead for the vagueness of that ethnographical term. Marshall explained resemblances between Indus and Sumerian religion through diffusion. He supposed that the Chalcolithic culture was likely represented at other sites in Sind and in Punjab, noting that explorations east of the Indus River were lacking. The focus had been on the western extents of the civilization. He made the case for investigations southeast from the Indus and towards the Ganges and encouraged the ASI to undertake that task, proposing that the identification of similar sites beyond Punjab would render the civilization an "Indian" one (Marshall 1931: 91).

Changing Priorities

The years leading up to World War II were marked by the Depression. The Crown faced increasing pressure over its possessions from competitors, particularly German and American. There was growing agitation amongst Indians whose interests were represented by neither the Government of India, nor the established political parties seeking *swaraj*. By the middle of the 1930s, elections were held for a spectrum of Indian political parties in the Provinces. This followed the Government of India Act 1935, which proposed a federated India through a coalition between Indian States and Provinces. Despite those proposals, the Government of India maintained key economic, defense, and foreign affairs offices into the War. It was in this context that the Government of India invited Leonard Woolley to visit Imperial India.

Out with the Old ...

By the 1930s, the ASI's archaeological and other works had lost both monetary and political support. This followed Marshall's departure from the ASI in 1928, and it was accompanied by Aurel Stein's retirement. The two men had worked together for almost three decades, since Stein was made Inspector General of Education and Archaeological Surveyor for the Northwest Frontier Province and Baluchistan. It was Stein who had declared that ancient sites in those sensitive territories belonged to an early Persian civilization (Stein 1905). In a career spanning over forty years, Stein recorded, collected, and mapped the furthest extent of "Indo-Aryan" culture. In so doing, he had not only charted the unknown "frontiers" of Imperial India but also defined what an Imperial possession was and what not.

ASI officers continued work at Mohenjodaro, with Mackay leading the excavations there. Excavations resumed at Harappa, as did explorations along the Indus under Hargreaves' direction. Mackay's publication built on those investigations and was intended as a follow-up to Marshall's volumes. Yet there were marked differences between the two works. Mackay's Mohenjodaro was scene to a weak government that was left defenseless to the will of persistent and aggressive tribal raiders from the mountains (Mackay 1938: 648). Skeletal evidence pointed to a "likeness" between inhabitants at Mohenjodaro, al-'Ubaid, and Kish. These similarities were due to an invasion or several invasions by Uruk people sometime in the recent past. The Uruk, Mackay explained, were races from Anatolia and North Syria (Mackay 1938: 668). This proposition had general agreement with Woolley's work at the sites of Carchemish and Ur.

Vats resumed excavations at Harappa in 1926. He published a comprehensive two-volume report on his work and the initial investigations in 1940, noting that Harappa made its own contributions to the "Proto-Indian" civilization (Vats 1940: i). He based this proposition on evidence from a cemetery that he had excavated. Pottery there was assigned a relatively late date of 2000 B.C. (Vats 1940: 11). He classified burials as belonging to early and late periods. Like Mackay, he recorded the height of artifacts below datum. Vats assigned complete and fractional skeletons with grave goods to the early period; the late period was designated by pot burials in which urns contained partial human remains (Vats 1940: 16). Vats assumed that similarities in material culture were evidence for a common origin. He interpreted contemporaneous sites east and southeast from Harappa using the same criteria.

Report on Indian Archaeology
With war looming, the Government of India invited Woolley to assess the state of Indian archaeology. The key task for Woolley was to develop a plan for the ASI. He was to identify promising sites for exploration and the best methods to attain those results, as well as the best methods for officer selection and training (*The Times* 1938). During his assignment, Woolley visited major sites in northern and southern India. He submitted his report to the Government of India and presented his assessment for the George Birdwood lecture at the Society of Arts in London in December 1939.

The Crown's reception and endorsement of Woolley's recommendations is evident in Robert Eric Mortimer Wheeler's appointment in 1944 as Director General of the ASI. The new hire had followed concerns over the Department's Directors. Sahni, who had trained under Marshall, served as Director after Hargreaves; another student of Marshall's, K. N. Dikshit, had taken over since and he was Director when Woolley's report was submitted.

As one of his many recommendations, Woolley suggested that a temporary European executive advisor be assigned to oversee ASI's reorganization. He named his protégé Wheeler for the job (Winstone 1990). Wheeler had previously founded the Institute of Archaeology in London. After his service in the Army, he was the Director of the ASI until India's independence in 1947. He went on to direct the newly created Pakistan Department of Archaeology. Wheeler was knighted for his contributions in 1952, and maintained an interest in the Indus civilization to the very end of his career in 1976 (Piggott 1977).

In addition to a thorough reorganization of the ASI, Woolley proposed a new research direction for the Department, intending to "fill" a blank era spanning 2000 years between Mohenjodaro and the Buddhist era. In particular, Woolley observed the lack of historical texts in India, with which archaeological material might be aligned, aiding the formation of a chronological sequence. He proposed that the Iron Age developed in the southern peninsula and that civilization subsequently spread north (Woolley 1940: 192). At the same time, Woolley suggested that the Classical influence be investigated at sites along the western and southern coasts, arguing for Western relations for Mohenjodaro and the Indus valley "cult." He rejected claims, especially Marshall's, that figurines pointed to later Indian deities and therefore suggested continuity in religious traditions. Woolley also rejected suggestions that the Indus script was antecedent to southern Indian scripts (Woolley 1940: 194). Woolley further questioned the geographical extent of the Indus "cult," arguing that more work was needed. It was within this context that Woolley recommended the appointment of three specialists in prehistory – one each for northern, central, and southern India.

Woolley called for changes in archaeological work conducted by both local universities and foreign institutions. This mirrored Marshall's amendments with one significant feature: Woolley sought changes in the legislation that would make archaeological work in India attractive for museums. He wanted a portion of excavated material for the excavating team. Until then, the 1904 Act and Marshall's amendments had authorized the ASI to keep *all* excavated antiquities. Indeed, Marshall's aims for the legislation were precisely to retain excavated material at local site museums. The Department, he had argued, could transfer select pieces to important museums in Imperial India. Woolley was averse to Marshall's methods and thought the protectionist Act was detrimental to Indian archaeology.

At the core of Woolley's report was the desire to popularize archaeology. Woolley saw that uncertainty in the Department's budget, especially during financial crises, resulted in small, ad hoc excavations (N.A.I. 1942). Without a fixed grant for excavation, immune to outside economic conditions, a long-term policy was not possible. Thus, in his report, Woolley urged that the Department be put on stable budgetary grounds, free from fluctuations in the Government of India's economy. Financial stability was to be augmented through greater co-operation with museums, particularly for financing excavations and for the display of select antiquities. He encouraged senior staff at the ASI to obtain part-time lectureships at Universities. These new responsibilities were to be introduced with greater specializations for staff (N.A.I. 1942). This also meant the reduction of (if not the elimination of), the *Annual*. Instead, Woolley proposed greater investment in low-priced, illustrative books for the wider public.

Woolley asserted that the ASI staff lacked proper training in scientific methods. To remedy this, two changes were proposed: two travel scholarships were to be offered to officers to gain training and experience in archaeological methods; secondly, a European advisor was to be appointed to oversee the administration of the scholarship and organization of excavations. Greater co-operation with the Geological Survey of India was advised, as was training for staff in Anthropology to assist the study of skeletal remains in the field. The extent to which Woolley's proposals were accepted is unclear. Yet it is possible to ascertain from Wheeler's appointment and the ASI's subsequent reorganization that Woolley's recommendations were not wholly rejected.

Conclusion

In this article's attempt to improve the understanding of perceptions about archaeological work in pre-1947 India and in what later became Pakistan and Independent India, readily available data on archaeological field studies is analyzed. The article presents evidence that questions commonly held views. In addition, careful examination of critical archaeological works that shaped civilization studies in Imperial India elucidates the relationship between archaeology and society.

Though often neglected, the changing relationship between archaeology and society in colonial settings has serious implications for theoretical developments in the discipline. By critically examining colonial histories, we can gain a better understanding of the complex relationships that shaped the practice of archaeology. In the case of Indian archaeology, competing aims of local and foreign excavators played out on a rapidly changing geopolitical scene. Understandings of the Indus civilization, the research aims, and the methods and tools employed by investigators at the time provide evidence of the relationship between society and archaeology prior to the professionalization of the discipline. When divorced from this history, incorrect generalizations are made about developments in the discipline.

Lawler's suggestion that Pakistan and India have been academically isolated is misleading. India has a tradition of working with foreign, particularly Anglo-American scholars. It was precisely a growing scholarly relationship between the United States and India that precipitated the formation of the American Institute of Indian Studies in New Delhi. Since its inception in 1961, the Institute has helped both American and Indian scholars further their research interests in archaeology and related disciplines. But the best evidence of collaboration between domestic and foreign institutions in Indus studies comes from Gregory Possehl's work on the West Pakistan Archaeological Project, based in Quetta in the mid-1960s, and his later work in India's Gujarat in the 1980s.

A frequently neglected aspect of Indus research is its formative period in colonial India. The Archaeological Survey's discovery of Mohenjodaro and Harappa amidst Imperial reorganization and nationalist movements in 1920s India was rousing to a growing class of Western-educated social elites that is commonly remembered in images of barristers Mohandas Gandhi, Muhammad Ali Jinnah (later the first Governor-General of Pakistan), and Jawaharlal Nehru (the first Prime Minister of India). This was the social milieu in which Indus studies were born, in which the nations of Pakistan and administratively independent India were created, and in which archaeology was practiced in both soon to be independent nations.

Propelled by competing excavators and Imperial interests, the Crown invested in the discovery of Indian heritage to gain and protect its territorial rights. Widespread archaeological remains of an unknown age materialized the Crown's imperial ambitions. Born amidst nationalist movements in Imperial India, the Indus civilization created frenzy amongst nationalists and Imperialists alike. For Western-educated Indian elites, the ancient society was evidence of an indigenous golden age and its recovery substantiated their claims for sovereignty. Meanwhile, excavators debated the identity of and by extension any claims on the ancient society.

In the years leading up to World War II, understandings of the ancient society in the Indus valley buoyed along the Crown's relationship with its widespread possessions. The Crown's efforts to secure territorial rights in the Kingdom of Iraq weighed heavily on the Government of India. Subsequent easing of its commitments abroad, both economic and military, ushered in increasing dissatisfaction within Imperial India and was accompanied by louder calls for a representative Indian government. It was precisely the rapidly changing geopolitical scenario that prompted critical amendments to the *Ancient Monuments Preservation Act*.

Marshall's research aim was to determine the origins of the Indo-Sumerian civilization. For him, the excavations presented the westward extension of one civilization towards the Kingdom of Iraq. The Indus civilization was born in the shadow of a declining empire that was facing fierce competition over territorial rights. His amendments protected Imperial rights and tightened its grip over antiquities and heritage. Marshall emphasized the long development of pre-Aryan civilization at Mohenjodaro. His conclusions presented heterodoxy to the accepted nationalist view that the Indus civilization was a northern Aryan accomplishment. It is unclear how the public at large received Marshall's unorthodox ideas. The shuffles in the ASI and its research direction as proposed by Woolley offer some clues.

In the wake of deepening economic woes, a thorough accounting of expenditures on archaeology and its return value were initiated. Ideas and practices that were not acceptable in times of plenty now found support. Woolley's initiatives aimed to open Indian archaeology to a wider audience. He wanted to make the field attractive to outside financiers, particularly public museums which would display excavated material. Publication of illustrative books for the public at large was meant to popularize the ASI's work. The senior staff was urged to take up lectureships at Universities. Most of these initiatives challenged not only Marshall's methods, but also the staff, many of whom suddenly found themselves "unqualified" for their jobs; opportunities consequently became available for the willing and able. With these changes came pointed research questions and new methods.

The importance of greater attention to critical works in archaeology and closely related disciplines in colonial settings must be stated. When its colonial history is ignored, Indian archaeology seems static and unchanging. But when examined in relation to society, the complexities of archaeological practice are made apparent. This chapter aimed to examine those complexities in the case of the Indus civilization.

References

Banerji, R. D. 1923. "Mohen-jo-daro." In: D. B. Spooner (ed.) *Annual Report of the Archaeological Survey of India, 1922-1923*. Calcutta: Government of India Central Publication Branch, 102-104.

Curtis, L. 1920. *Papers Relating to the Application of the Principle of Dyarchy to the Government of India*. Oxford: The Clarendon Press

Curzon, G. N. 1907. *Frontiers*. Oxford: Clarendon Press.

Curzon, G. N. 1911. "Archaeology in India. A Protest from Lord Curzon." *The Times of London*. Oct. 7, 4.

Gadd, C. J. & S. Smith 1924. "The new link between Indian and Babylonian civilizations." *Illustrated London News*, Oct. 4, 614-616.

Hargreaves, H. 1927. "Baluchistan." In: J. F. Blakiston (ed.) *Annual Report of the Archaeological Survey of India, 1924-1925*. Calcutta: Government of India Central Publication Branch, 50-60.

Hargreaves, H. 1928. "Baluchistan." In: J. F. Blakiston (ed.) *Annual Report of the Archaeological Survey of India, 1925-1926*. Calcutta: Government of India Central Publication Branch, 59-72.

Jeffery, K. 1984. *The British Army and the Crisis of Empire, 1918-22*. Manchester: Manchester University Press.

Lahiri, N. 2006. *Finding Forgotten Cities: how the Indus Civilization was Discovered*. New York: Seagull Books.

Lawler, A. 2008. "Boring no more, a trade-savvy Indus emerges." *Science*, 320(5881): 1276-1285.

Mackay, E. J. H. 1925. "Sumerian Connections with Ancient India." *Journal of the Royal Asiatic Society*, 697-701.

Mackay, E. J. H. 1930. "L' Area." In: J. Marshall (ed.) *Annual Report of the Archaeological Survey of India, 1926-1927*. Calcutta: Government of India Central Publication Branch, 89-97.

Mackay, E. J. H. 1931. "Plain and painted pottery, tabulation of painted pottery." In: J. Marshall (ed.) *Mohenjo-daro and the Indus Civilization*, New Delhi: Indological Book House, 287-337.

Mackay, E. J. H. (1938). *Further Excavations at Mohenjo-daro*. 2 Volumes. New Delhi: Manager of Publications.

Marshall, J. 1904. "Introduction." In: J. Marshall (ed.) *Annual Report of the Archaeological Survey of India, 1902-1903*. Simla: Government of India Press, 1-13.

Marshall, J. 1923. "Ancient cities of Sind: To the Editor of the Times." *The Times of London*, Jun. 29, 10.

Marshall, J. 1924. "First light on a long-forgotten civilization: new discoveries of an unknown prehistoric past in India." *Illustrated London News*, Sept. 20, 528-532, 548.

Marshall, J. 1926a. "Prehistoric India – Sind and Punjab Discoveries: A Limitless Field." *The Times of London*, Feb. 26, 15-16, 18.

Marshall, J. 1926b. "Harappa and Mohenjo-daro." In: J. Marshall (ed.) *Annual Report of the Archaeological Survey of India, 1923-1924.* Calcutta: Government of India Central Publication Branch, 47-51

Marshall, J. 1927. "The prehistoric civilization of the Indus." In: J. Marshall (ed.) *Annual Report of the Archaeological Survey of India, 1924-1925.* Calcutta: Government of India Central Publication Branch, 60-63.

Marshall, J. 1928. "India of 5000 years ago, Mohenjo-daro, the excavated cities." *The Times of London*. Jan. 4, 13.

Marshall, J. 1930a. "Introduction." In: J. Marshall (ed.) *Annual Report of the Archaeological Survey of India, 1926-1927.* Calcutta: Government of India Press, xv-xix.

Marshall, J. 1930b. "The Indus culture." In: J. Marshall (ed.) *Annual Report of the Archaeological Survey of India, 1926-1927.* Calcutta: Government of India Press, 51-60.

Marshall, J. (ed.) 1931 [1973]. *Mohenjo-daro and the Indus Civilization.* 3 Vols. New Delhi: Indological Book House.

Marshall, J. 1939. "The story of the Archaeological Department in India." In: J. Cumming (ed.) *Revealing India's Past.* London: The Indian Society, 1-33.

National Archives of India (N.A.I.) 1942. *Sir Leonard Woolley's Report on the work of the Archaeological Department – Considerations of Recommendations*, F 3-16/1942, Foreign Office and Lands Department.

Piggott, S. 1977. "Robert Eric Mortimer Wheeler. 10 September 1890 – 22 July 1976." *Biographical Memoirs of Fellows of the Royal Society*, 23: 623-642.

Possehl, G. 2002. "Archaeology of the Harappan Civilization: an annotated list of excavations and surveys." In S. Setter & R. Korisettar (eds.) *Indian Archaeology in Retrospect, Volume II Protohistory: Archaeology of the Harappan Civilization*, New Delhi: Manohar, 421-471.

Rai, L. L. 1929. "Europeanization and the Ancient Culture of India." *Annals of the American Academy of Political and Social Science*, 145(2): 188-195.

Sayce, A. H. 1924. "Remarkable discoveries in India." *Illustrated London News*, Sept. 27, 566.

Smith, G. 1884. "The history of Babylonia." In: A. Sayce (ed.) *Ancient History from the Monuments*. London: Society for Promoting Christian Knowledge. Retrieved July 9, 2010 from http://www.archive.org/details/historybabyloni00smitgoog

Smith, S. & C. J. Gadd 1931. "Sign-list of early Indus script." In: J. Marshall (ed.) *Mohenjo-daro and the Indus Civilization*, New Delhi: Indological Book House, 406-422.

Spooner, D. B. 1924. "Preface." In: D. B. Spooner (ed.) *Annual Report of the Archaeological Survey of India, 1921-1922*. Simla: Government of India Press, 1-3.

Stein, M. A. 1905. *Report of Archaeological Survey Work in the North-West Frontier Province and Baluchistan for the Period from January 2 1904 to March 31st 1905*. Peshawar: Government Press, North-west Frontier Province.

The Times of London 1922a. "The burden of Mesopotamia." August 31: 13.

The Times of London 1922b. "Indian financial outlook: facing big deficit." September 12: 9.

The Times of London 1922c. "Suspicious India – Will to mistrust: Charges against British." September 11: 4.

The Times of London 1923. "Ancient cities of Sind: Buddhist shrine excavated, remarkable finds." June 25: 13.

The Times of London 1924. "Important finds in India: Possible link with Sumerians." December 10: 13.

The Times of London 1925. "Indian Excavations." February 3: 11.

The Times of London 1938. "Survey of Indian Archaeology: Sir Leonard Woolley's Mission." June 16: 15.

Trigger, B. G. 1989. *A History of Archaeological Thought*. New York: Cambridge University Press.

Winstone, H. V. F. 1990. *Woolley of Ur: the life of Sir Leonard Woolley*. London: Secker & Warburg.

Woolley, C. L. 1924. "A second 'Tower of Babel': the 'house of the Mountain' at Ur." *Illustrated London News*, October 25: 780-781.

Woolley, C. L. 1927. "The excavations at Ur, 1926-7." *The Antiquaries Journal*, VII(4): 385-413.

Woolley, C. L. 1940. "Some aspects and problems of Indian archaeology." *Journal of Royal Society of Arts*, (1): 183-197.

Vats, M. S. 1974 [1940]. *Excavations at Harappa: being an Account of Archaeological Excavations at Harappa Carried out between the Years 1920-21 and 1933-34*. 2 Vols. New Delhi: Bhartiya Publishing House.

Esther Fihl

The Rolling Field Station
Danish Explorations of Central Asia in the late Nineteenth Century

In the late nineteenth century, the importance of scientific expeditions as the predominant way of knowing non-Western fields especially was challenged by new methodological inventions. In botany, ethnography, geography, and zoology, a growing number of scientists saw this way of collecting data, as they travelled long distances on expeditions in the Arctics, America, Africa, and Asia, as more and more problematic. New research questions had cropped up and brought into focus concepts such as holism, context, and habitual circumstances. These gave way to a new kind of fieldwork conducted during prolonged stays in one place, often in what came to be known as field stations.

In the following, I shall try to demonstrate how the challenges of the expedition became evident for travellers on two Danish expeditions led by lieutenant Ole Olufsen to Tsarist Central Asia from 1896-1899. At some point during their expeditions, the imperial saloon carriage belonging to the Russian Empress Dowager, the former Danish Princess Dagmar, became their field station, though it was a "rolling" one. The carriage came to function as their living and working quarters during their botanical, ethnographical, linguistic, and zoological investigations on their trip along the newly established Transcaspian railway. This railway was built by the Russians and ran from the Caspian Sea, through the Turkmen tribal area and, in 1896, as far as Samarkand, the famous Uzbek and Tajik city on the old Silk Road. From Samarkand the Danes followed the route of the Russian postal stage coach to Osh in the Ferghana Valley where they continued their expeditions on horseback to the Pamirs and adjoining areas, which was the main goal of both expeditions. On the second expedition in 1898-1899, they stayed for six winter months in a field station in Khorok, a small village in a deep valley bordering Badakshan in Afghanistan.

Functioning in certain ways as small scientific laboratories, both their carriage and their winter field station were rather narrow in space and fairly closed to outsiders except field assistants, servants, and invited local interlocutors or informants. As in many laboratories, the stay in both the carriage and the winter station meant intense social and scientific interaction between the experts, even though each had their own individual scientific assignments. Compared to the work they carried out while living in the open in tents among nomads or in peasant villages during their travels on horseback, the field stations entailed more mutual planning due to the fact that the experts had to share the same narrow physical space. Latent social conflicts and scientific contests between the experts came to the surface more easily in the carriage and in the winter station. Here leisure time interaction among the Danes was also more clearly influenced by the social conventions of the upper class bourgeois social life back in Copenhagen.

In my previous research on these two Danish expeditions, it has been the portrayals of Central Asian nature, society, and culture and the practical circumstances under which these representations were produced that have interested me. I have been especially concerned with the acquisition of material objects on the expeditions and with the cultural biographies of the objects, in order to analyze the social and cultural life of the collected items. In this way I have explored the circumstances under which the expedition members let items change from *objects of use* in Central Asia to objects designated as *ethnographica* in the National Museum of Denmark. In doing so, I have tried to bring into focus the complex interaction between the productions of representations of Central Asia in texts and collected objects on the one hand, and the social spheres and scientific practice of the four expedition members on the other (Fihl 2002).

The aim of this article is to illustrate how the expedition members got to know the field and how some of them felt trapped in a paradigmatic methodological shift from old-fashioned expeditions to modern concepts of fieldwork conducted in one place. This will be done by drawing on their diaries, reports, and published works and scrutinizing the guiding principles for their collecting of objects and information for Danish museums.

Preparations Caught in the Great Game
As Henrika Kuklick and Robert Kohler have also pointed out, scientific research in the field in contrast to in laboratories is possibly defined most clearly by the experiences of the natural places that the researchers are subject to as an integral part of their scientific work in the field (Kuklich & Kohler 1996: 14). What is evident in the material from the two late nineteenth century Danish expeditions to Central Asia is, however, that the academic, political, and colonial contexts at home and in

the field also played a considerable role and put distinct limits on their scientific endeavors. So before we embark on a description of the actual situations and events through which the members of the two expeditions through Central Asia came to know the field, a short outline of the Danish and Central Asian contexts of the expeditions should be given in order to illustrate these distinct limits.

In the late nineteenth century, the research milieu of several scientific disciplines in Denmark was very much focused on Greenland and the Arctic, and the Danish state had appointed a special Commission for Geological and Geographical Investigations in Greenland. Over the years, it equipped more than 50 expeditions, with questions of sovereignty and strategic natural resources playing a significant role in these decisions (Høiris 1986: 57 ff.).

In 1894, a young lieutenant in the Danish army named Ole Olufsen made his entrance on the expedition scene and showed up at the National Museum in Copenhagen. He was applying for means to equip an expedition to the Hindu Kush to investigate one of the few remaining white spots on the European maps of Central Asia. He wanted to visit a remote people called the Siaposh living in the area then known by the Persian name Kafiristan, or the land of non-believers. It was a name used by neighboring Muslims to designate a certain group of non-Muslim people who lived far up in the Hindu Kush and worshipped the sun and other natural phenomena. These people had not previously been visited by Western explorers and in the late nineteenth century the world knew very little about them; speculation ran that they were descendants of Alexander the Great, the Greek king, who around 330 B.C. had tried to conquer parts of Central Asia and whose troops went as far as Hindu Kush to control cities on the Silk Route. The book *The Kafirs of the Hindu-Kush* by the famous explorer Sir George Scott Robertson was published in London in 1896, only a couple of years after the official launching of the Danish expedition plan for Kafiristan.

In 1894, Olufsen asked for the support of the National Museum in Copenhagen, offering them the prospect of a collection of ethnographical objects from Kafiristan. He also contacted the University of Copenhagen to volunteer to produce the first maps of this remote area of Hindu Kush for the Department of Geography, since he had learned how to draw maps whilst in the army.

However, when his plans were made public in Danish newspapers, the English government delivered an official protest to the Danish government. Out of fear of espionage, the British would not allow him to travel in Kafiristan, on the borderland of British India towards Afghanistan and the Russian colonial empire.

Politically, all doors were thus closed for Olufsen in relation to Kafiristan. He was caught up in the biggest international conflict of the time called the Great Game, which was being played out between the three expanding superpowers of the world: the tsar-Russian empire was expanding in Central Asia, the British empire was expanding up into the Hindu Kush and also trying to gain control over Afghanistan, and finally China was expanding into Eastern Turkestan. The backdoors, so to speak, of these three expanding superpowers were slamming against each other in the Hindu Kush and the Pamirs and each of them feared being taken by surprise through these backdoors, and being exposed to military attacks by the others. This clash culminated in the Russian conquest of large parts of the Pamirs in 1891 and the corresponding British expansion up into the Hindu Kush (Fihl 2002: 50, 272f.).

One can get an impression of the difficult international conditions under which Olufsen prepared his first expedition from Denmark by reading the correspondence starting in 1894 between Olufsen and The Royal Danish Diplomatic Representation in St. Petersburg, the National Museum of Denmark, diverse ministries, and various departments of the University of Copenhagen.

Impediments also existed in the Danish academic context. Professor Løffler of the geography department, who later became Olufsen's supervisor, frankly stated in a letter to the Academic Council that Olufsen lacked almost all the prerequisites for carrying out a scientific exploration, since, having had an exclusively military education, he was entirely ignorant of geology, climatology, botany, zoology, and ethnography, as well as of the most recent developments within geography.[1]

Løffler's statement above demonstrates that the era of the amateur scientific traveller, collecting material and writing descriptions of folklore as he travelled along, was ending. Now, in the mid-1890s in Denmark, it took a solid scientific education and knowledge of the latest scientific developments to be trusted with state grants to equip expeditions. After two years of correspondences and more rounds of applications from Olufsen, the Ministry of Charity and Education received a statement from the Academic Council at the University of Copenhagen discouraging the use of university funding for Olufsen's expedition, which

> cannot be expected to be of immediate interest to our country, since it has no natural connection with Denmark's geographical position and national links

[1] Letter from the Faculty of Mathematics and Sciences at the University of Copenhagen, September 24, 1894: 517/94 25/9 4292, The Danish National Archives, Ministry of Charity and Education, 3rd Office, Office for Higher Education: Min:3 Expect. Matters dealing with educational and travel grants 1857-1915 (subsequently DNA-MCE-OHE).

such as the scientific expedition to the Arctic waters round Greenland that the Academic Council has previously approved.[2]

In this way, questions of Danish sovereignty were marked as important, and subsequently the answer to Olufsen's applications was negative. The final argument ran that the state could not support his plans, since his destination was of no strategic value to the welfare of the Danish nation, and since he did not have any academic credentials.

Olufsen did not, however, give up his plans of equipping an expedition. Through the support of private sponsors, he and his colleague Lieutenant Victor Philipsen raised enough money to travel and they spent a year on horseback, not in Kafiristan, since this was politically impossible because of the British protest, but in the Russian controlled parts of the Pamirs bordering Afghanistan and Kashgaria or Uighuristan, today the modern Xinjang Province of China. The good Danish diplomatic relations with Russia and the royal family connection to the Tsar smoothed the formal permissions for Olufsen's expedition to travel in the Russian controlled areas of the lowland steppe country of Central Asia as well as in the Russian parts of the highlands of the Pamirs.

The Pamir Treaty of 1895, which resulted in the drawing of international borders and for which deputations from Russia and Britain met to sign the papers in Pamirs itself, meant the settlement of a buffer zone between the two superpowers (Shahrani 1979). It was agreed that Russia was to keep the already conquered parts of the Pamirs, while Chitral, Kashmir, and Ladakh would remain within the British sphere of influence. The Vakhan valley had for thousands of years formed a corridor between East and West through the massif of the mountains. In the treaty, this valley was, however, divided along the Pandsh River. The right-hand side was to come under the Emirate of Bukhara, while the left-hand side was to remain part of Afghanistan, together maintaining a narrow buffer zone between the two major powers (Gvozdetsky 1974: 127ff, Dor & Neumann 1978, Grancy & Kostka 1978).

---

[2] Letter from the Academic Council, October 24, 1895: 517/94 30/10 1672 3644/95. Letter from Olufsen, August 22, 1895: 517/94 23/8 96 1190. Copy of a letter testifying that Olufsen has received teaching from R. Hammer, Commander of the Marines (Olufsen's handwriting), September 1, 1895. Letter to the Academic Council from the Head of Faculty C. Christensen, Faculty of Mathematics and Sciences, University of Copenhagen, September 20, 1895. All in DNA-MCE-OHE. See also: Copy of a statement from the Ministry of Charity and Education (Olufsen's handwriting), December 17, 1895: 651/95. The National Museum of Denmark, Ethnographical Collections, Correspondance concerning West and Cental Asia 1877-1952. See also: *Aarbog for Kjøbenhavns Universitet* 1895-96, 1897: 290. All Danish quotations have been translated into English by the author.

After having travelled in this buffer zone between 1896 and 1897, only a year after the Treaty, Olufsen returned to Denmark. Here he held a lecture at the Danish Geographical Society for an audience of learned academics and he succeeded in convincingly presenting his geographical and ethnographical results from his travels, especially in this area of inaccessible mountains bordering China and Afghanistan. In his lecture, and in an article, he described his stay in the Pamirs among Kyrgyz nomads who depended on herds of yaks and in the Vakhan Valley among farmers practicing transhumant pastoralism in the mountains along the Pandsh River (Olufsen 1897). In this valley, he had travelled on the northern side of the river, in what had only recently been declared the Bukharan part of the buffer zone by the Russian-British political settlement, but the sovereignty was not yet enforced by any military presence from either Bukhara or its ally, Russia. En route, Olufsen was fired at by Afghan troops and brigands and several times he witnessed how they had robbed villages in this newly Bukharan part of the valley. Olufsen claimed to be the first European to have travelled the difficult route along the northern side of the river Pandsh and to have made a map of the deep Vakhan Valley from where he also, in 1897, brought home a unique collection of ethnographical items to the National Museum of Denmark.

Soon after, he gained the support of the Carlsberg Foundation to equip his second expedition to Central Asia, which lasted from 1897-1899. As an amateur scientific traveller, Olufsen had learnt from bitter experience that he belonged to a dying era. He therefore tried to fulfill the scientific requirements for his second expedition by enlisting the assistance of two academically qualified persons. He engaged the botanical expert Ove Paulsen and the physician Anthon Hjuler, who was also trained in linguistics.

Both of Olufsen's expeditions had a multi-disciplinary approach, with the expedition members exploring the unknown, mapping and collecting information, and observing as they travelled along. Huge collections of flora, fauna, and ethnographical items were brought home to museums in Copenhagen (Fihl 2002).

Not least in their diaries, the two scientists, Paulsen and Hjuler, however, expressed frustrations with Olufsen, their expedition leader whose training had taken place in the military. They described incidents when they tried to challenge what they designated as his "old-fashioned" conceptions of the best way to get to know the field. They described how they tried to persuade him to stay longer at one place instead of always being on the move on horseback or in the imperial carriage.

On the Move

In the study of the history of anthropological field methods, attention has for many years been allotted to the fact that practices in the field depend heavily not only on the conditions of the place itself but also of the metropolitan places travelled through to get there. To acquire the right permissions, to reach the field itself, and finally to work there requires both improvisation and social skills in order to cope with inevitable challenges (Okely 1994, Frohlick 2006).

The above also applies to Olufsen's two expeditions. On their way to Central Asia, both expeditions made a stop in St. Petersburg to get the necessary permissions to enter the Russian controlled parts of Central Asia. Olufsen was fully aware that the Russian colonization of the area set the terms for travelling. Since his early youth he had read, and been inspired by, the prevalent travel literature on the lowland steppe country of Central Asia, which had emerged in the second half of the nineteenth century parallel to the Russian colonial conquest of large parts of the region and the gradual minimizing and transforming of the Emirate of Bukhara and the Khanate of Khiva into Russian vassal states. This also meant the establishment of a railway system and a postal coach system, as well as installments of the telegraph even in rather remote places, all controlled by strategically placed Russian military camps. This colonial infrastructure meant fairly safe travelling conditions for various European adventurers and scientific travellers who visited the area and subsequently told of their explorations in book form (Burnes 1975 [1834], Vambéry 1864, 1868, Wenjukow 1874, Ujfalvy 1879, Lansdell 1885, Moser 1885, Taylor 1892).

Immediately after arriving in St. Petersburg, the expedition leader Olufsen had an audience arranged for him and his crew at the court of the Empress Dowager, the former Danish Princess Dagmar. As the mother of the Russian Tsar, and also sister to the Queen of England, she was an influential lady at the Russian court. After the audience, all went rather smoothly obtaining the necessary passports and stamps from the Russian authorities. The Danish expedition eagerly accepted the terms given.

In the Pamirs, the Danes were given permission to travel on horseback and move freely as long as they did not cross any border to China, Afghanistan, or British India. They were offered permission, if necessary, to force Kyrgyz nomads to deliver them fresh yak milk, meat, shelter, and warm clothes in the severe climate in which they planned to stay for a whole year.

On their way to the Pamirs, they were issued free transport for themselves and their heavy equipment. East of St. Petersburg, they could use the trains free of charge. Through Georgia, the postal stage coach and the steamboat across the

Figure 1. Above the river Pandsh in the Vakhan Valley 1886. Part of the 1st Danish Pamir Expedition crossing a ledge gallery near Dashmak on paths "never before set foot on by White Man", as expressed by the expedition leader, lieutenant Ole Olufsen (Archives of the Ethnographical Collections, the National Museum of Denmark).

Caspian Sea were at their disposal and they were given free travel on the newly Russian-built Transcaspian railroad.

Outside the railway east of Samarkand, they gained permission to follow the routes of the tarantass, the stage coach system established by the Russians in the Ferghana Valley leading to the Pamirs. The Danish expedition members were given passports as diplomats, which meant that they would always have a right to fresh horses at the tarantass stations.

They could make stops wherever they wanted as long as they reported their arrival each time to the local Russian military camp leader. Papers were issued so that they could stay in these military camps for protection – and for control of their whereabouts.

Thus equipped with the necessary Russian permits and also introduction letters from the Tsar himself, they crossed the Caspian Sea and went ashore. On the first expedition in 1896, Olufsen and Philipsen immediately headed for the train station to buy tickets for the next train, but the station manager refused to issue tickets and ordered them instead to follow him outside. They were led to an imperial saloon wagon which was placed at their use, as was the Russian servant Alexander. Without their knowledge, this had been prearranged by the Empress Dowager who had sent a telegraph to the station master, instructing him that the two Danish expedition members were to be allowed to make use of the carriage anywhere on the railway line, for both outward and return journeys.

So on their trip to and from Samarkand, the carriage came to function as their living and working quarters during their investigations. While their saloon car was moved from one Russian goods train to the next, they would make a stop at the nearby oases. It became their rolling field station.

In the Merv Oasis

In the carriage, the expedition members kept note books in which they entered their observations and reflected upon the instructions they had picked up from scientific guidebooks or were equipped with from their home institutions. Neither Russians nor Central Asians, except the Russian servant Alexander, were allowed in the carriage; social life among the expedition members in the carriage consequently did not seem to have picked up any social conventions from the Central Asian field itself. In this sense, the rolling field station came to resemble a scientific laboratory fairly closed to outsiders (Kuklich & Kohler 1996: 3). The carriage also functioned as their home and so the limits between leisure and work became blurred; this is evident in the private diary entries that illustrate the cultural otherness of Central Asia and how they grappled with the Russian colonial setting in which they travelled. This situation was in some ways similar to the social conventions common back in the study chambers of their private Copenhagen homes.

The splendid furnishing of the imperial saloon wagon made fieldwork rather comfortable and relaxed for the expedition crew. Sitting on the rear platform of the imperial coach – which was also equipped with a separate bedroom, toilet, manservant's room and even an elegant saloon with upholstered sofas, armchairs, and a writing desk – the Danes made their journey through the land of the Turkmen, or Transcaspia, the provincial name allotted to the region by the Russians. Thus the expedition members transversed vast expanses of the wild shifting-sand desert of Kara Kum in great style, their Russian manservant Alexander punctuating their journey with *suchari* (a kind of rusk) and scalding hot tea from the samovar. As expedition leader, Olufsen could simply order when he wanted the saloon wagon hooked on and off trains along the Transcaspian rail, which followed the old Silk Road (Fihl 2002: 330 ff).

Visiting the Turkmen territory, they broke their journey at more railway stations and at some places they rented horses for excursions as a way to get to know the surrounding field, where the Turkmen nomads had been defeated by the Russian army in 1881 and their land included in the Russian controlled territory Transcaspia.

Two general aspects made a forceful impression on Olufsen and Philipsen in 1896 when they made a longer stay in the Turkmen oasis of Merv. The first was the unfamiliar landscape, with its contrasting areas of desert, steppe, and oasis. In their rolling field station, the two military trained persons tried to appropriate scientific practice as they entered descriptions of the relationship between man and landscape in their notebooks, inspired by what they had picked up from their private studies of the contemporary academic discourse of anthropo-geography (Ratzel 1882). They recorded how expanses of pure desert were interspersed with more congenial scenery made up of green steppe surrounding the fertile oasis. In May, grass grew high up the mountain slopes of the northern mountains of Kopet Dagh bordering Persia. Along the foot of the mountains there were cultivated fields that were irrigated with water from rivers descending Kopet Dagh that converted the loess soil into fertile farmland in a belt which varied in size. The large camps of the Turkmen nomads and semi-nomads were seen everywhere, sometimes comprising several hundred tents. The areas of steppe outside the cultivated sections of the oases were dotted with great herds of horses, oxen, camels, dromedaries, sheep, asses, and goats, which at this time of the year, according to the two expedition members, waded up to their bellies in a sea of grass and flowers – especially poppies. Using the theory of anthropo-geography they reflected on the degree of the "culture-friendliness" of the natural surroundings (Olufsen 1906: 272).[3]

The second feature which caught the two expedition members' attention, was the extent of the Russian activity, not only in the vicinity of the railway line but also in the form of expanding Russian settlements, based on the cultivation of the steppe especially around the Merv oasis. Colonies of Russian farmers, amongst whom the

---

3   See also Philipsen's private diary, 1896-1897 (in private ownership).

cultivation of cotton and tobacco was the main occupation, were to be found in almost all of the oases.

Being under the control, but also protection, of the Russian military, Olufsen's and Philipsen's stay in the oasis Merv, however, was to come to an abrupt end, as Philipsen described in an unpublished account of the journey.[4] One evening, just outside the Russian part of the town Merv, they paid a visit to a dance-hall "where a few Russians, but mostly Tekke-Turkomans, were dancing with Asian women to the piercing stains of Tekke Turkman music." Apparently the two foreigners did not appreciate the music, but unfortunately they gradually attracted the attention of both the musicians and the best of the female dancers – much to the resentment of the other male visitors who "began to shout and shower us with stones. The proprietor came and spoke to the crowd of people present, but that only made things worse – they began to storm up the staircase." Knives were drawn and revolvers loaded, with the two Danes holding the fort at the top of the stairs until a squadron of Russian Cossacks surrounded the building. Once escorted back to their imperial saloon car, the Danes prudently decided to have it connected to the first morning train.[5]

The Russian control of the Turkmen area also made an impression on Anthon Hjuler when he visited Merv with Ove Paulsen and Ole Olufsen in June 1899 on the second expedition. The Russians put limits on how the Danish expeditions got to know the field and they also controlled the investigated object, the Turkmen nomads and their symbiosis with the surrounding nature. The Russian control of the European travellers and their whereabouts in Central Asia was not only felt on the two Danish expeditions. This topic has also been an object of study in analyses of the collecting activities of other contemporary European travellers in the area (Balsiger & Kläy 1992, Dombrowski 1993, Burton 1997).

Although Hjuler confided his dissatisfaction with the Russian colonization of Central Asia and with the Russian bureaucracy in his notebooks and diary, he nevertheless credited the Russian imperialist mentality as being far less strict than that of other contemporary major European colonial powers:

> The Turkomans are now completely pacified; they are large and powerful in stature, with enormous fur caps. The Russians are expert at winning the confidence of these half-savage people. A number of the Turkomans are now Russian officers, bearing their distinctions on their kalats. Generally speaking,

---

4   Philipsen's recollections of his participation in the 1st Danish Pamir Expedition, 1942 (in private ownership).

5   Philipsen's private diary, 1896-1897 (in private ownership).

the Russians can be said to be ideal colonists. One is always hearing of scandals from the colonies of European nations in Africa. This applies to the French, German, and British alike. The Russians would certainly have handled the matter in a completely different manner – they do not feel themselves, as do other European peoples, to be vastly superior to the peoples among whom they live; they would understand the importance of mixing with them in a completely different way. The Cossacks would get on splendidly with the average black warriors, and the officers would not look down in contempt on the valiant Kaffir chieftains – they would sit together with them in the evenings exchanging war exploits (Hjuler 1945: 23).

According to the diaries belonging to Anthon Hjuler and Ove Paulsen, the stay in Merv during the second Danish expedition's return trip was quite short, lasting from Friday, June 2 to Thursday, June 8 1899. The three travellers arrived after a 14-hour train journey in their rolling field station from Bukhara to Merv at 9pm. In the intolerable heat, they had alighted from the train at the many stations where it made stops. Paulsen was busy at his botanist work, adding to his collection the valuable specimens of the local flora which were to be taken back to the Botanical Museum in Copenhagen. He also managed to add a large lizard, captured with the aid of a stranger, to the zoological collections he was also responsible for during the expedition.

During their stay in Merv, the expedition members were devoted to different assignments. While Paulsen went out in the landscape on his botanical excursions, Hjuler assisted Olufsen in drawing several maps of the area, and Olufsen collected approximately 100 ethnographical items from among the Turkmen nomads for the National Museum back in Copenhagen; back in the carriage, he made wrote notes on their nomadic way of life.

Paulsen kept notes in his diary of an excursion undertaken on one of the last days of their stay in Merv. In what was apparently a welcome change from his daily botanical work under the merciless sun, he rode out with Olufsen to visit a Turkmen chief, a friend of Olufsen from his previous visits. Paulsen's notes make clear that it was no easy task for Olufsen as a collector of ethnographical items to barter directly for items in the Turkmen camps, and he can hardly hide his sarcasm towards his expedition leader who had difficulties in purchasing women's clothes:

June 6, Tuesday. Trip alone with Olufsen. Drank tea with some Turkomans, who learned that I was 'Perengi' (=fereangi) and asked me if I then was an Afghan, if we had no railways – and the name of our King. Dreadfully hot, 38 C in the shade. Later with Olufsen in a Turkman camp with an old 'chan', who had been at Gök Tepé. The tranquillity and almost indifferent dignity he expressed! Two wives waited on him. But he absolutely refused to sell any women's clothes to Olufsen![6]

At this point in the expedition, after one and a half years' travelling, Paulsen is almost boiling over with frustration over Olufsen's authoritarian leadership and the latter's decision once again to stay only for a short time in Merv. He is annoyed and can hardly bear the sight of his expedition leader. In the lowland steppe country, the working conditions of the expedition members were marked by the brevity of the stops that the rolling field station made at selected locations, given the expedition form. For the botanist, who frequently expressed a desire to have intensive, more prolonged study periods in one location, these short – and hot – stays in Merv gave rise once more to frustrations at the decisions made by the leader of the expedition and it added to his exhaustion. On June 7, he entered the following in his diary:

> The day was miserable. I wanted to write a letter, but got no further than opening the ink pot. Went for a short walk with Hjuler, changed the papers between my collected plants, covered in sweat. I will never come to Merv of my own free will again![7]

Each time the train had stopped, sometimes for several hours at stations along the Transcaspian railway on their way to Merv, Paulsen had patiently gone out into the countryside in the hot sun, with his small spade, and carefully collected plants which were next sorted and stored neatly between papers in the carriage. However, this way of knowing the field was not at all in harmony with Paulsen's idea of modern botanical investigations. For him it was not, as it was for Olufsen, an option to bring home as many different plants as possible for other scientists to investigate and classify in the Botanical Museum of Copenhagen.

Paulsen stated that his interest was in investigating the botanical context of certain species of plants, to register their neighboring plants and their habitat and conditions for surviving (Paulsen 1912). This constituted research that had to be done on the spot and could not be left to writing desk or laboratory botanist colleagues at the Botanical Museum in Copenhagen. His research questions needed long-term

---

6   Philipsen's private diary, 1896-1897 (in private ownership).

7   Paulsen's private diary, 1898-1899 (in private ownership).

studies at one place and Olufsen would let him work only a few weeks, days, or even hours in one place before setting off again in the carriage.

The carriage was used on the way through the lowland steppe country on the way to and from the Pamirs. The main purpose of Olufsen´s two Danish expeditions to Central Asia was the exploration of the Pamirs and the adjacent valleys bordering Afghanistan, where the second expedition stayed six months in a standard winter field station. In several ways, the scientific and social life in this station was a continuation of the one experienced in the rolling carriage. It functioned as a place of retreat after work in the surrounding field as the collected data and material was systematized, described, and packed in the station. The field, in contrast to the laboratory, was an object of study in itself. It consisted of the plants, animals, natural phenomena, and human creatures that grew and lived in it (Kohler 2002: 6).

I find it important to underline, however, that the field was not just a natural place to be studied. On the contrary, it constituted an analytical object. The natural place studied was certainly particular, specific, and invariable and the various disciplines represented on these two Danish expeditions each had their own analytical definition of the field. It was defined according to the research questions asked, though these definitions were indirect. The research questions determined what natural and cultural contexts and kinds of symbioses were worthy of observation, registration, and eventually analysis in relation to the place. In that sense the field was studied from a certain empirical position depending on discipline and theoretical orientation (Fihl 2003). As I shall try to illustrate in the following, the field also consisted of other categories of actors who had come to conquer or explore the Pamirs and Hindu Kush regions and who also had great impact on the conditions under which it was possible for the two Danish expeditions to stay in the field.

The Winter Station in the Pandsh Valley
On account of the high altitude – 3,000-4,000 meters above sea level, with mountain peaks up to nearly 8,000 meters – the Pamirs were known for many years by the romantic name of the "Roof of the World." The two expeditions took place at a time in history when the enigma of the sources of the river Amu Darya had already been solved (Wood 1841, 1872, Yule 1872). Likewise, the romantic idea about the "Roof of the World," where the cradle of mankind was thought to be found, was already well on its way to becoming extinguished by the theodolite and the compass, and superseded by the accurate delineation of scientifically drawn-up maps as Lord George Curzon had phrased it in his 1896 book *The Pamirs and the Source of the Oxus* (Curzon 1978: 1, 83).

In the years when the Danish expeditions visited the area, there was a series of Western expeditions to the Pamirs and to what was called Chinese Turkestan or Kashgaria or Uighuristan, now the modern Xinjang Province of China. The end of the nineteenth century saw travellers with a wide range of interest arriving in the area. There were Westerners who were interested in geographical explorations and the collection of topographical data for the preparation of exact maps of the Pamirs and Chinese Turkestan. Adventurers also arrived on the scene, for whom hunting the horned big wild sheep *Ovis poli* and the wild goat *Copra sibirica* as well as mountain-climbing were the prime aims. There were also deputations sent out by the distant power centers in Russia and British India that wished to have the military-strategic and commercial potential of the areas assessed, for the defense, and the expansion, of their respective spheres of influence. The turn of the nineteenth century – until the closing of the borders after the communist revolutions – was characterized by intense competition between British and Russian scientific explorations of the Pamirs, Hindu Kush, and Chinese Turkestan (Shaw 1870, Morgan 1893, Hedin 1898, Alder 1963).

The Western scientists had to share the field of the Pamirs and the adjacent regions in the Hindu Kush and Chinese Turkestan with both military personnel and these different categories of travellers. However, for the scientists this field became a kind of laboratory for investigating a number of the issues that preoccupied different scientific disciplines at the time regarding climate, geography, archaeology, language, and race etc. England's Royal Geographical Society and Russia's Imperial Geographical Society in particular tried to outdo one another in publishing their results about the Pamirs and the Hindu Kush (Veniukof 1866, Yule 1872, Gordon 1876, Curzon 1878, Trotter 1878, Bonvalot 1889, Dunmore 1893a, 1893b, Younghusband 1896).

In the town of Osh in Ferghana, Ole Olufsen equipped both his first and second expeditions to the Pamirs, in 1896 and 1898 respectively. On both occasions, the expedition consisted of riders and of pack-horses with equipment, tents, and provisions. Nearly all the equipment needed had been brought along from Denmark, weighting a total of around 3,000 kg on the second expedition. All instruments and fragile items were packed in large luggage boxes that were hung balanced in pairs across the packsaddles of the horses. The rest of the luggage was packed in similar wooden boxes, or wrapped in felt, so that it could be transported on horseback. Apart from the extensive amount of equipment for climatological observations and cartographical measurements, cameras, glass photo plates, and a phonograph for recording sound, the three expedition members also brought with them 22 guns,

9,000 cartridges, a transportable canvas boat, tents, boxes of spirit containers for conserving fauna, medicines, candles, horse shoes, a transportable kitchen, flags, writing materials, a small library, and several other things (Fihl 2002: 199).

After nine months of exploration in the Pamirs proper, the second expedition went further south to the valley of Vakhan with the Pandsh River where they were to spend the winter in Khorok, a small town in Shugnan which Olufsen knew from his first visit. The valley of the Vakhan is, to quote the well-known British traveller Sir Aurel Stein, "nature's true highway between the Tarim basin and the Oxus region" (Stein 1925: 398). For at least the last couple of thousand years, trade, communication, and migrations between the east and the west – between China and the kingdoms of Central Asia and Persia – had followed this route and it was held to be the route of both Marco Polo and the troops of Alexander the Great.

However, because of the British protest against Olufsen's original plan to take his expedition to Kafiristan, the Danish expeditions had, prior to their departure from Denmark, been given strict orders by the Danish Ministry of Foreign Affairs not to "allow any members of the expeditions to intrude on any but absolutely Russian territory" (Olufsen 1904: 4). This meant that the expedition had to travel the almost impassable northern side of the river Pandsh. Along many of the stages of the route though Vakhan, Ishkashim, and Garan, the expedition members therefore had to content themselves with only observing that the Afghan side of the river was much more easily negotiable. Concerning the geography of Garan, Olufsen relates that the southern Pamir Mountains are like a sheer wall that descends to the river Pandsh, making them exceedingly difficult to pass and the water level of the river was, even in the summer, so high and full of boulders that it was impossible to ride down in the riverbed. The passage in the mountains "is along steep paths scarcely half a foot broad, along the border of the precipices that go sheer down into the foaming river that roars several hundred yards below" (Olufsen 1904: 34). At several points, the horses, donkeys, and luggage had to be lowered by rope from ledge to ledge; at other points it was impossible to pass with pack animals, so the luggage had to be transferred onto the men of the caravan or native Vakhani or Garani, who were kind enough to come to Olufsen's aid throughout the expeditions (Olufsen 1904: 22).

Through the valleys, first in Vakhan and later in other provinces, scientific investigations were carried out and also continued when the expedition members in November 1898 finally reached what came to function as their winter field station for six months in Khorok, a village with approximately 200 inhabitants. Here the Danes received accommodation in a house that had been placed at their disposal by the *beg* (governor), installed by the Emir of Bukhara under whose rule the area now was.

In the winter station, Anthon Hjuler's assignment was to continue his field study of the East Iranian languages spoken among the peasant population, who were Ismailis and practiced transhumant pastoralism. During the nine months the expeditions stayed in these remote valleys, a vocabulary and grammar of the Vakhi and Shugnani languages was noted, which, according to Hjuler, were unwritten languages (Hjuler 1912).

In relation to the topic of how he got to know the field, Hjuler stated that it was difficult for him to find suitable interlocutors or mediums, as he called them. Often they did not understand his intentions and abstractions sufficiently well to co-operate. Many wanted to deal with reality: "They answered instead of translating, they did not interpret my sentence as an example, but as a reality" (Hjuler 1912: 1). One of the mediums was difficult to work with, since, to begin with, he becomes

> half-offended when I ask him twice about the same thing. And then he has no sense of any subtle distinctions; he has, though, a great sense of the factual. When I asked him what 'The woman's knife is old' was in Shugnani, he said that this was impossible to say; it was inconceivable for a woman to wear a knife (Hjuler 1945: 148).

In his diary, Hjuler stated that a neighbor to the winter station, a man named Zalil, helped him with the Shugnani vocabulary and grammar, but this relationship gave rise to tension with the other two Danes in the station:

> December 16. I have got an appointment with our neighbor Zalil to come and visit me for an hour or so every day, for me to learn Shugnani from him. He is a handsome, tall man and I started to deal with him today. He seems quite intelligent and has a very clear pronunciation, but there is the drawback about him that his voice is so powerful that it makes the whole house resound, and the others are disturbed by this. Apart from this, Paulsen complains about him smelling.
>
> December 19. In the morning I had Zalil over for conversation. Shugnani appears extremely difficult, the inflections in particular are complicated; I begin to doubt that I will ever really get hold of the grammar.
>
> December 20. This morning, Zalil came for conversation; I think nevertheless that I will succeed with Shugnani.
>
> January 18. (…) Zalil came in the morning. I made the interesting discovery that the locative case exists in Shugnani (Hjuler 1945: 157 f., 170).

In the valleys, Paulsen systematically collected various wild plants that were typical of the area. He described the scenery in Vakhan as being like a paradise, with cultivated fields far up the mountain slopes. The fields had already been harvested in September but the threshing was still in full swing. The wild xerophytes of the mountain slopes were meticulously registered. The green vegetation of wild hydrophytes along the irrigation channels, which formed a delightful contrast to the brown color of the mountain slopes, was also an object of Paulsen's interest (Paulsen 1920). The collected plants were scrupulously entered into notebooks, dried and pressed, and then carefully placed in plant-folders, so that they could withstand long transportation in the luggage-boxes of the pack-animals.

The prospect of the winter station suited Paulsen very well. For months he had been looking forward to this, since his modern scientific ideals induced him to want to examine the flora of a limited geographical area, and now the expedition at last had a breathing-space. However, the six month stay in the winter field station did not, as it turned out, offer Paulsen any real botanical challenges as the snow soon arrived, blotting the landscape with a thick covering. In an article written for the Danish *Geografisk Tidsskrift* during his winter stay he described the various plants to be found around Khorok. Even so, one senses a deep-drawn sigh:

> Finally, I have to draw attention to the difficulty of describing the vegetation in a region where the expedition has only spent the late autumn months, after the harvest had been completed and most of the wild plants were either partially or totally withered. The plants cultivated are thus only known from their seeds, and information about their use and the time of the year when they are sown – as also the treatment of the soil – cannot, therefore, be based on first-hand evidence but only on what the natives say (Paulsen 1900: 22).

Paulsen's task also included the realm of zoology. In particular his unpublished diary from the winter field station contains many notes concerning prepared faunal specimens: foxes, hares, falcons, ermines, sparrows, round-worms, various species of mice, etc. These were added to the collection from the Pamirs of specimens of the wild sheep, goat, and marmot. The skin of the animals was removed from birds and the larger animals, and the skulls boiled; the smaller animals were preserved in alcohol. The latter method in particular caused Paulsen a number of troubles. The lack of alcohol was a problem: "Preparations, checked the animals to be preserved in alcohol – two more mice utterly decayed. The devil take zoology!"[8]

8   Paulsen's private diary, 1898-1899 (in private ownership).

The physical anthropological investigations of the population in Vakhan and Shugnan were also Paulsen's responsibility. The height and cephalic index of 98 adult men were measured. Paulsen was, however, unsuccessful in getting measurements of any females despite the persistent attempts of the *aksakal*s (village leaders) to press-gang the women. The average height of the men in the two valleys was 168.6 cm, although differences were observed between the populations. What was at stake was the current scientific, anthropological issue of the 1890s: to what extent were the people of isolated mountain valleys such as Vakhan and Shugnan in Central Asia, with their "Celtic and Greek" appearance, the key to the solution of the mystery of the origins and descent of the European race?

Alongside his linguistic studies, Hjuler also co-operated with Olufsen. They were responsible for the cartographical measurement of the terrain and they also carried out meteorological investigations as well as registering hot springs and collecting samples of soil and stones. The then current research question of the relationship between the electricity of the air and other meteorological factors in particular caught Hjuler's interest (Hjuler 1900, 1903, Olufsen 1900).

Olufsen's ethnographical observations and collections of artifacts took place while stationed in the winter quarter as well as in the villages along the Pandsh River on the route to Khorok. A number of houses were measured and drawn and objects of daily life were collected for the Ethnographical Collections at the National Museum back in Copenhagen.

Concrete facts about the artifacts that Olufsen collected in the Pandsh valley can be found in several of his publications as well as in his lists of items handed over to the museum in 1897 and 1899.

However, it is typical of Olufsen's books and articles that many of the sober observations and experiences about daily life among the agropastoralists of the Pandsh Valley recorded in his notes become transformed in the writing process into generalizations about the society and culture of the people (Olufsen 1911). The texts contain very few ethnographical examples of descriptions of actual observations concerning the harvest, preparation of food, manufacture of implements, religious ceremonies, etc. Likewise, there are practically no records of actual conversations with informants. The extensive expedition framework seems to have promoted an extensive acquisition of ethnographical knowledge as well. In relation to ethnography, for him the most important thing was to bring home a collection of artifacts for present and future generations of ethnographers at the National Museum to investigate, publish, and exhibit. According to his ethnographical mentor at the museum, Curator Bahne Kristian Bahnson, artifacts were also believed to represent the most objective and impartial means of studying culture (Bahnson 1887: 172).

Concerning the way he came to know the ethnographical field, Olufsen wrote as follows in the foreword to his book *Through the Unknown Pamirs:*

> The details of ethnographic interest which appear in this book have been gathered partly by autopsy and partly by questioning the most intelligent natives, such as Kasis (judges) and Aksakals (superintendents of the towns). I got much information from a Kasi from the town of Rang, in the province of Ishkashim called Khoda Dâ, who stayed with us at the winter station at Khorok for this purpose. The Kasi was a native of Vakhan, had studied at the Medressé in Badkakshan, was Kasi of Aghanistan, where he had met the Siaposh in Kafiristan, and had now settled down in Bokharan territory (Vakhan, Garan, Shughnan, &c.) where he had got the appointment of Kasi in the village of Rang (Olufsen 1904: XV f.).

Thus the winter field station was open to special informants like the loud voiced neighbor Zalil and the Kasi from Rang. Social life among the Danish expedition members in the winter station, however, was still different from what they experienced when living in tents while travelling on horseback.

By the order of the Emir of Bukhara, both Danish expedition crews were accompanied by the Bukharan attaché Mirza Abdul Kader Beg, since they travelled in new Bukharan territory in the buffer zone. In the winter station, there are pieces of evidence that Mirza adopted some of the habits of the Danes scientists. They spent their leisure time playing chess, reading and debating – often philosophical matters. Both Hjuler's and Paulsen's diaries contain several examples of Mirza's worldviews concerning such matters as the relations between heaven and earth, life and death, the organization of the human body, anecdotes from Bukhara, etc. Accordingly, Mirza apparently occasionally noted in his payer-book the essence of some of his discussions with the three "farangi" (foreigners), just as the Danes related some of the same discussions in their notebooks and diaries (Hjuler 1945: 125, 134).[9]

Thus, intercultural exchange took place in the winter station on various levels as a result of their shared accommodations and the narrow space. The Danish diaries are full of sympathetic understanding for the torments that the otherwise cheerful Mirza had to undergo when he had to refrain from eating during the daytime throughout the religious month of Ramadan.

The local scout and servant Shankar Mahamat (also called by the Russian name Benkoff) also stayed with the Danes in the winter station. Both he and Mirza

---

9   See also Paulsen's private diary, 1898-1899 (in private ownership).

learned something about the religious observations of the Danes as they celebrated Christmas in the winter station, trying their best to hold on to European manners:

> December 24. In the morning we were busy putting up the Christmas decorations. There was no way we could get hold of a fir-tree, since they do not exist here, so we used a different tree instead as our Christmas tree; we fixed lights on it and hung up apples, chocolate in silver paper, gold ribbon and various tin decorations which Paulsen had artistically cut out. We exchanged gifts: Olufsen gave me an elegant box of tobacco, and I received a Russian basilisk from Paulsen. Each of us gave gifts to Mirza and Benkoff. We were all in pretty good spirits. In the evening we had rice boiled in milk, goulash, sardines in oil and tomato, Mettwurst, pineapple, preserve, snaps, brandy and cigars. Mirza, Benkoff, and his younger brother were all exceedingly content with our way of celebrating (Hjuler 1945: 159).

In the winter station the Danes took it in turns to have a "housekeeping week," assisted by Benkoff. Sometimes, Mirza also helped to prepare the food, but he wanted it greasier:

> November 24. We baked cakes yesterday evening. Benkoff turned out to have considerable expertise within this area, being especially proficient at kneading dough. Mirza took care of the fire and also had plenty of good advice. He was most interested in the whole process but would, in our opinion, have folded too much sheep-lard into the dough. The cakes were excellent, tasting rather like wholemeal bread, although they are a bit pasty, and burnt on the one side (Hjuler 1945: 153).

After listing the individual research tasks of the Danish members of the expedition in the winter field station, Olufsen stated the following in one of his articles:

> … so here was enough to keep us busy night and day, and when in the evening we gathered around the samovar we had always plenty of subjects for conversation, either about what the day had produced or about scientific matters (Olufsen 1905: 181).

During this leisure time, Hjuler would often be wearing his Turkis fez, a nineteenth century European fashion for home-life relaxation, which indicated that life in the winter field station was once in a while informed by social conventions from upper-class bourgeois circles back in Copenhagen.

Perspectives

Analyzing the ways in which the two Danish expeditions to Central Asia in the 1890s got to know the field raises the importance of looking into the practical social situations in which the Danish explorers worked, collecting objects and information as they travelled along in what can be designated as social spaces where Danish, Russian, Turkmen, Tajik, Uzek, and Ismailis representatives with different cultural backgrounds met, clashed, and grappled with each other, often in highly asymmetrical relations of domination and subordination (Pratt 1992: 4, 6).

It points to the fact that in the daily work of the expedition members, many factors other than pure theory were steering their research. Craft, material surroundings, and practical reasoning played a considerable role; there was also a socially diverse array of agents to relate to. The method of producing knowledge during fieldwork is consequently qualitatively different from conducting the same activity in laboratories, with their closed and controlled workspace (Kuklich & Kohler 1996: 3).

For the expedition crew, the rolling carriage and the winter field station functioned more or less as social incubators, as they were fairly closed to outsiders. While travelling on horseback and living in tents among the people they were studying, however, more open social interactions with various agents would take place, as local people (informants, interlocutors, mediums) would constantly sit outside the tents or act as travel guides and social gatekeepers, influencing the scientific work in certain directions and thus having a great impact on the results. This inter-cultural contact zone between the expedition crew and the local people was characterized by friction between different kinds of cultural aspirations and dispositions, but it was actually the place of a great deal of the data production – as was also the case when locals visited the carriage and the winter station.

In his geographical descriptions, Olufsen sought to the best of his abilities to live up to the modern issues that dominated European physical geography in the 1890s. He tried to link topography to a precise, scientific mapping of geological, climatological, and biological conditions. From the amount of detail in his descriptions, it is clear that he focused on the subject of whether nature is culture-friendly or culture-hostile, as his supervisor Professor Ernst Løffler had stated that this should be the main concern of geography (Løffler 1876: iii ff, 1911: 84). In relation to ethnography, Olufsen struggled to live up to the scientific guidelines received from the ethnographic curator Kristian Bahnson at the National Museum as well as from Løffler. For both, geography as a discipline existed in close relation with ethnography and history. The anthropo-geographical approach, which saw nature, culture, and history holistically, implied the importance of the study of "cultural diffusionism" and "cultural development" over great geographical distances (Bahnson 1900: II:297, Olufsen 1911: 147 ff). As such, this scientific approach suited the expedition form fairly well.

Olufsen's Central Asian drawn maps, which in a European style were believed to see the spatial landscape from a bird's eye view, were like texts rather models for – not of – what they purported to represent. The zoological, botanical, mineralogical, and climatological investigations on the two expeditions were also carried out using knowledge based on distinctly Western ideas of objective natural scientific observations and classifications. No systematic investigations were undertaken as to how Central Asians themselves perceived the plants, animals, and geological phenomena of different kinds. The members of the expeditions neatly organized the physical world into units, classes, and hierarchies based on European conceptions of understanding. In the paradigm of their day, the Danish expedition members saw their observations as mirroring reality. Once classified, they ascribed the categories an objective existence. In accordance with their own European system of classification and worldview – which indicated that the physical world could be subject to objective measurements – they made their scientific abstractions of reality.

Olufsen, however, was of the scientific traveller or explorer type, understanding his journeys as taking place on behalf of others, in this case the institutions he was collecting for and to a certain extent the Danish nation. For him and Philipsen, the concept of a scientific traveller fitted in well with their aim of bringing back as much compiled material as possible relating to a wide range of different disciplines. Exploration and penetration was for these two expedition members the goal of the venture. For this, they defied wind and weather, heat and cold, as well as every kind of peril and threat posed by nature and dangerous unfamiliar people. Rivers, glaciers, mountains, and deserts were crossed with utmost difficulties. They expected that their talents, effort, and hardship would be recognized with medals and orders. They hoped that they would gain fame among not only other European travellers but also in relation to the Danish public.

The two Danish expeditions can be seen as examples of the fact that the importance of nineteenth-century scientific expeditions collecting data while travelling, often thousands of kilometres, as a predominant way to get to know the field (in botany, ethnography, geography, and zoology) was challenged by other understandings of the field. Paulsen's aim of conducting long-term fieldwork at one place particularly implied a different view on the space, context, and role of the scientist both in public and in academia. The rolling field station on the transcaspian railroad line and the winter station in Khorok were evidence of the emergence of these new methods.

References

Alder, G. J. 1963. *British India's Northern Frontier 1865-95*. London: Longmans.

Aarbog for Kjøbenhavns Universitet 1895-97. Copenhagen.

Bahnson, K. 1887. "Ethnografiske Museer i Udlandet." *Aarbøger for Nordisk Oldkyndighed og Historie*, 11: 2.

Bahnson, K. 1900. *Etnografien. Fremstillet i dens Hovedtræk, I-II*. Copenhagen: P. G Philipsen/Det Nordiske Forlag.

Balsiger, R. N. & E. J. Kläy 1992. *Bei Schah, Emir und Khan. Henri Moser, Charlottenfels 1844-1923*. Schaffhausen: Meier Verlag.

Bonvalot, G. 1889. *Through the Heart of Asia: Over the Pamirs to India*. London: Chapman and Hall.

Burnes, A. 1975 [1834]. *Travels into Bokhara. Together with a Narrative of a Voyage in the Indus*. Karachi: Oxford University Press.

Burton, A. 1997. *The Bukharans. A Dynastic, Diplomatic and Commercial History 1550-1702*. Richmond: Curzon.

Curzon, G. N. 1978 [1896]. *The Pamirs and the Source of the Oxus*. London and Nendeln/Liechtenstein: Krauss Reprint.

Dombrowski, G. 1993. "The Collector Willi Rickmer Rickmers 1873-1965." In: R. Pinner (ed.) *The Rickmers Collection. Turkoman Rugs in the Ethnographic Museum Berlin*. Berlin: Museum für Volkerkunde.

Dor, R. & C. M. Neumann. 1978. *Die Kirghisen des Afghanischen Pamir*. Graz: Akademische Druck- und Verlagsanstalt.

Dunmore, The Earl of 1893a. "Journeying in the Pamirs and Central Asia." *Geographical Journal*, 2.

Dunmore, The Earl of 1893b. *The Pamirs, I-II*. London.

Fihl, E. 2002. *Exploring Central Asia. Collecting Objects and Writing Cultures from the Steppes to the High Pamirs, I-II*. Copenhagen: Rhodos.

Fihl, E. 2003. "Samlingen. At forfølge tingenes biografi." In: K. Hastrup (ed.) *Ind i Verden. En grundbog i antropologisk metode*. København: Reitzels Forlag.

Frohlick, S. 2006. "Rendering and Gendering Mobile Subjects in a Globalized World of Mountaineering. Between Localizing Ethnography and Global Spaces." In: S. Coleman & P. Collins (eds.) *Locating the Field. Space, Place and Context in Anthropology*. Oxford/New York: Berg, 87-103.

Gordon, T. E. 1876. *The Roof of the World*. Edinburgh.

Grancy, R. S. de & R. Kostka (eds.) 1978. *Grosser Pamir. Österreichisches Forschungsunternehmen 1975 in den Wakhan-Pamir/Afghanistan*. Graz: Akademische Druck- und Verlagsanstalt.

Gvozdetsky, N. A. 1974. *Soviet Geographical Explorations and Discoveries*. Moscow: Progress.

Hedin, S. 1898. *En Färd genom Asien 1893-97, I-II*. Stockholm: Albert Bonniers förlag.

Hjuler, A. 1900. "Nogle Undersøgelser af Luftens elektriske Spændinger." In: O. Olufsen. "Den anden danske Pamirexpeditions Vinterstation 1898-99." *Geografisk Tidsskrift*, 15.

Hjuler, A. 1903. *Measurements of the Electric Tension of the Air. The Second Danish Pamir-Expediton*. Copenhagen.

Hjuler, A. 1912. *The Languages spoken in the Western Pamir Shugnan and Vakhan. The Second Danish Pamir-Expedition*. Copenhagen.

Hjuler, A. 1945. *Rejsen til Verdens Tag. Dagbog fra en Rejse gennem Rusland, Turkestan, Pamir og Persien*. Copenhagen: Nyt Nordisk Forlag.

Høiris, O. 1986. *Antropologien i Danmark. Museal etnografi og etnologi 1860-1960*. Copenhagen: Nationalmuseet.

Kohler, R. E. 2002. *Landscapes and Labscapes. Exploring the Lab-Field Border in Biology*. Chicago: University of Chicago Press.

Kuklick, H. & R. E. Kohler 1996. "Introduction." *Osiris*, 11: 1-14.

Lansdell, H. 1885. *Russian Central Asia, including Kuldja, Bokhara, Khiva and Merv, I-II*. London: Sampson Low & Co.

Løffler, E. 1876. *Haandbog i Geographien*. Copenhagen: Gyldendal.

Løffler, E. 1911. *Min Selvbiografi, En Geographs Levnedsløb*. Copenhagen.

Morgan, E. D. 1893. "On the Pevtsof Expedition and M. Bogdanovitch's Surveys." *Geographical Journal*, 2.

Moser, H. 1885. *A travers l'Asie centrale*. Paris.

Okely, J. 1994. "Thinking through fieldwork." In: A. Bryman & R. G. Burgess (eds.) *Analyzing Qualitative Data*. London: Routledge, 18-34.

Olufsen, O. 1897. "Über die dänische Pamir-Expedition im Jahr 1896." *Verhandlungen der Gesellschaft zu Erdkunde*, 6.

Olufsen, O. 1900. "Den anden danske Pamirexpeditions Vinterstation 1898-99." *Geografisk Tidsskrift*, 15.

Olufsen, O. 1904. *Through the Unknown Pamirs. Vakhan and Garan. The Second Danish Pamir Expedition 1898-99*. London: William Heinemann.

Olufsen, O. 1905. *Gennem Pamir: Rejser i Verdens mægtigste Bjærglande*. Copenhagen: Hagerup.

Olufsen, O. 1906. "Gennem Transkaspiens Stepper og Ørkener." *Geografisk Tidsskrift*, 18.

Olufsen, O. 1911. *The Emir of Bokhara and his Country. Journeys and Studies in Bokhara.* London: William Heinemann.

Paulsen, O. 1900. "Om Vegetationen ved Chorock." In: O. Olufsen (ed.) *Den anden danske Pamirexpeditions Vinterstation 1898-99. Geografisk Tidsskrift*, 15.

Paulsen, O. 1912. *Studies on the Vegetation of the Transcaspian Lowlands.* Copenhagen: Gyldendal.

Paulsen, O. 1920. *Studies in the Vegetation of Pamir. The Second Danish Pamir Expediton.* Copenhagen: Gyldendal.

Pratt, M. L. 1992. *Imperial Eyes. Travel Writing and Transculturation.* London: Routledge

Ratzel, F. 1882. *Antropogeographie.* Stuttgart.

Shahrani, M. N. M. 1979. *The Kirghiz and Wakhi of Afghanistan. Adaptation to closed Frontiers.* Seattle, WA: University of Washington Press.

Shaw, R. B. 1870. "A Visit to Yarkand and Kashgar." *Proceedings of the Royal Geographical Society*, 14.

Stein, A. 1925. "Innermost Asia: Its Geography as a Factor in History." *Geographical Journal*, 65: 5-6.

Taylor, B. 1892. *Central Asia. Travels in Cashmere, Little Thibet and Central Asia.* New York.

Trotter, H. 1878. "On the Geographical Results of the Mission to Kashghar, Under Sir T. Douglas Forsyth in 1873-74." *Journal of the Royal Geographical Society*, 48.

Ujfalvy, C. de 1879. *Expédition française en Russie, Sibérie et dans le Turkestan, I-VI.* Paris : E. Leroux.

Vámbéry, A.-H. 1864. *Travels in Central Asia.* London: John Murray.

Vámbéry, A.-H. 1868. *Sketches of Central Asia.* London: John Murray.

Veniukof, M. 1866. "The Pamir and the Sources of the Amu-Daria." *Journal of the Royal Geographical Society*, 36.

Wenjukow; M. 1874. *Die russisch-asiatischen Grenzlande.* Leipzig: Verlag von Fr. Wilh. Grunow.

Wood, J. 1841. *A Personal Narrative of a Journey to the Source of the River Oxus, by the Route of the Indus, Kabul, and Badakhshan.* London: John Murray.

Wood, J. 1872. *A Journey to the Source of the River Oxus. New edition with an essay on the Geography of the Valley of the Oxus by Colonial Henry Yule.* London: John Murray.

Younghusband, F. 1896. *The Heart of a Continent: A Narrative of Travels in Manchuria, Across the Gobi Desert, Through the Himalayas, the Pamirs and Chitral, 1884-1894*, 2nd edition. London.

Yule, H. 1872. "Papers connected with the Upper Oxus Regions." *Journal of the Royal Geographical Society*, XLII.

Serge Reubi

Exploring the Disciplinary Significance of Fieldwork Methods
A Case from the History of Swiss Anthropology

Should there be only one discipline to which fieldwork is central, it may well be anthropology. Not so much because of the location of its object, which requires the scholar to go into the field: the same rule stands for natural history, archaeology, and many other disciplines. Rather, it is because some anthropologists have promoted fieldwork as *the* landmark of the discipline (see for instance: Malinowski 1922: 6-25, Lowie 1937: 132, Mauss 1947: 13-21, Lenclud 1991: 470-472, Barth 2005: 18). Most presentist accounts of the history of anthropology indeed assert that "modern," "real," or "scientific" anthropology starts with Bronislaw Malinowski's experience in the Trobriand Islands during World War I. Before, then, it is said that anthropologists were armchair, i.e. amateurish, anthropologists: they used speculative methods and museum collections to produce highly disputable knowledge. Anthropology was not a field science yet; nor was it even a science, as fieldwork was central to its scientific status.

This conception of the discipline's history, which until today remains the most widely spread in textbooks and manuals, was challenged some thirty years ago by what one may call historicist historians of anthropology. James Urry had already questioned the reality of this "invention of fieldwork" in his Hocart Prize Essay of 1972. Ten years later, in the first volume of his series *History of Anthropology* dedicated to the question of fieldwork, the historian of Anglo-American anthropology George Stocking (1983) also tried to challenge its history (see also Kuper 1973). Malinowski, he wrote, was not so much the inventor of the fieldwork method than its mere promoter: fieldwork, as a method and as a practice, preceded his own work by far and Malinowski only popularized it. However, this interpretation of the discipline's past is not entirely incorporated yet. Testifying for the persistence of the presentist conception of the history of the discipline, young scholars still believe it to be necessary to focus their research on this question while examining the cases of other traditions, in France (Sibeud 2002, Laurière 2008), in Germany (Kraus 2004), or in Switzerland (Schmidt 1998, Reubi 2008).

Different from the presentist school as this historicist tradition may seem, it shares with its forerunner the idea that one may observe a very important or even revolutionary change around 1920. Fieldwork before World War I, it is believed, was very different from its post-Malinowski incarnation. To remain quite schematic, one may say that at first, fieldwork was extensive, conducted in large areas by scholars who did not master the native language and (therefore) focused on material culture and physical anthropology; it then became intensive, and concentrating on one single population, whose idiom the anthropologist can speak and understand. Hence he may focus on more complex issues (religion, social structure, mental representation) which require a more intimate relation to and an empathy with the natives. Now the question is not to decide if these two types of fieldwork are different or not – they obviously are. The question is rather to determine the nature of their relation: do they reflect two paradigms, or even two disciplines, as the proponents of the intensive method assert, or are they merely neutral tools at the anthropologists' disposal?

Most historians of anthropology support the paradigmatic or disciplinary significance of methods (see, among others Stocking 1983, Eriksen & Nielsen 2001, Kraus 2004, Debaene 2006). This change in fieldwork practices is thus generally seen as the sign of a more fundamental, cognitive change in the knowledge field, its aims and its methods, since expeditions are instruments (Kohler 2006: 137) and instruments are *théories matérialisées* (Bachelard 1934: 16). However, the fact that contemporary actors such as Alfred Reginald Radcliffe-Brown (1923) or Bronislav Malinowski (1922) were defending this very position in the 1920s and used it as an argument to establish the rule of functionalism in the anthropological field may be seen as an indication that one should not accept it without further investigation. It thus seems legitimate to re-open the case. Hence this paper will examine the nature of the relation between modes of inquiry and theoretical positions or, to use a more general formulation, the relation between tools and paradigms and disciplines.

My paper is structured in three parts. To begin with, I will give a short description of these two types of fieldwork, focusing on the work of two Swiss anthropologists: Felix Speiser (1880-1949) and Paul Wirz (1892-1955). Second, I will show that the diversity of fieldwork practices is synchronic, rather than diachronic. Third, I will examine the nature of the arguments put forward by Speiser and Wirz to explain their choices and thus determine the place of cognitive considerations. To finish, I want to clarify the consequences of all this regarding the major break, which is supposed to govern the birth of modern anthropology.

Extensive vs. Intensive Fieldwork

Felix Speiser and Paul Wirz form an interesting couple of examples. Indeed, they share most of their epistemic background and life-story. They come from a similar *milieu*, share their education, and work within the same paradigms. Felix Speiser was born in 1880 into a very wealthy family in Basel which ruled over the political, economic, scientific, and cultural life of the city. He received an education as a chemist at the University of Basel, before leaving for New York City where he worked in a Swiss chemical firm. As he confessed, life in the American metropolis and his occupation as a chemist did not satisfy him at all. Hence, to keep boredom at bay, he sometimes negotiated swaps between the Natural History Museum in New York and its counterpart in Basel. Besides, the career of his uncle Paul Sarasin, director of the Ethnographic Museum in Basel, surely set an example of a more thrilling life which led him to switch to ethnography and anthropology at the age of 28. After a short stay among the *Hopis* in 1908, he returned to Europe and settled in Berlin, where he studied under the supervision of Felix Von Luschan at the Ethnographical Museum. He was thus very much influenced by the moderate positivism of the school of Adolf Bastian, which also framed his uncle's conception of ethnography and still dominated the vast majority of German-speaking ethnography. After two years of study, he felt sufficiently self-confident to organize a two-year expedition to the New Hebrides during which he tried to implement the usual fieldwork methods. He undertook a second exploration in Amazonia in 1924, and returned to Oceania in 1929, following new aims which left aside the question of the *Urmensch* and focused on historical migration processes, in accordance with the aims of the *Kulturkreislehre* current towards which he had turned. Financially, like most of the Swiss ethnographers, he was entirely autonomous and may be seen as an atavism of the nineteenth century's gentlemen of science. He never had to work for a living and he sacrificed his entire fortune and his life to his ethnographic passion: he financed his field studies himself, supported the Ethnographical Museum of Basel, and was also an honorary professor at the university, where he taught for free. At the end of the day, he is, together with his uncle Paul Sarasin and his cousin Fritz Sarasin, among the founders and patrons of ethnography in Basel and in Switzerland.

Paul Wirz followed a similar track. He too came from a very wealthy family and received an education as an agronomist from the Federal University in Zürich. However, he never got his degree and turned to anthropology at the age of 24. He studied anthropology and ethnography in Zürich with Otto Schlaginhaufen and Alfred Steinmann for one year before leaving for New Guinea in 1914, where he studied the *Marind-anim* over four years. Back in Switzerland at the end of the war, he obtained his doctoral degree in ethnography under the supervision of Felix

Speiser. However, quite the opposite of Speiser or the Sarasin cousins, Wirz had no ties to the Basel world and was not integrated at all into the leading circles of the city. Added to the fact that he had a strong urge to visit "beautiful islands, jungle forest and free men" (Schmidt 1998: 28), this diminished his grounds for staying in Basel. Therefore, unlike Speiser who remained in Europe for twelve years before his next expedition, Wirz immediately left again for a new expedition. Slightly less well off than Speiser, he progressively changed from a scientific life to the life of collector in order to finance his passion for "the free life" of the natives.

Both wealthy and the recipients of a similar naturalist education, Speiser's and Wirz's careers nevertheless share more parallels. They were both trained as anthropologists in Bastian's moderate positivism tradition which ruled German ethnography before World War I. Moreover, understanding the limits of this tradition, they both turned towards *Kulturkreislehre* in the early 1920s, showing an interest in migration studies in Oceania. Finally, both felt quite unprepared for their first fieldwork expedition (Schmidt 1998: 39). Amidst all these similarities though, their fieldwork practices vary quite a lot: Felix Speiser valued extensive fieldwork while Wirz preferred intensive research. This builds up an interesting situation in which similar personal careers and identical theoretical orientations do not lead to the same practices.

Extensive Fieldwork

Extensive fieldwork is characterized by the rather superficial study of large areas. This is not only a scale issue, though; it has a considerable impact on research. Extensive fieldwork means multiple and short trips. These reduce the opportunities to study the populations and direct the scholar towards what is tangible: the study of material culture and anthropometry. These orientations require an important material infrastructure, both to bring back the collections of artifacts and to carry the measurement devices. To this end, the anthropologists have to organize bearers, food, and transportation facilities, clear security and health issues, and plan their journeys, which leave an important part of the decisional power in the hands of "the colonial world" – settlers, colonial officers, missionaries, and natives.

If this constitutes the methodological frame of extensive research, Speiser, when he arrived to the New Hebrides in 1910, complying with the newest Berliner trends (Kraus 2004: 253-265), was not inclined to use it. However, for reasons which we will examine now, he opted for extensive fieldwork. His conception of what the very object of anthropology may be helps us understand this shift.

Felix Speiser identified anthropology as the history of the human species. In this sense, every human group may be of some interest to the anthropologist.

However, the rapid vanishing of primitive populations – i.e. populations identified as immediately following the hominization process – promoted them to a privileged object of study in Speiser's eyes, while "accultured" populations did not interest him. Hence Speiser focused his attention on isolated populations or populations with little contact with the European settlers and neglected the natives who live near the colonial outposts,[1] in order to find tracks of "the old culture [without] external influence and degeneration"[2] or to study people who were "still pure and live isolated in the forest."[3] In his naturalist perspective, they were the most instructive, because they reflected a pure, natural state. They were "a part of Nature itself" (Speiser 1913:29).

This is highly instructive for understanding his research practice. Looking for primitive populations, Speiser had to move around, which had considerable consequences considering Speiser's convenience standards and his epistemic positions. Speiser could not imagine going into the field alone. It required at least a boy, a cook, an interpreter, and bearers. "Since I had no boy, I couldn't plan anything independently, and haven't seen many natives" (Speiser 1913: 34). If the cook and the boy were needed to match his demands of personal comfort, the interpreter and the bearers were a consequence of his choice for extensive research. The extensive, large-scale approach prevented Speiser from mastering every single local idiom and therefore required the help of an interpreter which, of course, reduced or hindered more intimate relations with the natives.

> This continual change had two disadvantages. In the first place, I was nowhere able to win the confidence of the natives and there could be no question of learning their language. The languages of the New Hebrides, of course, differ greatly from one another and consequently neighboring villages often speak quite different dialects. As a medium of communication I therefore used *biche de mer* or I employ an interpreter. Dealing with abstract matter in *biche de mer*, or even concrete questions of any subtlety, was impossible. As interpreters I could call on the services of my servants or the more intelligent people of the locality. My exchanges with them were also either in *biche de mer* or in a kind of

---

[1] Interestingly, Speiser never names the autochtonous populations which he studied, instead speaking of "natives" or of the "population" of this or that island. Thus favouring the generic appelation or the Portuguese toponyms against the native's emic ethnonyms, he attests his choice for an etic, objectifying description against an understanding of the native's point of view.

[2] Felix Speiser to Fritz Sarasin July 15, 1910, Staatsarchiv Basel, PA 212 (subsequently SB-PA-212). All translations by the author.

[3] Felix Speiser to Fritz Sarasin April 20, 1911, SB-PA-212.

English that was not very different from it. I was therefore deprived of the most important condition for profitable studies: intercourse in the native language (Speiser 1923: 3).

Aiming at an overview, this approach, when one considers the finances at stake, also limited the time that he could spend with each population. It therefore led him to focus on the study of material culture and anthropometry or, to put it in another way, the approach rested on the collection of objects and the measurement of body parts. This required the use of instruments and resulted in the collection of huge amounts of artifacts, which someone had to carry; hence the bearers. This had important consequences: in order to find the necessary bearers, Speiser needed the help (or coercion) of colonial officers and missionaries, thus himself becoming one of the colonial authorities, or at least related to them. This is precisely what happened in the following case.

While he was stuck on *Espiritu Santo*, Speiser finally had a chance to see natives when a French settler invited him to take part in an expedition to the other side of the island. On the verge of despair, Speiser accepted, even though the purpose of the expedition is to hire workers for the Frenchman's plantation, an activity which had long been considered as a "slavehunt" by the natives (Speiser 1924: 45). There is little wonder in such a case that the ethnographer was associated with colonial power, which of course explains the fact that the natives were somewhat reluctant to interact with him. Moreover, travelling with bearers requires organizing catering, since self-sufficient communities are not able to feed larger groups; it also hinders overnight stays in the middle of the small native villages. It therefore requires even more bearers in order to carry food supplies and camping gear. Above all, it forced Speiser to remain outside of the villages most of the time, which again prevented him from achieving any intimacy with the natives.

This lack of intimacy and the brevity of his stays complicated his relation to the natives. But it also put Speiser in a relation of dependence towards them. As a matter of fact, Speiser did not usually have enough time or knowledge to understand what was at stake. Of course, he studied the New Hebrides' material culture in European museums before leaving.[4] Making use of another usual method, he also brought along Parkinson's recent and well-illustrated monograph (1907) to help him recognize the interesting artifacts (or to have them recognized by the natives) in order to gather a representative and orthodox collection. Nevertheless, the natives were masters of the exchange: they decided what to sell, even if Speiser presented them with a wish

---

4   Felix Speiser to Fritz Sarasin August 2, 1909, SB-PA-212.

list. Indeed, his collections reflect the natives' choices much more than those of the anthropologist. When he arrived in a village, Speiser informed the inhabitants of his interest in buying any old stuff, and he was usually quickly surrounded by numerous crowds who found opportunity to sell old and out-of-use objects.

> As soon as people knew that I was buying objects, an unpleasant mob would gather. Excited men would arrive from neighboring villages, each of them with his junk which he would try and sell, screaming and shaking it under my nose. It was often impossible for me to rest and the struggle of the kulis and of the other boys to calm the mob would only raise the commotion and the hustle [...]. But the mob would not let me go and so I interrupted my visits and only bargain at M. Matthew's place, what calmed people down. Of course, people would still gather around me, clinging on to my sleeves or my trousers but the whole thing would become tolerable (Speiser 1913: 334).

Most of the time Speiser did not choose the artifacts which he bought: "I collect somehow almost everything that people bring me."[5] And when he made a choice, it was nothing more than a secondary choice from the native's corpus of artifacts. As Chris Gosden and Chantal Knowles have clearly demonstrated, natives are much more accustomed to conducting business with Europeans than the opposite: "The collection of objects was not unusual, as many whites did this too, including probably some of the patrol officers. Many patrol officers collected information on material culture [...]. Local people would have been far more experienced at dealing with whites than most anthropologists dealing with them and this must have smoothed many a transaction" (Gosden & Knowles 2001: 47). The making of ethnographical collections is hence within the competence of the natives. Of course, this is nothing new, nor is it prejudicial to research. It is simply important to underline that natives have much more experience with anthropologists than the other way round. Generally, they know best what is of interest for the anthropologists. To put it in a postmodern way: natives are the principal co-authors of anthropological knowledge.

Considering the difficulties in organizing the various trips to meet "untouched" populations, i.e. populations which were sufficiently colonized to be safe for the anthropologist but sufficiently strange to liven up his curiosity, and taking account of the difficulties in mastering the exchange and the impossibility of immediate communication, historians of anthropology should not be surprised to observe that, even while travelling on the other side of the world, Speiser, like other anthropologists,

---

5   Felix Speiser to Fritz Sarasin November 27, 1911, SB-PA-212.

only obtained a small part of his data from the natives. He bought important parts of his ethnographical collections from white settlers – from antiquarians, missionaries, or other anthropologists – and obtained the vast majority of his ethnographical data from his interactions with missionaries.

This should remind us of the capital influence of the colonial world on extensive research, influence which finds its roots in the necessity of moving around and concerns every dimension of fieldwork – from the definition of the field's limits to the gathering of information or the type of fieldwork. Thus, Speiser first wanted to experiment with intensive fieldwork, but "contrary to [his] original plan of staying put in one place for the greater part of [his] stay, external circumstances forced [him] to be continually on the move. Admittedly, [he] was, as a result, able to visit almost all the islands, but [he] was seldom able to stay longer than a few weeks in the various places, and even where [he] was in a position to prolong [his] stay, the place was about as unsuitable for [his] purposes as it could be" (Speiser 1923: 3). The external circumstances which he makes reference to are the exchanges which he had with settlers during his voyage to Oceania, and with officials upon arrival, which quickly led him to give up this project. The importance of Colonna, the French Governor of the New Hebrides, is particularly central in this regard. Upon arrival, Speiser paid the anthropologist's traditional visit to both the British and the French Governors of the *Condominium*, to introduce himself and to meet the political authorities. And, as one knows, such exchanges are more than a mere visit of *courtoisie* or an administrative obligation and must be seen as a part of the construction of scientific knowledge. As a matter of fact, Colona convinced Speiser not to start his intensive study right away but instead to come along with him on a survey trip of the archipelago. This is how Speiser embarked on his difficult and unplanned extensive fieldwork. Organizational and synchronization problems rapidly trapped him on the island of *Espiritu Santo* where his epistemological views prevented him from seeing the local natives as potential objects of study, and the lack of domestics hindered any plan of going on into the field. Therefore, Speiser lost several months, doing nothing but zoological collection and reading novels.

The impact of the colonial was not purely negative, though. Speiser had to thank the kindness of a missionary who was going on a survey of the nearby island of Vao for finally having the opportunity to meet some "real" natives. But his stay on Vao was short; shorter at least than what he needed, since he had to leave with the missionary some weeks later. With no domestic, no boy, and no interpreter, he did not see a reason to stay with the natives nor did he see that staying would profit him, which is certainly true since he obtained most of the information on the natives from the missionary himself. Distrustful and independent, the natives were reluctant to give him any information – even though money might have untied their tongues. Thus he gathered most of his data from the Père Bochu. "He gives me very precious information but is lazy and therefore I haven't managed to convince him to report to me systematically."[6] And this situation is rather exemplary. In the following twenty months of Speiser's stay in the archipelago, little changes. He is either stuck on some island or travelling to the next thanks to some settler or missionary, who remain his main informants.[7]

Even if the colonial impact is important for every type of research, the influence of natives and settlers seems to be much more consequent for extensive fieldwork. If this is the case, the contingencies of travelling are mainly to blame. Ignorant of the local specificities, anthropologists are told by the "colonial world" where to find the "most interesting" populations; vulnerable and isolated, they are helped with the logistic issues of the expedition; cognitively incompetent because of the size of their field, they benefit from the competencies of settlers, missionaries, and natives who pick out the artifacts which will form the anthropological collections or propose their interpretation of the observed facts. The focus on material culture and the perpetual moves from one population to the other which characterize this type of fieldwork strengthens these features of the colonial contribution to research.

Intensive Fieldwork
To some extent, intensive fieldwork is very similar to its extensive counterpart. They share the same object. In Wirz's eyes as in Speiser's, the only interesting natives are the "pure ones," i.e. those who have a non-corrupting relation to Europeans or those who are not acculturated (Schmidt 1998: 28 f.). This principle led Wirz, while he was stuck in the Dominican Republic during World War II, to rigorously ignore the *Taino*, the local aruak population, which he considered to be entirely degenerated (Schmidt 1998: 136-138). Moreover the influence of colonial officers, missionaries, etc. on the orientation and the nature of knowledge is as central in intensive as it is in extensive fieldwork. Paul Wirz left Europe for New Guinea in 1914 without any clear idea of his exact final destination. Like most anthropologists, he started to gather information on the New Guinean natives from his fellow travellers during his journey to Batavia (Schmidt 1998: 37). Various missionaries and settlers pointed out the *Marind-anim* of the south coast of New Guinea as a potentially interesting study case. This population, they argued, was in a perfect condition for an anthropological

---

6 Felix Speiser to Fritz Sarasin July 15, 1910, SB-PA-212.

7 Felix Speiser to Fritz Sarasin September 20, 1911, and Felix Speiser to Fritz Sarasin March 24, 1911, both SB-PA-212.

Figure 1. A Mise en scène of Wirz's relation to the Marind-anim, 1915-1919 (© Museum der Kulturen Basel, (F) Vb7045).

study: they were slightly touched by civilization, so that Wirz would be under no threat there, but most ancient customs were still in use – an argument that reminds one of what the apostolic vicar of Port Vila said to Speiser before sending him to the north of the island of *Fate*: "very primitive natives but not dangerous anymore."[8]

Extensive and intensive fieldwork practices present important differences, however. In particular, intensive fieldwork gives the anthropologist slightly more control over the research conditions. It gives him important assets: intensive fieldwork is conducted in small areas or in a single village. This spares the bother of travelling, which is the element of extensive fieldwork that makes anthropologists vulnerable. It therefore eliminates most of the difficulties which Speiser, for instance, was confronted with. Having delimited a small area, Wirz could stay in the same village for several months before moving to the next one on the coast. This allowed him enough time to go beyond the mere study of objects and bodies and to focus on the "psychic culture"; it also gave him the opportunity to master the local

---

8   Felix Speiser to Fritz Sarasin May 5, 1910, SB-PA-212.

idioms. As a consequence, Wirz needed no bearers, since he was not travelling much and because he brought no anthropometrical instrument and collected only a few objects. Neither did he need interpreters, which once again reduced his dependence on the settlers and missionaries. At the end of the day, he had a much closer relation to the *Marind*,[9] which had at least two consequences. Firstly, he was less dependent on their goodwill than Speiser since his knowledge about them was built not only on their disclosures, but also on his experience. He multiplied the points of view and the sources of knowledge and thus balanced their power and gained in autonomy. Secondly he had the complete trust of the *Marind-anim*, because they got to know him:

> In *Kumbe*, where we stayed for 14 days, we made acquaintance with the population. The people were very well tempered and it was possible to start something here; however it took about a week before they knew us and stopped running away when we tried to talk to them [...] I spent the following days from dawn till dusk with the natives who became my friends and showed interest in my work. [...] We became good friends with the *Papuas*, well, at least, this is how I perceived the situation [...] and they didn't seem to have any suspiciousness towards us" (Wirz 1928: 29, 264).

Their trust allowed him to enter deeper into an intimacy with the *Marind-anim*. And this was precisely his purpose:

> It was my aim to increase the intimacy of my relation to the natives. Everyone who engages in ethnographical research knows how important it is to live with the natives, in their village, in order to understand their language and culture. It is only possible to penetrate their soul if one has long and intimate relations with them and many things can only be known through observation. It is also important to be alone as much as possible with the natives (Wirz 1925: 50).

The decision to conduct local, intensive research obviously has more consequences than the mere spatial specificity which it seems to give. It has consequences on the anthropologist's relations to the natives as well as to the European settlers, missionaries, or officers. Moreover it has an important impact on the cognitive dimension of anthropology: the scale of the field determines which methods

---

9   Paul Wirz speaks indiscriminately of the "natives" (die Eingeborenen), the "Marind", the "Marind-anim", the "Marindinezen", and the "people" (die Leute).

are legitimate, which research programs make sense, which objects and which problematic are significant, and what relations one should have with the natives: in other words, everything that is at stake in anthropology.

Two Disciplines?

It is easy to see why these two field methods have been analyzed as two different paradigms in the history of the discipline, or even as belonging to two different disciplines. However they are merely two facets of one contemporary practice, as Wirz's case makes clear.

Unlike a famous contemporary Polish anthropologist, Wirz was not a fervent defender, nor promoter of his particular fieldwork practices – at least not before the 1940s. He did not believe that they derived from independent or separate disciplines, nor did he have any strategic or epistemic reason to defend his choice of procedures. And more importantly: Wirz practiced extensive fieldwork himself as well. While in New Guinea, between 1914 and 1918, he sometimes left the shores and the *Marind-anim* to travel inland and study unknown, savage, and shy tribes. Strangely enough, on these occasions, he acted like Speiser: practicing extensive fieldwork, and travelling with bearers, interpreters, and armed guards. And his results are also quite similar to Speiser's: "However there isn't much to see here. The more north you go, the shyer they get, and one cannot start anything with them […]. Women and children are not to be seen because they are immediately sent into the woods."[10] Sometimes the natives even fled upon his arrival: "As we finally reached the village, a couple of hours away from the river, I decide to stay for a couple of days. But when we arrived on the next day with our entire package, they had taken to their heels and there was nothing left to do but return."[11] Conversely, Speiser was not reluctant to adopt the intensive research method: as mentioned above, his initial plan was to start an intensive research project which, he believed, had a bigger heuristic potential. Feeling like an *"antiquaire"* while collecting objects in 1911,[12] Speiser, after meeting Alfred Radcliffe-Brown in Sydney in 1929, recognized the higher value of intensive research: "The thing is important indeed and a true understanding of cultures can only be found when one knows its social structure. Well, we'll see how far I can go with this secret science."[13]

10 Paul Wirz to Louise Wirz-Nidecker May 27, 1917 (Schmidt 1998: 42).

11 Paul Wirz to Louise Wirz-Nidecker December 7, 1916 (Schmidt 1998: 43).

12 Felix Speiser to Fritz Sarasin April 20, 1911, SB-PA-212.

13 Felix Speiser to Fritz Sarasin December 15, 1929, SB-PA-212.

In fact, both Wirz and Speiser believed that anthropological fieldwork comes in a variety of forms without losing its disciplinary essence. In the area of specialization which characterizes sciences embedded in the naturalistic paradigm at the end of the nineteenth century, the essence of a discipline rests in the segmentation of the real or an ontological specialization (Blanckaert 2006), rather than in a method or a question. To put it another way, the essence of anthropology lays in its object as a segment of the world, not in its method: anthropology is the study of primitive populations in contrast to, for instance, entomology as the study of insects. How, why, or with what means one should go about it remains unclear and unchallenged, because it is meaningless. Thus anthropology may have two (or even more) sides: it may study small-scale communities in depth or large-scale areas in a superficial manner; for Speiser or Wirz it nonetheless remains the same discipline. One method learns little about multiple populations, the other learns more about one isolated group, but this difference does not matter, since both strive towards the *inventory of the world*, which is the ultimate aim of the naturalist paradigm.

In this perspective, when Wirz travels in the New Guinean inlands, he is thus explicitly conscious that his research is superficial, just as Speiser is. And if they agree to do such research, it is because the superficiality of the analysis is made up for by the quality of the object studied under these conditions. The "untouched" character of these natives explains their willingness to work under these awful conditions, although they value in-depth studies much more. The two types of practice nevertheless equally constitute anthropological research, just as archaeologists may carry out salvage archaeology, surveys, or excavations, and still do archaeology. They are, as Robert Kohler says of explorations and natural history surveys, different scientific instruments. And anthropologists, like laboratory scientists, may choose which one fits each situation best while remaining real anthropologists.

Once again, Malinowski's invention of fieldwork and Radcliffe-Brown's brilliant symbolic *coup de force* that George Stocking unveiled twenty-five years ago are proven false (1983). But there is more to it than that. Speiser's and Wirz's choices also teach us that "functionalist fieldwork" is neither revolutionary, nor exclusive of any other type or method. They invented an inescapable and founding difference to replace continuity in order to promote their method, their conception of the discipline, and, finally, their careers. Using a strategy which has been analyzed by Pierre Bourdieu (2001: 72-77), they redefined the limits of the field to exclude their immediate concurrent (missionaries and scholars from the German tradition) and gain a dominant position.

Wirz's and Speiser's cases underline a second important point: the necessity for historians of science to focus on the analysis of practices rather than on programmatic and/or polemical statements to construct a chronology and understand the past. Practices have a rhythm and a history different from that of programs and ideas. If this is so, it is a consequence of the practices' particular *vis inertiae*. Revolutions do not exist here, but practices allow one to follow the numerous steps which change a scientific discipline. It forces one to concentrate on each one of them and to acknowledge the contingency of change. The study of practices finally weakens the power of ideas by demonstrating that they do not necessarily become real. Even when they are accepted by a community of scholars, they do not automatically become the new norm in practice.

Field Rationales

Hence, if principles do not give an orientation to practices, one has to wonder what does. It is therefore interesting to analyze the reasons which pushed Speiser or Wirz to choose one or the other, or both types of fieldwork.

First, of course, one has to note what one may name contingent reasons. The specificities and the context of the field have an impact on the research program. Rain, storms, drought, and any other difficult meteorological conditions, poor transport infrastructure, illnesses, dependence upon settlers or missionaries, political or colonial tensions, and costs of the expedition may all shorten researchers' stays or prevent them from travelling according to their original plans. Important as they may be in research, contingent reasons are not very interesting for the historian, however; what one may name "necessary reasons" are far more instructive. Four of them will be addressed here.

The Methodological Doubt

Even though Speiser recognizes the heuristic value of intensive studies based on the understanding of language, he is very careful about the significance of its results. He fears that it is impossible to determine the veracity of interview data, so he methodologically prefers the materiality of the artifacts. "I give up my own ethnographical studies on religion or the like. It would be presumptuous to believe that one may understand anything correctly in a couple of weeks or months. One may of course question all of them, annoy them and get bored and still receive wrong answers."[14] This is why Speiser prefers to concentrate on material culture: not only does it satisfy his positivistic desires thanks to its mere materiality, but it also has the advantage that it rests on personal observation rather than on the always suspect information given by the natives. "I collect of course everything that I can,

---

14 Felix Speiser to Fritz Sarasin September 2, 1910, SB-PA-212.

but I will be satisfied with the observation of the pure external life, with what you name, I believe, ergology. It has this important advantage that it rests on your own observations, and not on the native's information."[15]

Institutional Inscription
As a curator at the Basel Ethnographical Museum, Speiser had a particular interest in ethnographic artifacts and in the growth of collections *independently of their cognitive significance*. Even if he acknowledges the heuristic superiority of intensive fieldwork, collections of material culture suffice indeed for the needs of the museum. As a matter of fact, from a museological point of view, extensive fieldwork is always a better option: if one has the opportunity to organize one's trip, a couple of hours are enough to gather the complete series of a population's material production and the minimal information about its use and origin necessary to the museum's catalogue. The importance of the institutional inscription is confirmed *a contrario* by Wirz's example. As a freelance anthropologist, he financed his first fieldwork himself and hence was free to choose the method which he preferred. And this relation to the museum was again confirmed when Wirz, ruined by the postwar crisis, faced important financial problems. To support his research, he then began to work for several European museums and therefore turned more clearly towards extensive fieldwork.

Empathy and Freedom
Paul Wirz felt a sincere empathy for the *Marind*. Their life, which he perceived as a model of freedom and healthy relation to nature, resonated perfectly with his own *Freikörperkultur*'s aspiration for freedom and nature. In contrast, Speiser lacked Wirz's enthusiasm and empathy for the Bank or the New Hebrides islanders. He considered them with scorn and disdain, and attributed the failures of his research to their mindless state. However, it was actually his own position which was problematic. Despite being aware of the superficial character of extensive fieldwork, he could not even imagine spending the necessary one or two years with the inhabitants of each island. His situation was thus quite similar to what he observed in 1929 when he studied the Salomon Islands: "I always believed that I would have some problems to penetrate the spiritual culture and I was hoping to find a communicative old man and this was the case for every ethnographer who made in-depth studies within a short time. But this old man hasn't shown up yet and I still have hope for my next stay on *Arue*. [...] As one knows, it is an old story now: it is always easy to know the

---

15   Felix Speiser to Fritz Sarasin September 2, 1910, SB-PA-212.

natives superficially, but it requires one or two years to get underneath the surface. And this is something I don't like."[16]

Speiser didn't want to spend time with natives. Thus, while in Brazil in 1924, he travelled up the *Parù* on a steamer with *Aparai* Indians. Although he crossed half of the world to study these *Aparai* and although he was bored to death staring at the rainforest, he preferred to spend time on the upper deck drinking fresh lemonade. He occasionally went down to see the *Aparai* in the hold, but tried to avoid contact with them as much as possible. It may of course sound paradoxical to be misanthropic and dedicate oneself to the study of human beings. As a matter of fact, he was probably influenced by an atavism of his first education. He practiced anthropology as botany or zoology: as a naturalist, with no interaction with his object of study (Hacking 1999). In anthropology as in these disciplines, he wished simply to collect items without any of the personal commitment necessary to achieving intensive fieldwork.

Personal Inabilities
Speiser's choice of extensive fieldwork finally results from his personal inability to work with the indigenous populations, although at first he still blames destiny for his difficulties in understanding the spiritual culture of the natives. Later though, after his Brazilian fieldwork in 1924, he had to admit that the problem lay in his own hands. "There's nothing to do for anthropology, because I can't get anything from the people, because at the beginning, they were ready to co-operate and tried to answer, but didn't know a thing. I can't understand how travellers can collect any information about psychic life under those circumstances. If I blamed myself after the New Hebrides, this newest experience was unavoidable."[17] Speiser was merely unable to venture beyond ergology, understood as the material culture, because he was unable to direct an interview: "When I ask why the village was constructed here and not there, I don't get any satisfying answer. It is the custom, that's what they say, and this is what they answer to any question about their culture. Indians are indeed slaves to their traditions about which they don't think: it is simply the custom!" (Speiser 1926: 131-132).

But the problem cannot be explained simply with the fact that Speiser did not spend time enough with the natives: every evening, after the meal, he sat with them around the fire, but somehow it always ended with Speiser answering the natives' questions about Europe, rather than the other way round. At the end of the day, if these nights were very instructive for them, they were not for him. As he realised,

16 Felix Speiser to Fritz Sarasin March 31, 1930, SB-PA-212.
17 Felix Speiser to Fritz Sarasin November 19, 1924, SB-PA-212.

part of the problem was linguistic, since the Indians rapidly got tired of answering in their poor Portuguese. The principal problem, however, was that the natives distrusted him and did not like his questions; he did not have a trusting relationships with them. Thus such trustless conversations were standard:

> What happens when you die? – I'll be buried in the ground, in this and that way … – Good, and you never come back on earth? – No, I'm buried in the ground. – But you told me before that your dead father can see you. – Yes he can. – But how can he see you, if he is buried in the ground? – Yes, well, he is not buried in the ground. – Ok, so where is he then? – He is buried in the ground. – Ok, but, a.s.o … When I try again, later, sometimes I get better, sometimes (and much) worse answers, because the people are a bit tired by our questions and don't understand the meaning of it all (Speiser 1926: 212).

Hence, the best way to obtain information was by a stroke of good fortune; but this means that the research may have had to go on for a long time, which he could not afford.

This inability doesn't disqualify Speiser as an anthropologist. In his study of American survey naturalists, Robert Kohler (2006) also notes that not all field naturalists are good survey collectors: some lack the necessary competencies, some are too slow, or unorganized, but this, of course, doesn't disqualify them as naturalists. The same should go for anthropologists, but does it? Obviously not, if one reads Malinowski or Radcliffe-Brown. In the 1920s, they set a trend which linked the anthropological discipline to a particular fieldwork method and ethnography to another, which underlines the importance of method for these disciplines. It is all the more so in the case of programmatic articles where contingencies suddenly become constitutive of anthropology.

Conclusion

The cases of Wirz and Speiser help us to understand a number of issues. First with respect to the history of anthropology, it questions the total change, the "scientific revolution" which Radcliffe-Brown and Malinowski pretend to have brought about after World War I. For them, the study of human society suddenly divided into two disciplines. Ethnology, which rests upon extensive fieldwork and the comparative method, produces highly disputable, historical and ideographical knowledge. Social anthropology, which is based upon intensive fieldwork, defends the functionalist relation of the different elements of society and produces synchronic, verifiable, and nomothetic knowledge of practical value. This change is certainly true in the long term, when one compares scholars of the 1880s with scholars of the 1950s. But in a more detailed analysis, it appears to be more subtle and lacks its alleged revolutionary dimension. The history of anthropological field methods and practices does not fit too well in the canvas of the history of anthropological thought, which has traditionally framed the history of anthropology. One could have ignored this discrepancy but then one would be failing to acknowledge the disciplinary dimension of histories of disciplines, and the importance of methods and practices in promoting a field of knowledge from specialization to disciplinarization.

Second, it shows that changes in practices, field practices in particular, are slow; practices have much more persistency than programmatic articles or presentist histories of the discipline suggest. Hence they are extremely useful in order to reconsider chronologies and histories of the discipline, while not significant to a history of ideas of the old school which still has influence in the research field. To be sure, cognitive arguments are important in practice, as Kohler and Bachelard suggest. But this may well be true only at a pre-reflexive level, or for the definition of the very object of enquiry. Interestingly, they are never put forward by the actors as an explanation for their choices in favor of one or another method. Here personal, institutional, and contextual arguments play a more important role and explain the permanency of practices, putting individuals and institutions back at the center of the understanding of sciences.

Finally, Wirz's and Speiser's positions question Bachelard's concept of instruments as materialized theories and force us to reconsider its significance. As a matter of fact, it may well be effective in the case of a disciplined science. As long as a research field remains at the level of specialization, i.e. remains a simple ontological division of the world, it does not rest on specific questions or methods. These are of secondary importance, and thus irrelevant in the age of specialization.

References

Bachelard, G. 1934 [1973]. *Le nouvel esprit scientifique*. Paris: PUF.

Barth, F. 2005. "Britain and the Commonwealth." In: F. Barth, A. Gingrich, R. Parkin & S. Silverman. *One Discipline, Four Ways: British, German, French, and American Anthropology*. Chicago: Chicago University Press, 3-57.

Blanckaert, C. 2006. "La discipline en perspective. Le système des sciences à l'heure du spécialisme (19ème-20ème siècle)." In: J. Boutier, J. C. Passeron & J. Revel (eds.) *Qu'est-ce qu'une discipline?* Paris: Editions de l'Ehess, 117-148.

Bourdieu P. 2001. *Science de la science et réflexivité. Cours au Collège de France 2000-2001*. Paris: Raisons d'agir.

Debaene, V. 2006. "'Etudier les états de conscience': La réinvention du terrain par l'ethnologie, 1925-1939." *L'homme*, 179: 7-62.

Eriksen, T. H. & F. S. Nielsen 2001. *A History of Anthropology*. London: Pluto Press.

Gosden, C. & C. Knowles 2001. *Collecting Colonialism. Material Culture and Colonial Change*. Oxford, New York: Berg.

Hacking, I. 1999. *The Social Construction of what?* Harvard: Harvard University Press.

Kohler, R. 2006. *All Creatures. Naturalists, Collectors, and Biodiversity, 1850-1950*. Princeton: Princeton University Press.

Kraus, M. 2004. *Bildungsbürger im Urwald. Die deutsche ethnologische Amazonienforschung (1884-1929)*. Marburg: Curupira.

Kuper, A. 1973. *Anthropologists and Anthropology: The British School 1922-1972*. London: Routledge.

Laurière, C. 2008. *Paul Rivet, le savant et le politique*. Paris: Muséum national d'histoire naturelle.

Lenclud, G. 1991. "Méthode ethnographique." In: P. Bonte & M. Izard (eds.) *Dictionnaire de l'ethnologie et de l'anthropologie*. Paris: PUF, 470-475.

Lowie, R. 1937 [1971]. *Histoire de l'ethnologie classique*. Paris: Payot.

Malinowski, B. 1922 [1963]. *Les argonauts du Pacifique occidental*. Paris: Gallimard.

Mauss, M. 1947 [1967]. *Manuel d'ethnographie*. Paris: Payot.

Parkinson, R. & R. Heinrich 1907. *Dreissig Jahre in der Südsee: Land und Leute, Sitte und Gebräuche im Bismarckarchipel und auf den Deutschen Salomoninseln*. Stuttgart: Strecker und Schröder.

Radcliffe-Brown, A. R. 1923. "The Methods of Ethnology and Social Anthropology." *South African Journal of Science*, 20: 124-147.

Reubi, S. 2008. *Gentlemen, prolétaires et primitifs. Institutionnalisation, pratiques de collection et choix muséographiques dans l'ethnographie suisse, 1880-1950*. Thèse de doctorat, EHESS-Université de Neuchâtel, 2 vols.

Schmidt, A. E. 1998. *Paul Wirz. Ein Wanderer auf der Suche nach der "wahren Natur."* Balser Beiträge zur Ethnologie 39, Basel: Museum für Vökerkunde.

Sibeud, E. 2002. *Une science impériale pour l'Afrique ? La construction des savoirs africanistes en France, 1878-1930.* Paris: Ehess.

Speiser, F. 1923 [1990]. *Ethnology of the Vanuatu.* Bathurst: Crawford House Press.

Speiser, F. 1913 [1924]. *Südsee, Urwald, Kannibalen.* Stuttgart: Strecker und Schröder.

Speiser, F. 1926. *Im Düster des brasilianischen Urwalds.* Stuttgart: Strecker und Schröder.

Stocking, G. W. 1983. "The Ethnographer's Magic: Fieldwork in British Anthropology from Tylor to Malinowski." In: G. W. Stocking (ed.) *Observers Observed. Essays on Ethnographic Fieldwork.* Madison, WI: University of Wisconsin Press, 70-120.

Urry J. 1972. "'Notes and Queries in Anthropology' and the Development of Field Methods in British Anthropology, 1870-1920." *Proceedings of the Royal Anthropological Institute,* 45-57.

Wirz, P. 1925. *Im Herzen Neuguineas Tagebuch einer Reise ins Innere von Holländisch Neuguinea.* Stuttgart: Strecker und Schröder.

Wirz, P. 1928. *Dämonen und Wilde in Neuguinea.* Stuttgart: Strecker und Schröder.

Christopher J. Ries

Armchairs, Dogsleds, Ships, and Airplanes
Field Access, Scientific Credibility, and Geological Mapping in Northern and North-Eastern Greenland 1900-1939

Science is like a tree – continuously growing. Every new result begets new questions, new investigations. And we can be certain that if a branch is deprived of the possibility to put out new shoots, other branches will soon smother it and take its place. (Lauge Koch, speech given in the Commission for Geological and Geographical Investigations in Greenland, 1924).[1]

In the Second International Polar Year 1932-1933, a controversy about the sovereignty over East Greenland between Denmark and Norway culminated in a trial at the International Court in The Hague. Settling the issue in Denmark's favor, the trial marked the end of a decade of escalating imperialistic rivalry between the two Scandinavian countries, and a competitive increase in the scientific exploration of the region by both nations. In the very year of the final verdict, Denmark's commitment to securing total scientific and political dominance over the last remains of its colonial empire was expressed in the most intensive season of the numerously and internationally crewed Three Year Expedition 1931-34 to East Greenland. That year 109 expedition participants descended upon the East Greenland coast under the leadership of the Danish geologist and Arctic explorer Lauge Koch. Armed with new field technologies, such as permanent wintering stations, airplanes, and radios, international teams of highly specialized scientists inaugurated a new era in Arctic research under Danish leadership.

The technological and institutional innovations that accompanied the launching of this expedition completely transformed the mode of producing and mapping geological knowledge about Greenland in Denmark. In this article, I will discuss these changes as they affected institutional and technological networks for

---

1   All translations by the author.

production of knowledge about the Arctic field during the early twentieth century. I will also demonstrate how, in one particular instance, this development undermined the social and epistemological basis for speaking with authority about Greenland geology to such a degree that all attempts at scientific debate and reasoning became futile.

My argument hinges on two fundamental premises. Firstly, that the different technologies of travel with which Danish scientists and explorers accessed the field in the early twentieth century – be it the proverbial armchair, the dogsled, the ship, or the airplane – were intimately tied to a number of semi-independent social and institutional networks, each wielding the authority to ascribe credibility to claims to knowledge about Greenland according to different social and epistemological standards. Secondly, that the common agreement on a progressively clearer understanding of Greenland geology was inherently dependent on these networks adhering to a well-defined division of labor, which ensured that the knowledge emerging from each network could be welded into a common body of knowledge without challenging the social and epistemological authority of the others.

During the late 1920s and early 1930s the foundations of this division of labor were increasingly strained by massive and far-reaching institutional and technological changes, and in 1935 they collapsed completely, when a group of Denmark's most influential geologists accused Lauge Koch of scientific fraud and misconduct. Seriously questioning a wide range of scientific results claimed by Koch over the course of his career, the criticism was widely published in Denmark and abroad, challenging Koch's status as an Arctic explorer and one of the leading international authorities on Greenland geology. In the end, Koch filed a libel charge against his attackers, initiating a two and a half year court debacle, which would shatter the foundations of the Danish geo-scientific community and leave it divided for more than twenty years (Ries 2002, 2003).

We shall return to some of the details of the ensuing debate later in this article. Suffice it here to say that the final ruling of the High court in 1938 was a "split decision" which allowed both parties to boast of victory. In 1939 the outbreak of World War 2 put an end to Danish Greenland exploration for six years, but even after the war, the conflict was never resolved. From 1947 until Koch's retirement in 1958, two independent Danish state-sponsored geological surveys were operating in Greenland, with little or no co-operation. Under the auspices of the newly established Greenland Geological Survey, Koch's university-based opponents occupied the west coast, while Koch continued his operations on the east coast as an independent advisor to the Greenland Administration, with almost no participation from Danish geologists (Ries 2007).

In this article, I will direct attention to the scientific dispute that sparked the Lauge Koch Case at a meeting of the Danish Geological Society in December 1935. For, although issues of academic rivalry and personal animosity played an important part in this controversy, it was not merely a socially generated crisis that polluted scientific debate and diverted Danish geologists from their chosen subject of interest. Rather, confused and overly agitated as the debate may seem today, the Lauge Koch Case raised crucial and necessary questions about the basis for scientific credibility: questions which had to be addressed at a time when changes in fieldwork methodology challenged traditional bases of scientific authority, ultimately threatening to dismantle the foundation for ordered scientific debate and common progress.

More specifically, I will home in on four specific points raised against Koch, and on the way the conflicting parties carried their argument on these points throughout the trial. For while the discussion of these four points brought into question various scientific results claimed by Lauge Koch, it was not so much the facts of the geology itself that set the parties apart. Rather, as the scientific combatants began to debate issues of scientific priority, they inadvertently moved into a much more fundamental discussion of the scientific authority of various modes of field observation. Essentially unable to disagree about what had been observed, the discussion began to revolve around questions about when these observations had been made, by whom, and by what means. As we shall see, the debate pitted different technologies of travel and fieldwork against each other, and in so doing, revealed a corrosion of the traditional epistemological division of labor in the geological exploration and mapping of Greenland which had been brewing for decades.

1878 to 1925: Armchair Geologists, Navy Captains, and Dogsled Pioneers

For most of the nineteenth century Danish geological and geographical exploration of Greenland was mainly centered around the colonial settlements on the west coast, where doctors, priests, and colonial administrators with little or no specialized knowledge carried out their investigations by boat or dogsled on the side, along with their more official and administrative duties. Navy officers contributed to the mapping of the coast between the colonies by ship, but the northern and eastern parts of Greenland remained largely unexplored or only visited by foreign expeditions.

In 1870 an increase in foreign activities in Greenland caused growing concerns about the lack of planning, scope, and co-ordination for the Danish exploration of Greenland (Bistrup 1943: 11-13, Garboe 1961: 298-305), and in 1878 The Danish Commission for Geological and Geographical Investigations in Greenland was established with a board of three, to further a more consistent and co-ordinated exploration of the great Danish colony. Head of the commission was Frederik

Johnstrup (1818-1894), professor of geology at Copenhagen University, supported by Hinrich Rink (1819-1893) director of the Royal Greenland Trade and Niels Frederik Ravn (1826-1910), Member of Parliament, and captain of the Danish navy.

Although the number and names of the members changed over time, the mixture of geo-scientific, political, and military representatives on the board remained a significant feature of the Commission until it was dissolved, reflecting not only the scientific profile indicated in its name, but also the national importance attributed to the exploration of Denmark's great northern colony. From its founding until it was closed down in 1925, this Commission advised, sanctioned, and supported Danish exploration efforts in Greenland, while editing the international scientific journal *Meddelelser om Grønland* (*Messages on Greenland*) in which all expeditions to the colony were expected to publish their results.

As was intended, the founding of the Commission was accompanied by an intensification of Danish geological investigations in Greenland. Nevertheless, although a small handful of trained geologists visited selected locations on the West Greenland coast over the next three or four decades, geological specialists remained few and far between in the field until after the turn of the twentieth century. The most important exception in this period was K. J. V. Steenstrup, who worked extensively in West Greenland, traversing the region by boat or dogsled on a number of expeditions between 1870 and 1880 (Noe-Nygaard 1984: 23); yet even Steenstrup conformed to the general picture in that his impressive fieldwork activities were not matched by his scientific publications. While Steenstrup managed to add considerably to the collections, it would largely be up to others to process, publish, claim, and certify the scientific results (Noe-Nygaard 1979: 307-8).

As a rule then, Danish field geologists in Greenland in the last quarter of the nineteenth century, regardless of their scientific qualifications, were mainly that. The transformation of their observations and collections into published certifiable scientific results was generally performed in laboratories and offices in Copenhagen. Here university specialists accessed the field by proxy, preparing fieldwork instructions and coordinating the activities of individual travellers while establishing scientific truth on the basis of the available literature and museum collections. The university armchair was the geological expert's vehicle for accessing the field, by way of the collections brought to him by others. His credibility stemmed much more from his institutional affiliation and scientific education than from his personal experience and observations in the field.

After a flurry of Danish geological expeditions to West Greenland in the 1870s – most of them conducted by Steenstrup – focus shifted to the geographical exploration of the largely uncharted East Greenland coast in the 1880s. Of course there were large and unexplored regions of potential scientific interest in East Greenland, but Danish interest in the area was also sharpened by growing Norwegian activity in the region during these years. Until 1814 Greenland, Iceland, and the Faroe Islands had been counted among the Norwegian possessions of the Danish King. After the Napoleonic wars, Denmark signed two peace treaties – one with Sweden and one with England – ceding Norway to Sweden in the Swedish treaty, while maintaining supremacy over all overseas possessions with the English. Since then, Denmark and Norway had maintained a strained but peaceful co-existence in the North Atlantic, with Norwegian hunters and fishermen mostly restricting their activities to the Norwegian Sea.

In the final decades of the nineteenth century, the Norwegians moved their activities closer to the East Greenland sea-ice, and Denmark responded by an increase in expeditions to the region. These were all conducted by navy personnel travelling by boat or ship combining assertions of national sovereignty with marine biological investigations and topographical surveys along the coast. No trained geologists participated in these expeditions, and geological observations and collections were only made sparingly and at random to be later handed in to the experts in Copenhagen (Garboe 1961: 360-361).

Until the end of the nineteenth century then, the establishing of geological and geographical truth about Greenland was carried out by representatives of the political and cultural elite of government officials that had dominated public life in Denmark throughout the 1800s. Speaking from the privileged positions of authority of the university armchair and the navy exploration vessel, the scientist and the navy captain both commanded the respect and admiration of their countrymen. United in the board of the Commission for Geological and Geographical Investigations in Greenland, the professional experts at the university ensured the credibility of scientific results while the navy provided national esteem and importance as well as regional logistical support.

At the turn of the twentieth century, a new type of authority entered the Arctic exploration scene to perform alongside the traditional scientific and military authorities: the sled-driving heroic adventurer. Of course, the Navy, which protected and demarcated Danish national borders, had always been able to generate national celebration. Similarly, during the later decades of the nineteenth century even culturally refined urban scientists – in many respects the counter-image of the sturdy and rugged navy captain – were increasingly celebrated as national heroes, selflessly and tirelessly working in the service of progress for humanity in general and the Danish nation in particular (Kragh et al. 2008: 329-339). In that sense, it is possible to discern the contours of a Danish Arctic hero in the East Greenland

Figure 1. Lauge Koch at the drawing board on Bicentennary Jubilee Expedition, North Greenland, 1922 (Arctic Institute, Copenhagen).

expeditions of the late decades of the nineteenth century, as ministers and celebrities attended the ceremonies of arrival and departure in the harbor of Copenhagen. Still, the university scientist and the navy captain gained their authority mainly as representatives of the traditional cultural elite. Respected and admired for their knowledge and social standing, they were not, however, embraced and idolized by a general public able to see in them a "fellow man" with whom they could identify.

The new Arctic hero was a modern man of the people. As a truly "public man," the Arctic explorer would become an Arctic incarnation of what has been termed "the Danish hero of the 1920s": the autodidact humanitarian who "clad only in his honesty" threw off the ties of civilization and devoted himself to experiencing his own existence in the world (Andersen 1998: 77). The young journalist Ludvig Mylius-Erichsen (1872-1907) was such a man. Strongly critical of Danish colonial policy in Greenland, it was he who sketched out the plans for the Danish literary Greenland expedition 1902-04. Members of the expedition, apart from Myius-Erichsen himself, were the artist Harald Moltke, the physician Alfred Bertelsen, and the young Danish-Greenlander Knud Rasmussen as an interpreter, as well as two West Greenlanders, Rasmussen's childhood friend Jørgen Brønlund and a hunter from Upernavik, Gabriel Olsen. Stressing the absence of natural scientists among the crew as an advantage, Mylius-Erichsen wished to approach and describe Greenland and its people in a new and different way. More specifically, he wanted to visit an isolated tribe of Inuit in the remote north-western corner of Greenland at Cape York, to experience "people clad in bear skin and living off raw meat" (Gilberg & Taagholt 2000: 164).

In fact, the Cape York Inuit had not been entirely isolated. Between 1891 and 1902 the American explorer Robert E. Peary had spent 4 winters in Greenland in close contact with the Cape York Inuit, who served as sled drivers and hunters on his expeditions in Northern and North-Western Greenland as part of his quest for the North Pole. In 1892, Peary reached the northern border of the inland ice at Academy Glacier and, from there, observed geographic formations which suggested the existence of a channel between Greenland and present-day Peary Land. The investigation of the Peary Channel and the establishment of Greenland's northern boundary became central to the Danish exploration efforts.

In 1902, the year that Peary terminated his activities in the Cape York District, "The Danish Literary Expedition" became the first in a series of Danish dogsled expeditions to Northern Greenland in the first three decades of the twentieth century, which aimed specifically to manifest Danish sovereignty in Northern Greenland. The Danmark-Expedition 1906-08, the Alabama Expedition 1909-12, the 1st and 2nd Thule expeditions 1912-13 and 1916-18, and the Bi-centenary Jubilee Expedition 1921-23: these were the expeditions that fleshed out the dogsled pioneer as an independent authority on North Greenland.

Mylius-Erichsen, Einar Mikkelsen, Knud Rasmussen, and Peter Freuchen were among those who both soothed and aroused the yearning for death-defying heroism and Arctic adventure, which grew in the Danish public during the first three decades of the twentieth century. Through a concerted effort, explorers such as these solved the last geographical mysteries of the northernmost parts of Greenland, proving Peary's channel to be a forgery and demonstrating Danish sovereignty over all of Greenland to Denmark and to the rest of the world. Their credibility stemmed not from their long education or the stripes on their uniform, but from the personal sacrifices they made in the conquest of the farthest unexplored shores of Denmark's great Arctic colony, and from the media coverage they got from the rapidly growing market of affordable newspapers, accompanied by a surge in literacy in the Danish population.

In summary, between 1875 and 1925, the geological and geographical investigation of Greenland was carried out within three distinct "regimes of exploration" each with their own expertise, their own claim to authority, their own geographical focus, and their own way of accessing the field. The exploration of the North was in the hands of heroic dogsled explorers with little or no scientific qualification, but with the will and the wish to challenge themselves and the hostile Arctic environment in pursuit of greatness and respect for Denmark as a colonial power, and for themselves. Patrolling the Greenland coastal areas by ship, navy captains had special access to the largely unpopulated east coast, where they combined oceanographic research and topographical coastal surveys with military flag-waving. Focusing on the colonized west coast, the scientific experts rarely ventured far beyond the colonial settlements. In fact, they rarely entered the field themselves, preferring instead to work in their labs and libraries in Copenhagen describing and processing collections and observations made by others.

Although a number of the larger expeditions were funded by private investors and foundations and arranged by independent committees, the Commission for Geological and Geographical Investigations in Greenland still played a role by supplying committee members, and by editing the journal *Meddelelser om Grønland* (Ries 2003: 133-134).

New Strategies, New Institutions, New Technologies, and New Controversies 1925-1935

Between 1875 and 1925 then, representatives of each of the three "regimes of exploration" could, under the common auspices of the Commission, stake claims to knowledge about Greenland according to their own standards, in practice supplementing and supporting rather than challenging the authority of the other two. But towards the end of the 1920s this division of labor began to corrode; this corrosion process saw its culmination in December 1935, when the publication of Koch's book *Geologie von Grönland* caused pent-up frustrations in the Danish geo-scientific community to explode in what would arguably become the greatest public scandal in the history of Danish science (Ries 2002, 2003).

Lauge Koch's career until then had been as unusual as it had been impressive, thoroughly integrating adventurous and life-threatening field explorations with exceptional and internationally acclaimed scientific accomplishments. At the age of 23 Koch participated in Knud Rasmussen's 2nd Thule Expedition in 1916-1918, charged with the responsibility of all geological and topographical mapping on the expedition. After graduation in 1920, Koch led his own Bi-centenary Jubilee expedition to North Greenland 1920-1923, during which he traversed North Greenland on dogsled for two and a half years, alone or with Inuit assistants, covering a distance of about 14,000 kilometers. The longest of these trips was a 200-day journey along the north coast of Greenland, round the coast of Peary Land, and back over the inland ice, covering a distance of nearly 2,900 km. By 1924 Koch's

Figure 2. Lauge Koch (2nd line, fourth from left) surrounded by crew members of the Three Year Expedition 1934. Among them Arne Noe-Nygaard (front line, center), Hans Frebold (third line, second from left) and Curt Teichert (third line, fourth from left) who all joined in the attack on Koch in 1935. Sitting third from left, front row is Gunnar Seidenfaden, Koch's then right hand, logistics manager and botanist (courtesy of the Archive of the Danish Geological Museum).

geological and topographical surveys had covered almost a quarter of Greenland's exposed land surface, and over the following two years he proceeded to strengthen his international scientific reputation in a flurry of maps and books and articles, while conducting a number of lecture tours in Europe and the USA.

In 1926 Koch was employed as geological consultant to the Greenland Administration, and charged with the responsibility of planning the geological investigation of East Greenland north of 70°N. In 1926-27 he led a geological dogsled expedition in the coastal areas between Scoresby Sound and Danmarkshavn, followed by geological investigations on Disko Island in West Greenland in the summer of 1928. In 1929 he received his doctoral degree on his treatise *Stratigraphy of Greenland*, summarizing his previous field results in a general account of the geological development of Greenland (Koch 1929), and in the summers of 1929 and 1930 he returned to East Greenland, leading teams of geologists on ship-based expeditions.

In 1930 Koch was commissioned by the Danish state to develop a plan for an exhaustive scientific investigation of East Greenland between Scoresby Sound and Danmarkshavn in response to increased Norwegian activity in East Greenland. The built-up tension between Denmark and Norway culminated in 1931, when Norway claimed sovereignty of part of the East Greenland Coast, and Denmark responded by summoning Norway to the International Court in The Hague. Boosted by the political urgency of the time, not only the Danish state but also a wealth of larger and smaller private contributors generously invested in the Danish expedition efforts realized in the massive and technologically advanced Three Year Expedition to East Greenland 1931-34, under the leadership of Lauge Koch. Pioneering the use of airplanes in Arctic geographical and geological survey, it was the largest expedition that had ever been sent out from Denmark, counting no less than 109 participants in the summer of 1933 alone. That year Denmark won the case against Norway in The Hague, retaining full sovereignty over all of Greenland.

After the end of the Three Year Expedition in 1934, Koch stayed in Copenhagen through 1935, preparing the manuscript for his *Geologie von Grönland*. It was this publication which would spark The Lauge Koch Case in December 1935, as eleven of the most influential geologists in Denmark published a damning review of the book in the *Bulletin of the Geological Society of Denmark* (Bøggild et al. 1935). "The Eleven," as they were immediately dubbed by the press, counted among them both former teachers, colleagues, and students of Lauge Koch, and their charges were serious indeed, ranging from theft of scientific property to misleading argumentation and sloppy scholarship. The review was followed by a call for an extraordinary general meeting of the Danish Geological Society at which a resolution to the effect of excluding Lauge Koch would be presented. Furthermore, 600 offprints of an English version of the review were distributed to geologists throughout the world.

When Koch responded by suing his attackers for slander, the press quickly took advantage of the situation. Over the next two and a half years The Lauge Koch Case haunted Danish newspaper headlines as it moved through the Danish Court System. During that time, what had at first seemed like a particularly fierce but fairly straightforward controversy over scientific priority proceeded to expose a number of fault lines in the institutional, technological, and epistemological foundations of Danish Greenland geology which had been developing since the beginning of the twentieth century.

Originally written in Danish, Koch's book had been translated into German for publication, and as it turned out, the proper interpretation of the German word "nachweisen" became a fundamental issue in the whole controversy. Even after a lengthy etymological expert treatise had been produced in court, the word "nachweisen," constantly oscillating between the meanings "suggest," "show," "determine," "prove" and "discover," remained a linguistic loophole in the debate until the very end.[2] For all the apparent comedy of this, it was in fact a direct and serious reflection of the epistemological crisis, which had been brought about by changes in the institutional and technological foundations for the acquisition of knowledge about the geology of Greenland over the past 20 years – with Koch himself as a pivotal point in that process.

It would be impossible in the limited space available here to discuss more than a few of the criticisms which Koch's book met with in Denmark. In the following, I will discuss four specific points of criticism put forth by "The Eleven" at the fateful meeting in the Danish Geological Society in December 1935, which each in their own way shed light on the ways in which Lauge Koch brought about an epistemological break-down in Danish Greenland geology by blending, challenging, and transcending the traditional practices and authorities at play in Danish exploration of Greenland in the early decades of the twentieth century.

Transcending the Navy Captain: The Trans-Atlantic Basalt Formation

Koch had conducted fieldwork in the Greenland basalts both east (1926-27) and west (1920 and 1928) around 70°N, and in *Geologie von Grønland* he claimed to have shown or "nach-gewiesen" these basalt formations and their relation to a larger trans-Atlantic formation running from Scotland, The Faroe Islands, and Iceland to Baffin Island in Canada (Anonymous 1935b: 554).

As pointed out by The Eleven, the relation of the Greenland basalts to a trans-Atlantic formation had already been discussed not only by Bøggild in his *Handbuch der Regionalen Geologie* (1917), but also by the English geologist Arthur Holmes (1890–1965), who in 1918 had mentioned a titanium bearing formation stretching from West Greenland to the Faroe Islands. Interpreting Koch's use of "nachweisen" here as a claim to "discovery," The Eleven felt compelled to counter Koch's presentation on this point. Koch on his part expressed his astonishment that The Eleven could insist that he had intended to claim to have "discovered" the Greenland basalt, especially since the basalt outcrops in East Greenland had been first noted by the English explorer William Scoresby (1789-1857) on a visit to these regions back in 1822 – seventy years before Koch had even been born. According to

---

2   Hammerich, L.L.: Responsum, January 15, 1937. Parcel: Retssagerne 1936-37. Lauge Koch's Archive, Danish National Archive (subsequently LKA-DNA).

Koch, The Eleven must have willfully misinterpreted the word "nachweisen," which must in this case obviously take on the meaning "point out" or "draw attention to" (cf. Anonymous 1935b: 563-564, Hammerich 1937). But there was a further detail to this: neither Bøggild, nor Holmes, had mentioned the continuation of the West Greenland basalts in Baffin Island, pointed out by Koch, and this would call into action the Danish naval captain Eigil Riis-Carstensen (1892-1953).

In 1927 Riis-Carstensen had laid out a plan for a comprehensive oceanographic exploration of the Davis Strait and Baffin Bay and in the spring of 1928 the Danish state – aided by the Carlsberg foundation – decided to fund Riis-Carstensen's expedition, transferring custody of the Greenland Administration ship *SS Godthaab* to the Ministry of Naval Affairs. With this ship, a navy crew, and five civilian scientists under his command, Riis-Carstensen investigated the waters between Greenland and Canada in the summer of 1928 (Riis-Carstensen 1929). En route to Baffin Land, Riis-Carstensen called in at Godhavn on Disko Island, where Koch was studying the basalts. Knowing that the ship would continue to Baffin Land, Koch visited the ship to ask if some of the crew could be tasked with looking for basalts on the other side of the bay. According to the young botany student Gunnar Seidenfaden (1908-2001) who was on the ship, neither Riis-Carstensen, nor any of his crew were really qualified for such a task, so Koch briefly explained to Seidenfaden and another crew member what to look for:

> The two of us ... who were the most interested in geology, had to admit that we were not sure that we knew basalt all that well. So Koch took us onto the deck, as we lay there in Disko Bay, and told us to look at the basalt terraces on Disko Island, and then we got a piece of basalt from Disko. So when we came over to Baffin Land, the commander had a boat lowered into the water ... Then he rowed ashore, and we went around looking for those stripes that we saw, and we damn well couldn't make out whether there was any basalt, so we picked up some rocks, which commander Riis-Carstensen gave to Lauge Koch, and then he could claim that he had shown that there was basalt there too.[3]

During the Lauge Koch Case seven years later, Seidenfaden, by then serving as Koch's private secretary and provisions officer on his East Greenland expeditions, was asked to testify on the episode in court, as Riis-Carstensen decided to stake claims of his own to the basalts in Baffin Land. As captain of the ship, Riis-Carstensen argued, he had been in charge of the investigation on Baffin Land, and should rightfully have

---

3   Gunnar Seidenfaden, taped interview conducted by author, February 20, 2000.

been credited for his effort in bringing about results to confirm Koch's theory.[4] Riis-Carstensen had not previously made any claim regarding the basalts. It therefore seems likely that Riis-Carstensen's attack on Koch in 1935 was spurred on by anxieties about the changing role and status of the navy in Greenland exploration in the 1930s, brought about by recent developments in Koch's field practices and his position within Danish Arctic research.

The foundations of the strained relationship between Riis-Carstensen and Koch had been laid in 1930. In that year, Danish concerns with growing Norwegian activity in parts of East Greenland materialized in the plan for a Danish navy inspection and hydrographical inspection of the East Greenland waters. Mimicking the 1928 expedition, Riis-Carstensen was to be put in charge of the navy crew and the civilian scientists on board the Greenland Administration ship *SS Godthaab*. Eyeing the possibility of continued geological explorations in East Greenland, Lauge Koch asked the Greenland Administration permission to use the ship as a base for a geological expedition to the region. The captain of the ship was Riis-Carstensen, and perhaps it was the seemingly unproblematic co-operation between him and Koch in 1928 that led to the decision to let Koch and Riis-Carstensen develop a common plan for the expedition.

After a meeting with potential scientific personnel and professors within the scientific disciplines concerned, Koch developed a plan which "as far as possible would allow the individual scientists to perform the tasks expected of them according to their abilities, on the basic assumption that I would be the scientific leader while the ship travelled along the coast north of Scoresby Sound."[5] But as his negotiations with Riis-Carstensen commenced, it became clear that Koch had reckoned without his host. In February 1930, Riis-Carstensen, without Koch's consent, submitted his own plan to the Marine Ministry, which in turn forwarded it to the Greenland Administration. Riis-Carstensen also forwarded a copy of the plan to Koch, but from the accompanying letter it was clear that Riis-Carstensen considered matters settled:

> Of course we have not talked about the last point in my plan – the chain of command – but that would really have been unnecessary. As the ship is fitted as a navy ship, there can only be one commander on board – I have really just put it in writing in order to prevent any possible misunderstanding up front.[6]

4   Gunnar Seidenfaden, taped interview conducted by author, February 20, 2000.

5   Lauge Koch to Daugaard-Jensen, March 18, 1930. Parcel 3, Archives of the Editors of *Meddelelser om Grønland*, Danish National Archive (subsequently AEMoG-DNA).

6   Lauge Koch to Daugaard-Jensen, March 3, 1930. Parcel 3, AEMoG-DNA.

Koch immediately took up the issue with the Director of the Greenland Administration, Jens Daugaard-Jensen, directly: "I am sorry that R-C. has submitted a plan without my knowledge, for the contents of which I cannot take any responsibility," he wrote.[7] Koch found Riis-Carstensen's mode of conduct completely unacceptable. Even worse, he considered the plan itself amateurish and completely inadequate as a response to the Norwegian activities, on several accounts. Firstly, Koch had very little faith in Riis-Carstensen's abilities to navigate in ice-packed waters, and was very concerned that his plan did not include a professional ice pilot on board the ship. During their negotiations, Riis-Carstensen, having complete faith in his own abilities in this respect, had in fact downright refused to discuss this matter, bluntly stating, that " … if an ice pilot was brought along, the ship would not sail at all."[8]

Moreover, according to Koch, Riis-Carstensen's plan revealed a complete lack of understanding of the importance of proper food and lodging for the scientists' ability to fulfill their tasks. While Koch himself was awarded a cabin of his own, the rest of the scientists would be crowded into the forecastle of the ship and expected to prepare their own meals and clean the ship. Furthermore, Koch was concerned with Riis-Carstensen's decision to rely mainly on pemmican for emergency winter provisions. All this, Koch argued, would increase the chance of scurvy considerably:

> The first condition for avoiding this ailment is of course a proper diet, but according to my experience, this will not do it alone. Scurvy can only be avoided if the spirit and emotional wellbeing of the expedition members is completely in order. I cannot define this more closely, but I know the importance of good spirits during a wintering. I fail to see how bearable wintering conditions can be maintained when the personnel is to be moved abaft from the orlop deck, and the young scientists are placed in hammocks in the salon.[9]

In one sense, the conflict between Riis-Carstensen and Koch can be interpreted as one of competing regimes of Arctic competency – of two different breeds of Greenland travellers. Both parties realized the political implications of the expedition, but while Riis-Carstensen reasoned mainly as the captain of a military inspection, Koch thought mainly as the leader of a scientific expedition:

---

7   Lauge Koch to Daugaard-Jensen, March 3, 1930. Parcel 3, AEMoG-DNA.

8   Lauge Koch to Daugaard-Jensen, March 3, 1930. Parcel 3, AEMoG-DNA.

9   Lauge Koch to Daugaard-Jensen, March 3, 1930. Parcel 3, AEMoG-DNA.

> R-C. mentions in his plan the importance of showing the Danish national flag in places where it has not been shown before ... I have always felt that the best way to "show the flag" was to carry out real work up there; work which should certainly not be inferior to that carried out by the Norwegians.[10]

Faced with Riis-Carstensen's plan and attitude, Koch informed the Greenland Administration that he and his scientific personnel would not be able to co-operate with Riis-Carstensen. Riis-Carstensen on his part openly declared that the navy had no interest in adding complications to the trip by bringing scientists on board.[11]

Under these circumstances, the fact that Koch, Riis-Carstensen, and their respective crews were aboard the ship when it departed in June 1930 can only be ascribed to serious pressure from the Greenland Administration. The Norwegian threat made it paramount that Denmark made its presence known in East Greenland, and while the navy was indispensable for logistical purposes, Koch's argument about the political importance of scientific activities seems to have swayed the final arrangement in his favor. The transport costs were still upheld by the navy, but Koch got his ice pilot and command of the ship in the East Greenland coastal waters (Koch 1955: 26). As it turned out, on all of Koch's subsequent expeditions before the outbreak of World War II, the Greenland Administration placed not only SS *Godthaab*, but also its other ship, the SS *Gustav Holm* directly under Koch's command, staffed by civilian captains and crew (Koch 1955: 249).

The subservience of the navy to Koch's scientific agenda was underlined by the fact that the air survey and topographic mapping on these expeditions was carried out with the use of airplanes and personnel from the Navy Air Command and the Geodetic Survey under the supreme command of Koch himself. In 1928 the staff of the Military Topographical Department had been put under civilian leadership in the newly established Geodetic Survey under the Ministry of Defense. And just as the transition from a military organization to a civilian administration throughout the 1930s gave rise to several smaller and larger conflicts within the Geodetic Survey itself (Poulsen 1997: 11), the subordination of military geodesists and navy air pilots to Koch's civilian leadership in Greenland during that same period did not pass without clashes of authority.[12]

---

10 Lauge Koch to Daugaard-Jensen, March 3, 1930. Parcel 3, AEMoG-DNA.

11 Lauge Koch to Daugaard-Jensen, March 18, 1930. Parcel 3, AEMoG-DNA.

12 "3-årsekspeditionen 31-34, Nordøst Grønland, Obl. Bruhn, flyvefoto – luftfartøj, Alm. Skrivelser." Archive of the Danish National Survey and Cadastre, General Archive.

Since 1875 representatives of the navy had at any one time taken up no less than half the chairs in the Commission for Geological and Geographical Investigations in Greenland. Thus, while the scientific disciplines mentioned in the Commission's name were each represented by one university scientist, the navy had still played a very important role in the planning and execution of the exploration of Greenland – and East Greenland in particular – combining military inspection and coastal topographical surveys with hydrographical and marine biological investigations. During the inter-war years, the role of the navy as an important authority in the scientific investigation of Greenland was gradually undermined by some of the very developments that accompanied Lauge Koch's movement to the center stage of Danish Arctic exploration.

To a certain extent then, the incompatibility between Riis-Carstensen and Koch might be seen as a reflection of institutional tensions that related to much broader changes in the scientific exploration of Greenland. In 1924 an increase in foreign exploration activity in Greenland had prompted Koch to suggest that the Commission presented the Ministry of the Interior with a 5-year plan for the exploration of Greenland, applying at the same time for a tripling of the yearly grants for the Commission. The submission of this proposal, in conjunction with recurring complaints about the narrow scientific scope of the Commission from a wide range of scientific disciplines, led to a complete revaluation of Danish Greenland research policy during the mid-1920s (Ries 2003: 137-154).

In 1926 a new and larger Commission for Scientific Investigations in Greenland was established under the Ministry of the Interior. Chair of the new Commission was the Minister of the Interior himself, seconded by the Director of the Greenland Administration. Out of the fifteen remaining members, twelve were scientists and scholars representing different scientific disciplines: geology, geography, archaeology, botany, meteorology, physics, geophysics, mathematics, ethnology, and medicine (Bistrup, 1943: 20-21). The remaining three seats were still taken up by navy officers, but reckoned by membership ratio alone, the influence of the navy on the Danish agenda for Arctic exploration had been greatly diminished. The original Commission for the geological and geographical Investigation of Greenland was not dissolved, but since its only remaining task would be to complete the already planned issues of *Meddelelser om Grønland*, it had in reality played out its part. From then on, the initiative for the dispatching of expeditions lay with private individuals, the scientific institutions, and first and foremost with the Greenland Administration (Jørgensen 1978: 128, Secher 1992: 25-27).

Transcending the Armchair Scientist: The North Greenland Caledonian Fold Zone

Koch was an extraordinarily bright and rising star on the firmament of Danish geology in the 1920s, but his way to fame and power passed through a scientific landscape disrupted by political and institutional change which would ultimately leave the Danish geological community transformed, and undermine the social and epistemological authority of the traditional scientific establishment. Broadly speaking, the criticism of The Eleven must be seen as the attempt of the Danish geo-scientific community to meet and counter that challenge. In order to address this situation more directly, we may take our departure in the criticism raised by The Eleven against Koch's claims regarding the determination of the North Greenland Caledonian Fold Zone. This point of discussion directly pitted Koch against his own former teacher and mentor Ove Balthasar Bøggild (1872-1956), member of the Commission for Scientific Investigation of Greenland and the only full professor of geology in Denmark. With only a single field season in Greenland in Ivigtut in 1900, Bøggild was not only an extremely skillful and careful crystallographer, but also primarily a laboratory scientist, who considered the museum collections his most valuable research tool (Noe-Nygaard 1979: 311). In 1905 he had made his name as a leading international expert on the minerals of Greenland with the publication of his 625-page *Mineralogia Groenlandica* (Bøggild 1905).

In 1917 Bøggild had published a brief summary of the geology of Greenland for the German handbook series *Handbuch der Regionalen Geologie* (Bøggild 1917). When Koch's *Geologie von Grönland* appeared in 1935, it was internationally welcomed as a much-needed update and summary of the vast increase in knowledge about Greenland geology brought about since then – largely through the efforts of Koch himself (see, e.g., Backlund 1936, Hobbs 1936). According to Koch's critics however, the fact that he completely omitted dealing with the geology of South Greenland – previously described by well-known geologists such as Bøggild, K. J. V. Steenstrup, and the Swedish explorer-scientist Adolf Erik Nordenskiöld (1832-1901) – was in itself a serious flaw in what purported to be a general summary of Greenland's geology. But even more serious was Koch's apparent failure to acknowledge what other people had accomplished in Northern and Eastern Greenland – among them Professor O. B. Bøggild.

As pointed out by The Eleven, Koch had written in *Geologie von Grönland* that he, in 1917, had shown – or "nachgeweisen" – the existence of a fault zone spanning the entire North Greenland coast, continuing westwards to Ellesmere Land and Grant Land (Højesterets Protokolsekretærer 1938: 493). According to The Eleven, this could only mean that Koch claimed to have discovered the North Greenland

Caledonian Fold Zone single-handedly, although the existence of such a formation had already been pointed out by Bøggild in his *Handbuch der Regionalen Geologie* (1917). Thus, Bøggild on the basis of the available literature and collections had suggested a possible connection between the metamorphosed and folded formations on Canadian Ellesmere Land, which had been known since 1878, and similar formations noted by the Danmark Expedition (1906-08) in the Northern and Eastern parts of Peary Land. According to The Eleven, Koch could therefore not rightfully claim the establishment of the North Greenland Fault Zone as the result of his own research, since all he had done was to confirm what had already been established by Bøggild.

As Koch's professor and as a member of the Commission for Geological and Geographical Investigations in Greenland since 1913, Bøggild had almost certainly presented Koch with his idea about the existence of a fault zone in North Greenland before Koch went away on the 2nd Thule-expedition 1916-18. Even so, Koch was unable to take Bøggild's critique seriously. Escaping death by only the narrowest of margins in 1917, Koch travelled along the north coast of Greenland on Knud Rasmussen's 2nd Thule Expedition, making first hand observations of the mountain range in question. In Koch's opinion, the fact that Bøggild's article had been published before Koch was able to publish his own results by no means deducted from the importance of Koch's field observations, which for the first time had established the existence of the fault zone as a proven fact, rather than a mere speculation on Bøggild's part.

Bøggild's claim to establishing the North Greenland Fault Zone rested upon the traditional division of epistemological labor, according to which trained geologists working in offices and laboratories were in charge of establishing reliable scientific knowledge. Synthesizing and extrapolating from available literature and museum collections, the armchair geologist meticulously studied any new material brought home by travelling amateurs, piece by piece adding to the comprehensive and ordered geological framework that he was composing. Koch, on the other hand, defied the traditional field/lab division of labor, adhered to by Bøggild. Not only a uniquely seasoned fieldworker, he was also an accomplished scientist, with an impressive international track record of highly respected publications. Excelling across the whole spectrum of skills and practices involved in the investigation and determination of the geology of Greenland, Koch did not have to rely on the intervention of others in his scientific practice. More "complete" in his geological practices, he would always be able to challenge the "incomplete" practices of other practitioners, who primarily worked at either end of that spectrum – such as Bøggild, the armchair specialist.

Of course, this epistemological challenge was weighted further by the fact that the status of the Geological Museum had been increasingly undermined as the central institution within Danish geology, as Koch's expeditions grew in scope and political importance during the late 1920s and early 1930s. Originally, Lauge Koch was trained as a geologist at the Geological Museum in Copenhagen. In his early career, this was where he found his daily professional network and where he recruited most of the geologists that took part in his expeditions, and it was to here that the collections from his early expeditions were returned for storage and investigation or for exchange with universities and museums abroad.

With the Three Year Expedition, the productivity of Danish Arctic exploration simply exploded. The establishment of an actual infrastructure in the field enabled a fast mobilization of enormous amounts of data, followed by a correspondingly urgent need for storage space as well as for manpower and laboratory facilities for registration and working up of the material. The geologists at the Geological Museum in Copenhagen might have been justified in expecting the dawning of a golden age for Danish Arctic geology – and for Danish geologists. However, as Koch's international reputation began to soar, he also became increasingly dissatisfied with the productivity and professional skills of his Danish colleagues, and began looking for more proficient colleagues in Sweden, England, Germany, and Switzerland. Also, after 1931, the collections from Koch's expeditions were kept locked up in a storage building in the harbor of Copenhagen. From here the different foreign experts employed by Koch would require selected samples to be delivered to universities and research facilities around Europe. In short, the Danish geologists were not invited to participate in Koch's expeditions, could not gain access to material, and repeated requests to have the collections handed over to the Geological Museum were turned down.

By 1935 then, the geopolitical urgency attached to the Danish exploration of the East Greenland coast had given Koch almost complete control of Danish Greenland exploration. Koch's unique fieldwork experience, his unquestionable scientific qualifications, his political skills, and the introduction of modern technology into geological fieldwork had allowed him to undermine the Geological Museum as the cornerstone of the Danish geo-scientific establishment, thereby challenging the institutional and epistemological authority of his own former mentor O.B. Bøggild – the incarnation of the scientific establishment and armchair geologist *par excellence*.

Transcending the Hero Explorer: The Peary Channel
Throughout the 1930s, Koch's modernization of expedition practices had transformed field logistics, enabling the well-trained scientific expert to supplant the death-

defying dogsled explorer as the ideal Arctic traveller. The heroism that characterized the early decades of the twentieth century was in decline, and Koch, while having earned his own Arctic spurs in the classic explorer tradition, increasingly found himself in opposition to the old school of heroic travellers (Ries 2003).

During the Lauge Koch Case, Koch's ambiguous position within the general school of Danish Arctic adventurers was brought into question by the attack of The Eleven. For, although the initial critique had revolved around his 1935 publication *Geologie von Grönland*, the debate quickly extended to include some of Koch's earlier publications as well. Thus, in the resolution presented on the fateful meeting in the Danish Geological Society in December 1935, Koch was accused of having exaggerated his cartographic accomplishments on the Jubilee Expedition to North Greenland 1921-23 – not in *Geologie von Grönland*, but in his popular book on the expedition: *Nord om Grønland* (1925a).

The critique focused on two maps intended to display the knowledge about the Peary Channel that existed before and after Koch's exploration of it. According to The Eleven, these maps imposed upon the reader the impression that the non-existence of the Peary Channel had been established by Koch (Anonymous 1935b: 566). This, The Eleven argued, was directly contradicted by the actual course of events. As early as 1907, members of the Danmark Expedition had found the bottom of Independence Fjord closed off to the West, and this had been publicly known since the publication of Mylius-Erichsen's account of this in 1913. Furthermore, in 1912 the same discovery had been made independently by Knud Rasmussen and Peter Freuchen on Rasmussen's 1st Thule Expedition 1912-13, and displayed in the maps published from that expedition in 1915. Therefore, The Eleven argued, Koch had violated the code of proper scientific conduct by ascribing to himself the honor of disproving the existence of the Peary Channel in the two maps accompanying his book.

In court, Koch countered this accusation on four main grounds. Firstly, *Nord om Grønland* was a popular book, which did not pretend to account for the discoveries of earlier expeditions. Secondly, the book actually did contain a word-for-word quotation of the Cairn report in which Mylius-Erichsen had originally published his observation. Thirdly, Koch had already published a treatise in the scientific journal *Geographical Review* in 1925, presenting a correct and exhaustive account of the history of the exploration of the Peary Channel to an international audience (Koch 1925b). And fourthly, and most importantly, the maps themselves did in fact portray the situation quite accurately.

As Koch pointed out, the map of "North Greenland before Lauge Koch's mapping" did show Peary Land connected to mainland Greenland by a thin line at the bottom of Independence Bay. The most important thing, he argued, was the fact that it was Koch himself who, as cartographer on Knud Rasmussen's 2nd Thule expedition in 1917, had first noted that the northern part of the Peary Channel was closed off:

> Before 1917, no-one had been able to observe what the conditions were along the remaining part of the channel shown on Peary's map. It is therefore quite appropriate to base a map purporting to show the knowledge of these regions before 1917 on the observations made by Peary, the error of which was only discovered in 1917 by the claimant on the 2nd Thule Expedition (Højesterets Protokolsekretærer 1938: 254).

The Eleven were not satisfied with the way Mylius-Erichsen's and especially Freuchen's discoveries around Independence Bay were marked on Koch's map. Arriving at Independence Bay from the west across the Inland Ice, Freuchen had been able to determine that Peary Land was connected to Greenland by "a considerable stretch of land" between Independence Bay and Adam Biering's Land. According to The Eleven, Koch's marking of that discovery with "a hardly visible line" was tantamount to him claiming the discovery of the non-existence of the Peary Channel for himself.

But Koch maintained his position. On the Danmark Expedition Mylius-Erichsen had only seen the bottom of Independence Bay closed off standing in the bay itself, and was unable to remark upon the situation further inland. On the 1st Thule Expedition, Freuchen had become snow blind and his instruments were damaged by salt water, rendering his map both unclear and partly in conflict with observations made by Freuchen himself as well as by the Danmark Expedition. Koch's map of the region, based on observations made on his Jubilee Expedition 1921-23, was in Koch's opinion therefore the first comprehensive and reasonably accurate depiction of the geography there, and marking the mapping results of previous explorers with a symbolic connecting line was an entirely appropriate acknowledgment of their accomplishments (Højesterets Protokolsekretærer 1938: 265-266).

At first glance, we may be surprised that The Eleven, representing the very scientific expertise which Koch had so determinedly sought to infuse into Danish Greenland exploration, would side with the heroic amateur explorer in their attack on Koch. But we must remember that the scientific production and position of the scientific experts rested in part on their close co-operation with the travelling adventurer amateur, who saw his justification and credibility in the production of scientific knowledge challenged during these years. In this respect, the uncompromising modernization of Arctic research carried out by Koch during the

1930s posed a painful challenge to a particular set of cultural values that had lain at the heart of Danish Arctic exploration since the turn of the twentieth century.

The great Knud Rasmussen was the archetypical incarnation of a particularly Danish spirit of Arctic conquest combined with romantic primitivism and cultural bonding with "the aboriginal Greenlander" – the Inuit. Rasmussen was of mixed Greenlandic-Danish descent, raised in Greenland, and endowed with an outstanding knowledge of Greenlandic language and culture. Intimately familiar with Arctic life conditions, he had been able to continue, develop, and embody the systematic use of native experts and traditional Inuit travel and survival techniques introduced by Peary (Harbsmeier 2002: 34-35). In a Danish context, this co-operative strategy carried a new symbolic meaning, conveying as it did a natural and historical bond between colonizers and colonized in the willingness with which the Inuit accepted and assisted the rightful and benevolent Danish conquest of their land.

Yet by the mid-1930s, the gradual introduction of modern technology into Arctic research carried out by Koch had left little room for romantic heroism, let alone empathic veneration for the traditional Inuit way of life. In the words of Gunnar Seidenfaden: "Just as Arctic research has learned to use the machine from industrial enterprise, it has also adopted the modern industrial form, – its organization and consistent rationalization" (Seidenfaden 1938: 27-28). In the Danish imagination of the 1930s, Greenland, formerly a space of freedom, adventure and heroic spectacle was falling under the spell of modern industrialized society: " ... the great polar explorer is dead, and the monument on his grave is the modern expedition" (Seidenfaden 1938: 25).

When The Eleven launched their attack on Koch in 1935, the great Arctic explorer was indeed dead in a very tangible way. In December 1933, the demise of Knud Rasmussen had thrown the Danish people into national mourning. As described by the Danish newspaper *Nationaltidende*:

> He was personal charm from head to heels. He was ever surrounded by an aura of light and festivity, and one got the impression of a man who lived his life free of all complexity. Easily and playfully he conquered, the youth gave him torchlight processions, the university made him an honorary doctor, the entire country called him by his first name (Anonymous 1935a).

Cast in the press as "Greenland's Aladdin," the great Knud was everyone's friend – and everyone's hero – linking history with the present, and forging the ties between Greenland and Denmark in a way that corresponded well with the Danish self-image as a rightful, benevolent, and welcomed colonial power.

To the Danes, Rasmussen was great in life, but even greater in death, and in December 1933 the newspapers were flooded with articles, notes, and obituaries in his praise. Peter Freuchen, polar hero in his own right and Rasmussen's faithful Arctic companion, freely expressed the Danish emotion as he mourned the loss of a close and beloved friend: "With Knud we witness the death of the last polar explorer who lived the life of the natives, and went where no-one had gone before him. Life is poorer without him. Words cannot express my pain" (Freuchen 1933). It seems only natural that Koch, having travelled with Rasmussen on the 2nd Thule Expedition 1916-18, was asked to make his contribution. Koch complied with this request, praising him as "our only polar explorer of world format" and as a charming and wonderful person. But he also took the liberty to make some critical remarks about Rasmussen's scientific achievements:

> Knud wrote what he believed to be science, but from time to time it came closer to fiction … From his earliest youth Knud stood in awe and admiration of science and he wished to be its servant. As a young man Knud was well aware of his limitations. To receive an honorary doctorate was the greatest dream of his life, which he hardly dared believe would be fulfilled. It is questionable whether he should have been awarded that title. He considered it to be something different from and more than a title. He saw it as a stamp of scientific probity and misunderstood it (Koch 1933b).

As might be expected, the patronizing tone in Koch's remarks was received with widespread resentment among Rasmussen's friends, co-workers, and assistants, as well as within the scientific establishment which had yielded members – such as O. B. Bøggild – to the different committees that had supported Rasmussen's many exploits. But Koch maintained his standpoint. As a polar explorer, Koch embodied completely different cultural values. Just as Rasmussen was cast in the press as "Greenland's Aladdin," Koch was cast as "Greenland's Nureddin":

> It never occurred to anyone to call Lauge Koch by his first name. As for his doctoral title, he had to earn that himself on a geological treatise. Heavily and stubbornly he worked his way to the name he wished to make for himself as an explorer … Born in steel armor … he never felt any desire to ingratiate himself with his fellows. He reserved for himself the right to express his opinion straight out, at any time and about anybody (Anonymous 1935a).

In Koch's opinion, Arctic heroism must be weighed on a different scale than Arctic science, and past observations superseded by more contemporary, more solidly documented, and better results. As Koch's attorney sarcastically argued in court, he found the continuing and uncritical worship of past exploits pitiful, sterile, and counterproductive to the modernization of Arctic research that was Koch's great ambition:

> But perhaps the geologist's rules of proper conduct require more than ordinary mention of previous researchers in dealing with a specific area. Perhaps one must repeat the names every time one comes across a piece of rock or a fragment that others have dealt with over the years? ... It is extraordinarily far-fetched and artificial to pin defamations on a man just because he does not continuously go around repeating his compliments to the men of the past (Anonymous 1937).

Transcending the Field Specialist: Central East Greenland Sediment Zones
The discussion about the exploration of the Peary Channel was only one of several disputes about Koch's way of assessing the observations of previous explorers – armchair or otherwise – which arose in the wake of his debacle with The Eleven (see Ries 2002: 209-213, 2003: 297-304). In general, Koch was able to defend his position with reference to the innovative combination of extensive fieldwork and internationally recognized scientific expertise that had been his hallmark throughout his career. In fact, it was precisely his ability to muster ship-loads of scientific experts to remote field sites in East Greenland during the early 1930s that had made his geological exploration program so successful, effective, and distinctly modern. But among the original points of criticism raised against Koch, there was one that pitted him against one of his very recent East Greenland collaborators, who seemed to comply perfectly with the very standards set up by Koch: the German geologist Curt Teichert (1905-1991). The debate about this point reveals particularly well the nature of the collapse in the social organization of epistemological authority that lay at the heart of the Lauge Koch Case.

After receiving his PhD degree from Albertus University in Königsberg in 1928, Theichert's efforts to build a professional career had unavoidably been complicated by the turbulent political and economic conditions in Europe. In 1928 he married a Jewish woman, and after completing an assistantship at Freiberg, Teichert received a Rockefeller Foundation award in 1930 for paleontology studies in Washington, D.C., New York City, and Albany. This one-year experience brought Teichert international recognition for his research on cephalopods and earned him a position as geologist

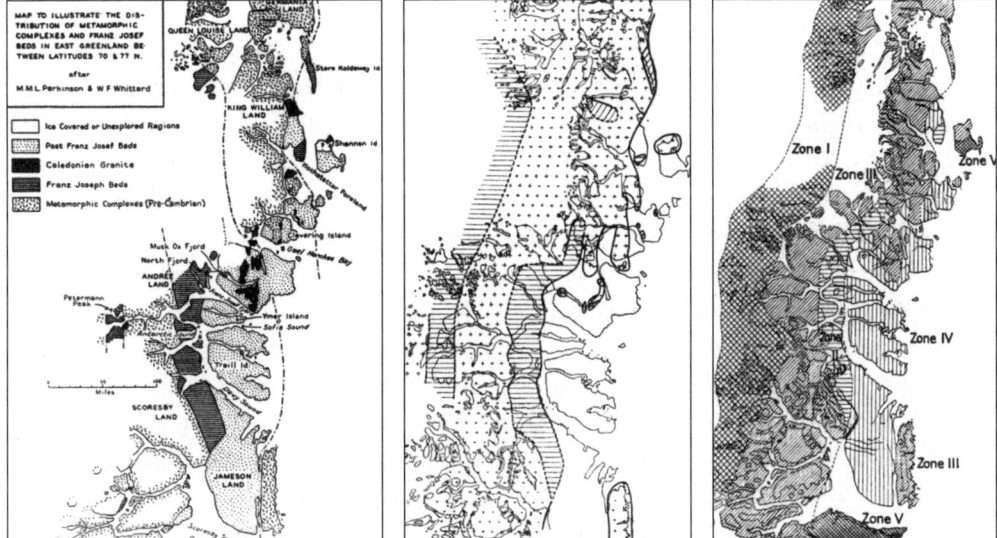

Figure 3. Maps of East Greenland sediment zones published in: Parkinson & Whittard 1931 (right), Teichert 1933 (Center), Koch 1933a (left).

on Lauge Koch's Three Year expedition, overwintering in 1931–1932. After his return from Greenland, Teichert and his wife had found political conditions in Germany so unfavorable that they moved to Copenhagen, where he received a small temporary stipend as a research paleontologist at the Geological Museum in Copenhagen (Reinemund, 1997: 39). It was from this position that Teichert entered the feud against Koch as one of The Eleven in December 1935, accusing him of having used and published his maps and results without permission, and without naming his source. To illustrate their point, The Eleven presented three maps of East Greenland between 70°N and 77°N produced by Parkinson & Whittard (1931), Teichert (1933), and Lauge Koch (1933a) (see Figure 3).

The situation was this: during the winter 1931-32 Teichert had wintered in East Greenland as part of a small expedition crew in the field station *Eskimonæs* on Clavering Island at 74°N. In spring 1932, Teichert had conducted geological investigations in the surrounding landscape, in April and May making a longer dogsled journey northwards along the edge of the Inland Ice to Bessel Fiord at 76°N and back along the coast. In June 1932 Koch – who had spent the winter in Copenhagen – arrived at *Eskimonæs* to partake in the expedition during the summer. During the first couple of days, Koch conducted a series of interviews with the overwintering expedition members, including Teichert, about their activities during the past nine months. Immediately after, Koch embarked on a series of geological air surveys covering the region visited by Teichert. Both Teichert and Koch returned to Copenhagen with the expedition ships in the summer of 1932, and in October Koch received a comprehensive report of Teichert's fieldwork including a number of maps and drafts. Koch immediately returned the material asking Teichert to prepare it for publication in the volumes of *Meddelelser om Grønland* that had been especially reserved for results from Koch's expeditions and for which Koch had exclusive editorial rights. In May 1933 Teichert sent in the finished manuscript and a map, which was then published in December 1933.

So far, so good. But according to The Eleven there was one big problem. In October 1933 the American journal *Geographical Review* had published an article by Koch, giving a short summary of the geology of East Greenland based on the results achieved on the Three Year Expedition so far; that article contained a map prepared by Koch, which, according to The Eleven, must have been based on Teichert's results. The fact that Teichert's name was not mentioned in the article was all the more serious, as Teichert had laid out what The Eleven described as "an entirely new idea about the geology of that area" (Anonymous 1935b: 568). By way of comparison, The Eleven presented an earlier map of the same region, prepared by the geologists Parkinson and Whittard in 1931, and pointed to important differences in Teichert's map: differences which Koch had apparently incorporated in his own map.

In the opinion of The Eleven then, Koch had ignored Teichert's unshakeable results, proposing his own but essentially similar speculation, for they found it highly unlikely that Koch's airborne observations alone would have provided him with a sufficient basis for proposing such a hypothesis. In their original presentation of their critique in the Danish Geological Society, they had described the concept of geological survey from the air as "sheer madness" and "a completely hopeless idea" (Anonymous 1935b: 562). Unable to accept that Koch could draw any authoritative geological conclusions on the basis of airborne observations alone, they asked him to explain "how you can determine from the air, that a red eruptive is acidic, and that a black eruptive cannot be acidic. Surely one cannot look solely at the color" (Anonymous 1935b: 561).

Since none of Koch's attackers had any experience in airborne geological survey, he dismissed them as completely incompetent in assessing results achieved in this manner. His answer was as arrogant as it was evasive, but it was also revealing of the different opposite epistemological positions from which the parties argued: "Only such persons who have not seen eruptives in nature, and know them only in the form of museum pieces, would ask a question so naive" (Anonymous 1935b: 553). Without question Koch, on his early dogsled expeditions, had developed and practiced skills of geological field observation and regional mapping far beyond those of any his teachers and colleagues. Summarizing this experience in 1929, he wrote in the preface to his doctoral thesis *Stratigraphy of Greenland*:

> If a geologist who travels by sledge aims to obtain a general impression of the conditions, and is not satisfied with making a series of casual observations in the places where he happens to camp, he must get into the habit of taking a broad view of things. I may, perhaps, illustrate this by an example. In civilized countries a magnifying glass is an instrument necessary for making geological studies. On geological reconnaissance by sledge the magnifying glass has to be replaced by a telescope, and sometimes the time left for geological investigations is so limited that one is tempted to turn the telescope around so as to get a diminished picture of the landscape and thus in an easier way obtain a general view of the conditions (Koch 1929: 8).

On Koch's dogsled expeditions in the north, very few had been in a position to question the validity of his observations, and his unique results were received with much admiration and little discussion. Since the late 1920s, however, this situation had been gradually changing as Koch began bringing other geologists with him on his expeditions. And while Koch's use of the airplane in the 1930s did still provide him with a privileged observational position in the field, the vast increase in the number of geologists on his expeditions made his results ever more susceptible to questioning and debate among colleagues exploring the same regions by different means.

Teichert, unable to able to get a full regional eagle-eye view of the geology, had arrived at his results by deduction and extrapolation from ground-based observations and samples collected along the route he had travelled. Koch, on the other hand, had been able to get a more general view of the conditions from the air, and produce what he considered to be a reliable scientific result based on direct observation. True, by working from an airplane, Koch had achieved his results in a way that made them less open to evaluation and critique by his colleagues. In essence, however, that situation was not much different from what had been the case on his early dogsled expedition in North Greenland. As Teichert and Koch had arrived at very similar results the only issue that remained open for debate was therefore – in this as in the previously mentioned examples – which of the results that carried the more authority.

Conclusion: The Power to Know

In relation to Teichert's mapping, Koch had already put forth an argument at the meeting of the Danish Geological Society in December 1935 that addressed the problem of epistemological authority from a very different but also very revealing perspective: Even if he had based his own results on Teichert's – which he maintained that he had not – Koch as leader of the Three Year Expedition was entirely licensed to cite the results of his expedition members, even without informing them or asking for their permission. As one of Koch's few supporters among the Danish geologists, vice president of the Danish Geological Society Valdemar Nordmann (1872-1962), put it: "We have yet to establish whether or not Koch might have the same right as f. ex. the state geologists of the Danish Geological Survey, who can readily publish the results and observations of lower ranking geologists and assistants participating in their surveys" (Anonymous 1935b: 572).

During the 1930s, Koch had indeed obtained a very ambiguous yet overwhelmingly powerful position in a geo-scientific milieu steeped in international geopolitical interest and institutional upset. The territorial conflict with Norway made Greenland a main priority in Danish research policy during these years, and as leader of the Three Year Expedition 1931-34, Koch had been directly appointed by the Danish government to lead the field campaign for scientific and political dominance to secure a Danish victory at the international court in The Hague. Although Koch did not formally hold a position as state geologist, and although there was no formal existence of a Greenland Geological Survey, Koch, with solid governmental backing, was in fact acting as both – even to the point of stamping "Greenland Geological Survey" on everything from geologist tools and supplies boxes to letters and official documents.

At the same time, traditional institutions of fieldwork organization and generation and distribution of authority in Danish Greenland exploration were crumbling. In 1925 the honorable Commission for Geological and Geographical Investigations in Greenland was superseded by the much larger and scientifically more diverse Commission for Scientific Investigations in Greenland under the Ministry of the Interior, and with the formation of a special Three Year Expedition Committee around Lauge Koch under the new Commission, Koch took up a powerful and unique position at the political center of Danish Arctic science. In the dawning age of modern high-tech expedition, the heroics of the dogsled amateur-pioneer lost their claim to scientific legitimacy. For the navy, ambitions of independent scientific agency in Greenland were stifled, as ships, planes, and personnel were drafted for logistical services in Lauge Koch's operations.

At the university, Danish geological expertise was humbled by Koch's predilection for working with international experts. In this situation, Danish geologists were faced with the uneasy choice between drafting themselves to Koch's expeditions, to work for a man who in the words of Koch's then private secretary Gunnar Seidenfaden "knew how to exploit and prey on his co-workers and their results, and then recklessly discharge them to look for new blood,"[13] or sitting idly by and watching substantial governmental and private funds as well public and professional fame flow in Koch's direction. They chose the third option: attack.

Throughout this article I have interpreted the "Lauge Koch Case" as an expression of a social and institutional crisis in the Danish polar exploration community, brought on by the massive changes in the practices of Danish polar exploration that converged around Lauge Koch during the interwar years. I have, however, tried to avoid the mistake of writing off the ensuing scientific debate as a mere "pretend discussion" to smoke screen more base motives of professional envy or personal animosity. On the contrary, by accepting the scientific arguments posed by the conflicting parties as sincere, I have attempted to approach an understanding of the ways in which the epistemological basis of scientific knowledge production in Danish Arctic exploration was directly dependent on a particular set of technological, social, political, and institutional conditions. In a period of rapid cultural, institutional, and technological change, the "Lauge Koch Case" was a direct reflection of an upset in modes of knowledge production and authority distribution that spawned new epistemological questions, for which the social institutions necessary for a resolution were not yet in place.

Epilogue

Koch's book *Geologie von Grönland* was, as Koch readily admitted to Peter Freuchen, "poorly translated and encumbered with careless mistakes."[14] Whether this in itself justified a critique of Koch's entire life's work as radical as that presented by The Eleven remains a matter for discussion. Likewise, Koch's decision to take The Eleven to court might at first sight look like a violent overreaction, but it was advised and supported by the Danish government, who would not see its scientific ace in the East Greenland sovereignty conflict discredited so shortly after the victory at the Hague in 1933 (Ries 2003: 44).

---

13 Gunnar Seidenfaden to Peter Freuchen, July 19, 1936. Peter Freuchen's Archive, Danish Royal Library.
14 Lauge Koch to Peter Freuchen, January 23, 1936. Folder: Correspondance 7 (1935-36), LKA-DNA.

With Koch's lawsuit against his attackers, all hope of ordered scientific debate was essentially abandoned and the fight was transported to the legal arena of the courtroom. When in March 1937 The Eleven were proclaimed free of all charges in High Court, Koch appealed and in June 1938, the Supreme Court reached a verdict that enabled both parties to express their satisfaction to the press: while the original critique presented in the book review was considered within the limits of the law, the subsequent covering letters and the proposed resolution to exclude Koch from the Danish Geological Society were judged excessive. Even the costs of the trial were split in half between the parties, and although the extended hostilities had fouled relations between Koch and his opponents beyond repair, things otherwise quickly went back to normal.

One week after the final verdict, Koch left Copenhagen for East Greenland to round off his Two-Year Expedition in its last summer, after which he went on a four-month lecture tour in the USA. Even Koch's support from the Danish government seemed unfailing when in January 1939 the chief of the Danish Greenland Administration proposed the establishment of a permanent institution for conducting a systematic geological survey of all of Greenland, under the leadership of Lauge Koch. Meanwhile, prominent members of The Eleven regrouped to propose their own plan for continued geological investigations in Greenland in a permanent collaborative effort between the Danish Geological Survey, the Geodetic Institute, and the Meteorological Institute under the auspices of the Greenland Administration – and without Koch's participation.

On April 9, 1940 German Forces invaded Denmark, terminating all exploration of Greenland for almost six years. Stifled under German occupation, the fractured Danish geological community settled in a new form, the conflicting parties emerging from the cover of war even more set in their respective positions than before. In 1946, Koch's opponents formed the Greenland Geological Survey (GGU), specifically designed to start operations in West Greenland. Koch, on his part, was able to continue his East Greenland activities as a freelance geologist for the Greenland Administration, relying almost completely on foreign geological expertise. Unable to reach a common understanding, these two Danish state-subsidized geological surveys worked side by side for more than a decade, with little or no co-operation. But that is another story.

References

Andersen, J. 1998. *Vildmanden – Sandemose og animalismen i mellemkrigstidens litteratur.* Copenhagen: Gyldendal.

Backlund, H. G. 1936. "Geologie von Spitzbergen, der Băreninsel, des König Karl- und Franz-Josef-Landes by Hans Frebold; Geologie von Grönland by Lauge Koch." *The Geographical Journal*, 87(4): 358-360.

Anonymous 1935a. "En mand i rustning. Lauge Koch, den voldsomt angrebne, i lynskud." *Nationaltidende*, December 4.

Anonymous 1935b. "Referat af ekstraordinær generalforsamling i Dansk Geologisk Forening d. 9/12 1935." *Meddelelser fra Dansk Geologisk Forening*, 8.

Anonymous 1937. "Videnskabelige angreb er hidtil blevet mødt med åndens våben." *Berlingske Aftenavis*, January 26.

Bøggild, O. B. 1917. "Grönland." In: G. Steinmann & O. Wilckens (eds.) *Handbuch der regionalen Geologie*. Volume 4. Heidelberg: Carl Winter, 1-37.

Bøggild, O. B., R. Bøgvad, et al. 1935. "Remarks upon Lauge Koch: Geologie von Grönland." *Meddelelser fra Dansk Geologisk Forening*, 8(5): 497-512.

Bistrup, H. 1943. "Kommissionens Historie 1878-1943." *Meddelelser om Grønland*, 134(1).

Dawes, P. R. 1992. "Lauge Koch som polarforsker: 50 år i Grønlands tjeneste." *Tusaat/Forskning i Grønland*, 3: 9-15.

Freuchen, P. 1933. "Nu savnes den ånd de alle levede i." *Politiken*, December 22.

Garboe, A. 1961. *Geologiens Historie i Danmark*. Volume II. Copenhagen: Gyldendal.

Gilberg, R. & J. Taagholt. 2000. *Thule og polareskimoerne*. Copenhagen: Reitzel.

Harbsmeier, M. 2002. "Bodies and voices from Ultima Thule: Inuit Explorations of the Kablunat from Christian IV to Knud Rasmussen." In: M. Bravo & S. Sörlin (eds.) *Narrating the Arctic: A Cultural History of the Nordic Scientific Practices*. Canton, MA: Science History Publications, 33-71.

Hobbs, W. H. 1936. "Geologie von Grönland by Lauge Koch." *Journal of Geology*, 44(6): 758-760.

Højesterets Protokolsekretærer 1938. "Dr. phil. Lauge Koch mod Professor O. B. Bøggild m. fl." *Højesteretstidende*, 82(14-15): 249-81.

Jørgensen, J. 1978. "Kommissionen for Videnskabelige Undersøgelser i Grønland." *Grønland*, (10): 307-312.

Koch, L. 1925a. *Nord om Grønland*. Copenhagen: Levin & Munksgaard.

Koch, L. 1925b. "The geology of North Greenland." *American Journal of Science*, 9: 271-285.

Koch, L. 1929. *Stratigraphy of Greenland*. Copenhagen: Levin & Munksgaard.

Koch, L. 1933a. "The Danish Three-Year Expedition to King Christian X Land." *Geographical Review*, 23(4): 599-607.

Koch, L. 1933b. "Knud Rasmussen." *Politiken*, December 22, 1933.

Koch, L. 1935. *Geologie von Grönland*. Berlin: Gebrüder Bornträger.

Koch, L. 1955. "Report on the expeditions to East Greenland 1926-39 conducted by Lauge Koch." *Meddelelser om Grønland*, 143(2).

Kragh, H., P. C. Kjærgaard, H. Nielsen & K. H. Nielsen (eds.) 2008. *Science in Denmark: A Thousand Year History*. Aarhus: Aarhus University Press.

Noe-Nygaard, A. 1979. "Geologi." In: T. Wolff (ed.) *Københavns Universitet 1479-1979. Bind 13. 2. del*. Copenhagen: Gads Forlag, 261-375.

Noe-Nygaard, A. 1984. *Et geologisk specialmuseum tager form (Mineralogisk Museum 1870-1970)*. Volume I. Copenhagen: Geologisk Museum.

Noe-Nygaard, A. 1986. *Et geologisk specialmuseum tager form (Mineralogisk Museum 1870-1970)*. Volume II. Copenhagen: Geologisk Museum.

Parkinson, M. M. L. & Whittard, W. F. 1931. "The geological work of the Cambridge expedition to East Greenland in 1929." *Quarterly Journal of the Geological Society of London*, 87: 650-674.

Poulsen, K. 1997. "En epoke er slut." In: K. K. Sørensen (ed.) *Topograf historier, topografisk kortfremstilling og kortlægning af Grønland, Island, Færøerne og Danmark, 1928-1992*. Copenhagen: Geodætisk Institut, 6-20.

Rasmussen, L. B. 1988. *En jordisk krønike – træk af DGU's historie 1888-1988*. Copenhagen: DGU, Miljøministeriet.

Reinemund, J. A. 1997. "Memorial to Curt Teichert, 1905-1996." *Geological Society of America, Memorials*, 28: 39-42.

Ries, C. J. 2002. "Lauge Koch and the Mapping of North East Greenland: Tradition and Modernity in Danish Arctic Research, 1920-1940." In: M. Bravo & S. Sörlin (eds.) *Narrating the Arctic: A Cultural History of the Nordic Scientific Practices*. Canton, MA: Science History Publications, 199-231.

Ries, C. J. 2003. *Retten, magten og æren. Lauge Koch Sagen. En strid om Grønlands geologiske udforskning*. Copenhagen: Lindhardt & Ringhof.

Ries, C. J. 2007. "Inventing the four-legged fish: palaeontology, politics and popularization of the Devonian tetrapod Ichthyostega 1931-1955." *Ideas in History*, II(1): 37-78.

Riis-Carstensen, E. 1929. "Nogle Bemærkninger vedrørende Godthaab ekspeditionen 1928." *Geografisk Tidsskrift*, 32(1): 128-136.

Secher, K. 1992. "Lauge Koch og Kommissionen." *Tusaat/Forskning i Grønland*, 3: 25-34.

Seidenfaden, G. 1938. *Moderne Arktisk Forskning*. Copenhagen: Jespersen & Pio.

Teichert, C. 1933. "Untersuchungen zum Bau des kaledonischen Gebirges in Ostgronland." *Meddelelser om Grønland*, 95(1): 121 pp.

Kristian H. Nielsen

Expedition "Live"
Science, Media, and Politics on the Galathea 3 Expedition 2006-2007

On August 12, 2006, the Galathea 3 Expedition departed from Copenhagen for its eight-month round the world journey. The expedition was a multifaceted and multidisciplinary multimedia event. Managed by the Danish Expedition Foundation, a business foundation established by the Danish Ministry of Science, Technology, and Innovation, the expedition formed part of a larger government initiative supporting research and research communication with the aim of consolidating Denmark's competitiveness in the emerging global culture of knowledge societies. The expedition included more than fifty research projects within the fields of biology, geology, culture and history, and climate and environment. Four independent media partners were involved in communicating the expedition to the general public using print and electronic media. Adding to the organizational complexity of the expedition was the fact that the expedition was conducted on a naval ship called *Vædderen* with naval officers in charge of the operation and with a full naval crew on board (Galathea 3 2006).

Owing to the great expectations on behalf of scientists, politicians, naval officers, private funders, and others, but also because of the objective of sailing around the world, the Galathea 3 Expedition was an extraordinary venue for scientific exploration. Today, most scientific research projects at sea are carried out on a routine basis on much shorter expeditions. Expedition ships are fitted with standard scientific instruments used to obtain substantial amounts of data to be processed in laboratories on board or at home. Ships are expensive and more or less standard technologies of scientific exploration (Sorrenson 1996). They also are confined social spaces in which people have to get along and work together until the job is done. And expedition ships are usually employed in places far away from public attention and at more than an arm's length from policy-making.

More than a scientific expedition, the Galathea 3 Expedition departed in quest of scientific results, publicity, the interest of young people, and adventure. The patron of the expedition was HRH The Crown Prince, known to the Danish public as an adventurer, mostly because of his travels with the sledge patrol Sirius, a unique Danish navy unit conducting long-range reconnaissance patrolling and enforcing Danish sovereignty in Northern and Eastern Greenland. At the ceremony of departure, he emphasized that, even in the twenty-first century, scientific exploration means adventure, and, moreover, that this expedition was a public adventure:

> A white globe is attached at the top of the mast. More than anything, this globe separates Galathea 3 from all the other expeditions we have dispatched throughout history. This satellite antenna will make it possible for all of us to follow the expedition closely. We can look over the shoulders of the researchers while sitting in our own living rooms and in schools. We are present when they drill the ocean floor and when the trawl and net is being pulled in. In a way, we can all take part in the adventure of travelling around the world. It still is an adventure, even in our time (Boelskifte 2006).[1]

Pointing out a vital communication device of the expedition while also invoking some of the places in which narratives and images of the expedition would be received, The Crown Prince touched upon important issues relating to the history and public understanding of science. These are issues relating to science as a form of communicative action, to the publics of science, to science politics, and the media of science and science communication. Much like its predecessor, the Galathea Deep Sea Expedition 1950-52, the Galathea 3 Expedition ventured to combine scientific research with intensive media coverage and national identity (K. H. Nielsen 2006, 2008). From the beginning, therefore, the emerging relationships between scientists, science communicators, politicians, the general public, and others played an important role in defining what the expedition was all about. The expedition was not a two-faced Janus, as science is in Bruno Latour's portrayal (Latour 1987), but had many different faces. The live coverage of the expedition in several Danish news media presented those various aspects of the expedition to the general public on a daily basis.

For example, as the expedition took off, Morten Meldgaard, director of the Danish Expedition Foundation, had to respond to accusations of the expedition being all about PR. To the Danish newspaper *Politiken*, one of the four media partners of the expedition, he said:

---

1   All translations by the author.

First and foremost, it is a research expedition providing the researchers with the opportunity to work in areas they normally have no access to. And, because we will be going through all climate zones and most of the oceans, there are projects which will be joining us all the way. This is unique.

But, on top of this, there is an educational and communicative aspect. And here the ship is important as it provides a good frame for telling stories about research to all who are interested. *Vædderen* in itself will become a small world, which will be broadcasted to the larger world. It ought to appeal to curiosity so that in the future we will be able to secure recruitment for education and maintain a high level of knowledge at home (Bergsagel 2006).

Meldgaard defined the expedition in terms of research being the top priority (which later caused controversies between the expedition management and *Jyllands-Posten*). He also considered the multifaceted character of the expedition, expressing some concern for its public image. So far, getting so many people from different ministries, research institutions, disciplines, media, and work cultures to co-operate had been one of his most difficult tasks as expedition manager. The staff of the navy was used to strict schedules and fixed boundaries, while typically researchers were more anarchistic, demanding full autonomy. Even though most problems had been solved in a peaceful manner, he foresaw that with the addition of the press with its clear interest in conflict and sensation, tensions such as these were bound to be widely exposed to the public. Communicating the expedition online and in print, sometimes even in real time, would not only create excitement and arouse curiosity, but potentially also raise more pressing issues about what constitutes expedition life in the twenty-first century (Bergsagel 2006): issues pertaining to the nature of expedition research, the communication of research to a wider public, the funding and the politics of research, and the interaction between researchers, naval personnel, journalists, politicians, and expedition management.

My aim in this article is to explore such issues by looking at a variety of different ways of understanding the Galathea 3 Expedition. With a particular emphasis on a few critical incidents, I offer several stories about the expedition from its conception through to its evaluation. I rely more or less exclusively on published material, in particular the expedition homepage, reports, and newspaper articles, most of which was published in real time during the lifetime of the expedition. This is a choice of convenience, but also a matter of methodology. By providing the reader with a reconstruction of some of the main events of the expedition, I wish to give an idea of how the expedition could have been perceived by someone who followed the expedition in the media. In short, I am particularly interested in finding out how the expedition was presented to a larger public, and, as far as possible, I will refrain from making my own judgments about the true nature of the expedition. This is not an essay about expedition life as it really took place on the expedition; nor is it about expedition life as such. Rather, it is about the live coverage of expedition life and its consequences for the public perception of scientific expeditions. I also have no intentions of addressing the numerous research projects undertaken during the expedition, although there is little doubt that most people involved saw research as the dominant aspect of the expedition (Jørgensen 2008). In contrast, I would like to delve into other ways of seeing the expedition, thus somewhat downplaying the research aspect. More specifically, I intend to investigate how people in Denmark – had they followed the reporting of the Galathea 3 Expedition as intensively as I and others did – would have been able to acquire different kinds of meaning about the life of this quite extraordinary expedition: meanings which were being communicated live by different people, institutions, and media, and, thus, with little chance of concurrently reconstructing the expedition to fit particular purposes.

Narratives Making the Expedition Possible
Scientific expeditions need more than research questions, ships, instruments, and money to travel. In the case of the Galathea 3 Expedition, national support and media interest were important to the legitimization of the expedition. Narratives that link scientific exploration with national identity, with national policy-making, and with the reality of the mass media, not to mention narratives that lend the scientific aspects of the expedition plausibility and legitimacy, played a crucial part in making the expedition possible. Indeed, it might be argued, as historians of Arctic exploration Michael Bravo and Sverker Sörlin have done, that it makes "sense to speak of narratives in the same breath as precision instruments, as technologies of travel" (Bravo & Sörlin 2002: 21). The narratives explored in their book about the Arctic include scientific, religious, nationalist, imperial, evolutionary, indigenous, geological types, among others.

Mobilized by networks of communication, narratives are grounded in human experiences and institutional practices. Scientific narratives travel along highly institutionalized chains of communication, where peer review and publication constraints play an important role. Narratives about science, however, also appear in the public domain mediated by the media and other channels of communication. Here, they interact with other kinds of narratives to produce hybrid, heterogeneous narratives about particular aspects of scientific practices. In this sense, expedition science and the field sciences are no different from the laboratory sciences. Narratives enable certain collaborations, mobilizing particular actors in relation to specific projects.

In the following, I wish to trace three interrelated narratives about the Galathea 3 Expedition that were pivotal in getting the expedition going, namely those presenting the expedition as an adventurous undertaking, as a people's project, and as a way of popularizing the role of the researcher. These three narratives were useful in aligning the interests of journalists and editors with those of politicians and researchers.

Expedition Research as Adventure
In the spring of 2000, chief editor of the Danish newspaper *Jyllands-Posten* Henrik Thomsen, known as a successful visionary and a persistent organizer (Mikkelsen 2004), came up with the idea of launching a third Danish Galathea expedition. Conceived on the fifty-year anniversary of the launch of the second Galathea expedition, the new expedition was to provide a multi-faceted platform for Danish researchers and research communication to the Danish public. The main purpose of the expedition was to foster public understanding and public appreciation of science using new, primarily web-based, media. Besides research and communication, Thomsen's idea also included a cultural element involving Danish and foreign artists, and a business element that created an international display window for Danish businesses (Thomsen 2008).

Thomsen built on the newspaper's experiences with launching journalistic expeditions, the celebrated JP Explorer concept which today forms a permanent part of *Jyllands-Posten*. The first JP Explorer expedition, which also was Thomsen's idea, took off in the summer of 1998, travelling round the world in 500 days and arriving in Copenhagen on 31 December 1999. It was the largest venture of the newspaper, a countdown to the new millennium, and also a way of demonstrating the potential uses of new media in journalism. The expedition was presented as the first ever interactive expedition. Anyone with access to the Internet could retrieve logbooks, photos, roadmaps, informational material, and ask the participating journalists questions. Also, a dedicated web-based school service was established. On several occasions, the team of journalists held online meetings with school children and others, providing the journalists with the opportunity to engage with their readers as well as the other way around. Each day, the printed newspaper featured the logbook of the journalists on the back page of section one. The JP Explorer concept was hugely popular among the readers of *Jyllands-Posten* and has won international acclaim (JP Explorer 2009).

Moreover, Thomsen referred to the ongoing debate about young people's (lack of) interest in science and technology. One of the reasons why, at the time, this issue was particularly on the agenda was the launch of the Danish government's industrial policy, known as *.dk21*, in February 2000. Devised in close collaboration with 300 executives of the largest corporations in Denmark, *.dk21* focused primarily on entrepreneurship, innovation, and increased competency development and competition (Skovgaard 2000). Although widely received as important and timely, the initiative, aimed at securing the national foundations of growth and welfare, was criticized by the director of the Academy of the Technical Sciences, Torben Klein, for failing to deal with two general problems: the lack of public appreciation of science and the failure to recruit young people for science and engineering educations (Klein 2000). Later, in April 2000, the Danish Villum Kahn Rasmussen Foundation launched an investigation into the causes of the declining interest in science and technology among young people. Professor of education Niels Egelund, the leader of the investigation, asked why it is that young people today would rather study the humanities, or become TV hosts or journalists (Dannemand 2000).

In an article describing the creation of the Galathea 3 Expedition, Thomsen specifically mentioned an interview with a Danish executive that had had an impact on his idea. The executive had predicted that the disregard for science education would eventually lead to massive problems for research-based industries and for society. The shortage of qualified manpower would endanger continued economic growth in Denmark. In his article, Thomsen then recalled the times when things were quite different: when, in the 1970s, Jacques Costeau, wearing a red hat, would communicate ocean research, gluing everybody to the TV screen; and when, in the early 1950s, on the second Galathea Expedition, Hakon Mielche's reports of scientific exploration would capture the imagination of the entire nation. Why did they succeed? What was their recipe? The answer that Thomsen came up with was that they combined research communication with adventure. With the recent success of the first adventurous JP Explorer expedition in mind, he concluded:

> Consequently, the wildest idea emerged – to launch a national research expedition that, by joining research with adventure, would recreate public interest in the natural sciences. A third Galathea Expedition, which would transform the Galathea tradition in an age of digital media, where young people would be able to follow the expedition online, looking over the shoulders of the researchers and sharing their curiosity. In this way, a new window to natural science could be built (Thomsen 2006a).

The promise of adventure remained an important element in pitching the Galathea 3 Expedition, not the least because adventure was seen as key to promoting public appreciation of science and stimulating young people's interest in science. The idea of adventure was also flexible enough to accommodate different interpretations. Upon departure, the editorial of the Danish trade journal for engineers, *Ingeniøren*, remarked that the touch of adventure and exploration was needed "to light the bonfire of interest, which will become manifest when one identifies oneself with the thought of travelling as part of an expedition and as an adventurer in a hidden and secret world" (Editorial 2006b). In a more modest vein, the editorial team of *Jyllands-Posten* simply stated that "every expedition is an adventure" (Editorial 2006a), while the reporter of *Berlingske Tidende* used bombastic terms like "the wildest Danish adventure in decades" (L. H. Nielsen 2006). Acknowledging that, in 2006, the other end of the globe is just a click on the mouse or a phone call away, the reporter insisted on the Galathea 3 Expedition being a "genuine adventure" in several ways: "Not just a journey, but a Grand Tour. Not just another experience, but discovery and exploration. Not just for now, but for the future" (L. H. Nielsen 2006).

Most writers and commentators seemed to find enough adventure in the combined idea of circumnavigating the globe while conducting research. Although the point was seldom made explicit, research, like exploration, was depicted as a venture into the unknown. Research and adventure seemed to be two sides of the same coin. Research was adventurous, and adventure brought about new discoveries and thus research. However, there were also a few critical voices to be heard for whom the equation of adventure and research sounded a bit too romantic. Under the heading "science and nostalgia," the editor of *Politiken* stated:

> With a lot of show, the idea of curious Danish researchers venturing round the world in quest of breakthroughs in the name of science is being conjured up. You would almost believe that we were in a different age: when the world map still contained many white spots and even larger mollusks might be undiscovered. (...) However, since truth is the highest value of science, another truth about Galathea has to be told. Sailing round the world in order to make scientific discoveries – and even across a variety of scientific disciplines – makes little sense in the year 2006. Today, research is highly specialized, and the world is connected in ways that have reduced most shipping to transportation of containers or tourism (Editorial 2006c).

The People's Expedition
Despite the fact that adventure not only invoked exciting images of expedition life, but also gave the expedition a somewhat quaint appearance, the affiliation of Galathea 3 with adventure, as we shall see, stuck with the expedition right to the

end. However, this is getting ahead of the events. The Galathea 3 Expedition needed more than vague notions about adventure to support it. The road from Thomsen's initial idea to actually embarking on the expedition was long and winding, involving considerations about research policy-making, quests for scientific knowledge, and media competition.

At first, *Jyllands-Posten* had little success in promoting the concept of the Galathea 3 Expedition, except to Danish oceanographers who generally responded with no or little reservation to the idea of an ocean-going expedition round the world. Sympathy also was to be found in private foundations, firms, the government, and various organizations; all of which, however, were reluctant to put their money into the expedition. Then, in the spring of 2004, at a meeting with the editorial team of *Jyllands-Posten*, Prime Minister Anders Fogh Rasmussen developed a keen interest in the idea. Headed by the Ministry of Science, and with the purpose of realizing the expedition, a task force consisting of government officers representing several ministries was set up. The Ministry of Science also established the Danish Expedition Foundation to raise funds for and organize the expedition (Thomsen 2006a, 2008). The foundation's board of trustees consisted of senior administration officials from the Ministry of the State of Denmark, the Ministry of Science, the Ministry of Education, and the Ministry of Defense, as well as scientists representing the humanities and the natural sciences. The board also included one person from an independent media production company (Danish Expedition Foundation 2006).

When the government entered the equation, the expedition became embedded in national politics, or, more specifically, research and educational policies. With its double focus on research and research communication, the expedition fitted well with some of the government's new initiatives. In 2003, a new Act on Universities had been introduced. The Act lists public communication of science as a third obligation of the universities, on an equal footing with research and teaching. In fact, the obligation to communicate concerns not only the more traditional, if not canonical mode of science communication whereby new scientific results are being disseminated to the general public, but an interactive, two-way mode of communication allowing for knowledge and competencies to be exchanged with society and encouraging university employees to take part in public debate (Nielsen, 2005, Nielsen et al. 2007).

An independent think tank on the public appreciation of research was also set up, publishing its final report in May 2004. In the report, the think tank assessed current activities within research communication, while also making recommendations with regard to the future of research communication in Denmark. The report made clear that there was a growing interest in research in the population. Although many science communication activities could be identified, the report recommended further promotion of public understanding of science. From the recommendations, it was evident that the think tank wanted the research councils and universities to spend more money on communication. Among other things, the think tank suggested allocating up to 2% of the total public research budget to public communication. Although the exact translation of 2% into how many millions of Danish kroner depended on the interpretation of what amounts to the total research budget, the message was clear: we need to promote research communication in various ways both inside and outside of the research system (Tænketanken for forståelse for forskning 2004). A subsequent survey showed that the majority of Danish scientists approved of these recommendations (Nielsen et al. 2007).

When in August 2004 the expedition plans were launched to the general public, minister of science Helge Sander stressed the need to mobilize public goodwill in order to support an expedition like the proposed Galathea 3 Expedition:

> We have allocated DKK 2 million of the Budget to make the necessary preparations. Then the huge expenses for such a huge project come. It's part of our agreement that we will try and procure the necessary means from institutions, businesses, and foundations so it will become a common project. Not just a *Jyllands-Posten* project, or a ministerial project, but a people's project (Boelskifte & Grund 2004).

The expedition as a project of and for the people: such were the ambitions presented by the minister. Presenting the expedition in this way, the minister was able to justify the inclusion of the media, but also the investment on behalf of the government in what was originally *Jylland-Posten*'s idea. The media took part in order to communicate the expedition as widely as possible, and the involvement of the ministry or government was a guarantee of parliamentary or even democratic representativity. In Sander's view, the expedition fully delivered what the people needed in terms of research policy: increased funding for scientific research and more public appreciation of science. According to Sander's think tank, the people wanted to become more informed about and even more involved in scientific research. Finally, the expedition was a means of producing and distributing good stories about ongoing scientific research.

Cool to Be a Researcher
The abovementioned think tank on the public appreciation of research also stressed the need to think more carefully about recruiting young people to the sciences.

Some seven months later, on the explicit recommendation of the think tank, a working group on science communication targeted at young people and children was established by the Ministry of Science, the Ministry of Education, and the Ministry of Culture. The working group concluded that the challenges of globalized knowledge society with its increased international competition for growth and innovation forced Denmark to step up its priority of knowledge in terms of scientific research, technological innovation, and science and engineering education.

Together, the two reports testified to the political attention given at the time to science communication as a means of promoting science education. Specifically, in full agreement with the emphasis on adventure, both reports concluded that in order to attract young persons to science and technology education, initiatives showing the excitement and interest of research were needed. The report on science communication argued that "we have to take on the image of research as dusty and being inhabited by nerds and, instead, we need to develop new images and new role models, which, together, make it "cool" to be a researcher" (Tænketanken for forståelse for forskning 2004).

At the launch of the expedition plans, the call for promoting young persons' science careers received particular attention. The research executive of the Confederation of Danish Industries Charlotte Rønhoff said: "Bearing in mind the low position of Denmark within the EU when it comes to the natural sciences, we have every reason to foster excitement about and interest in scientific research." Minister of science Helge Sander concurred: "The Galathea 3 Expedition will strengthen young people's interest in science." Vice-president of the Royal Danish Academy of Sciences and Letters Christian Berg responded to the news of the Galathea 3 Expedition using terms like "sensation," explaining that the expedition would have an enormous impact on Danish research and, in particular, would push the natural sciences forward. Chief Editor of *Jyllands-Posten* Henrik Thomsen, originator of the expedition, added to this discourse of excitement:

> A lot of young people think that only nerds take up natural science. This is a serious societal problem since it might lead to research-based industries lacking qualified manpower. The Galathea 3 Expedition continues what we already do at our newspaper: Promoting interest in and appreciation of science. It has to be cool to be a researcher (Boelskifte & Grund 2004).

Together with the notion of adventure and the "live" inclusion of all the Danish people, the aim of promoting young people's interest by making it cool to be a researcher was one of the leading narratives that enabled the realization of the expedition as part media, part government, and part research project. However, due to disagreements between scientists and journalists with respect to the best and most appropriate way of representing research and researchers to the public (see below), it is highly questionable whether the expedition did much in the way of making it cool to be a researcher. In fact, in their study of web-based news articles, Nielsen & Autzen (2011) found that the journalists described researchers primarily as dedicated, knowledgeable, and hardworking experts. Researchers are presented as being special because of their specialized and highly esoteric knowledge, and they are different from other people in that they devote their entire life and most hours of the day to their work. Rather than making it cool to be a researcher, the evidence seems to suggest that the Galathea 3 Expedition confirmed the idea of researchers being fully devoted to their work and entirely absorbed by their identity as researchers.

Conflict and Co-operation
Even though Helge Sander and others proclaimed that they wanted it otherwise, the expedition became to a large extent a government project. Government officials were more or less in control of the Expedition Foundation, and the Danish Agency of Science, Technology, and Innovation assisted the Foundation in the selection of research projects. In late 2004 the agency conducted a consultation, during which scientists and research institutions were invited to submit proposals for destinations and possible research projects to be included in the expedition. About 120 proposals came in, all of which were screened by the Danish Councils for Independent Research covering the medical, technical, natural, social and human sciences. The preliminary survey was primarily a tool to help the Expedition Foundation determine the route of the expedition. In June 2005 the Expedition Foundation distributed a request for specific research proposals primarily within the natural sciences, but projects from other disciplines would also be considered. By the deadline of September 2, 2005, 124 proposals had been submitted. The independent research councils reviewed all projects according to their scientific merits to ensure that the expedition only included relevant research projects of high scientific quality. On the basis of this review process, but also with due consideration of the projects' communication and educational potentials (see below), the Expedition Foundation approved about 70 projects, about half of which were projects within the natural sciences and one quarter within the social and the human sciences (Boelskifte 2005).

The Expedition Foundation was unable to provide funding for most of the accepted projects, but did offer its recommendations to independent grant applications submitted to the research councils and private foundations such as the Carlsberg Foundation. By November 2005, the Finance Committee had agreed to allocate 60 million DKK for the expedition, most of which was used for modifications and the manning of the expedition ship. About 70 million DKK for the research projects was still lacking. Director of the Expedition Foundation Morten Meldgaard said that, based on positive indications, he was confident that the rest of the money would be found (Larsen 2005). As it turned out, Meldgaard was right in his prediction. The total budget of the expedition amounted to about 200 million DKK, about half of which came from private foundations (65 million DKK) and public research councils (35 million DKK); the remaining half was paid by the state (Haslund-Christensen & Meldgaard 2008).

Negotiating Media Rights

Because *Jyllands-Posten* had developed the original idea, the newspaper was granted special rights with respect to covering the expedition. The first outline of regulations for the Expedition Foundation, drafted by the Junior Council to the Treasury (in Danish: Kammeradvokaten) and presented to the Finance Committee on October 11, 2004, stated that the foundation and *Jyllands-Posten* had entered into a co-operation agreement. The agreement gave *Jyllands-Posten* exclusive rights to the "newspaper related coverage of the expedition, so that the journalists of the newspaper as the only print media would be able to take part in the whole of the expedition. Other media will be able to cover the expedition according to the decision of the board of trustees" (Kammeradvokaten 2005).

The Finance Committee was hesitant about granting exclusive rights to *Jyllands-Posten*. In a speech to the committee on November 10, 2004, minister of science Helge Sander responded that, normally, he would have preferred a bidding round as recommended by the committee. However, he also believed that it would be unfair to *Jyllands-Posten*, who had come up with the idea in the first place, to leave without them. *Jyllands-Posten* also already owned some important copyrights for the Galathea 3 Expedition, which could be violated. The Expedition Foundation had to make the final decision regarding which media groups were to be included in the expedition. The minister urged the foundation to include news media other than *Jyllands-Posten*, such as, for example, the two newspapers *Politiken* and *Berlingske Tidende*, both of which had already taken an interest in the coverage of the expedition (Sander 2004).

Later in the process *Jyllands-Posten*'s exclusive rights were reconsidered. On June 22, 2005 the Ministry of Science submitted a confidential document to the Finance Committee, revealing that the Junior Council to the Treasury had begun negotiating media contracts on the following conditions:

1. None of the media on board will be granted juridical exclusiveness.
2. In consideration of the physical conditions, the aim is to secure as broad media coverage as possible.
3. The conditioned agreement with *Jyllands-Posten* will be renegotiated in consequence of the changed conditions with respect to the ship platform. (Kammeradvokaten 2005)

The new ship platform referred to in the quote was the naval ship *HDMS* (Her Danish Majesty's Ship) *Vædderen*. In the beginning, the expedition ship was expected to be the research vessel *Dana*, which could accommodate a total of 38 scientists and crew in single cabins (DTU Aqua 2009). Partly because of the overwhelming interest in the expedition on behalf of researchers and journalists, it was decided to employ *HDMS Vædderen* with a complement of 61 persons (Wikipedia 2009). Choosing *Vædderen* gave room for more researchers and more journalists on board. The terms of the partnership between the expedition and the news media followed lengthy and intense negotiations between, on the one hand, the Expedition Foundation and the minister of science, assisted by the Junior Council to the Treasury, and, on the other hand, *Jyllands-Posten* and other Danish news media. On October 14, 2005, four co-operation agreements were in the bag. According to the agreements, the three Danish newspapers that had shown an interest in the expedition, i.e. *Jyllands-Posten*, *Politiken* (in a joint venture with the national TV weather service TV2 Vejret), and *Berlingske Tidende*, were allocated four, three, and two places, respectively, while the trade journal for engineers, *Ingeniøren*, received one place during two periods of approximately one month's duration each. In the agreements, the four participating media were all referred to as "media partners."

In the agreement with *Jyllands-Posten*, the special status of this newspaper was explicated: *Jyllands-Posten* would retain its right to appoint one member of the Expedition Foundation's Board of Trustees; *Jyllands-Posten* would have access to a total of four places on the expedition (two of which were to be used for electronic media production); *Jyllands-Posten* would receive compensation for handing over to the Expedition Foundation a number of immaterial rights (in particular the homepage www.galathea3.dk); *Jyllands-Posten* would hand over to the Expedition Foundation the rights for the Galathea 3 concept and project description at no cost; and, in all official publications and on the homepage of the Expedition Foundation, *Jyllands-Posten* would be mentioned as the originator of the Galathea 3 Expedition (Kammeradvokaten 2005).

The agreements between the four media partners and the Expedition Foundation explicated mutual rights and responsibilities. Planning and carrying out the expedition, the Foundation had bound itself to ensure variety in a way that would accommodate the needs of the media. This included taking into consideration the communication and educational potentials of the research projects that were to be included in the expedition. The Foundation was also to make sure that about half of the total time frame of the expedition would be used for coastal sailing, in harbors, or anchoring of the ship. The media needed to be able to disembark regularly in order to cover not only the expedition itself, but also the many countries and regions visited. The agreement also made clear that the media would have full editorial freedom, which meant that the Foundation would not be able to demand replacement of particular journalists based on arguments pertaining to content produced or opinions expressed by the journalists. Finally, it was agreed that, at least eight months prior to departure, the Foundation would provide the media partners with a route plan. Changes to the plan could only be enforced because of force majeure or research-related considerations (Kammeradvokaten 2005). As it turned out, the events that followed the publication of the route plan did in fact necessitate changes.

The Intervention of the Cartoon Crisis

On the basis of the preliminary consultation performed in late 2004, the Expedition Foundation published a first route plan for the expedition in June 2005. The preliminary route was intended as a guideline for all applicants who wanted to have their research project included in the expedition. Originally, the route went from the Azores in the Atlantic Ocean to Australia via the Mediterranean, the Suez Canal, the Arabian Sea, and the Bay of Bengal with planned visits to several Islamic countries.

On September 30, 2005, *Jyllands-Posten* published twelve editorial cartoons, most of which depicted the Islamic prophet Muhammad. The cartoons were commissioned by the newspaper's cultural editor, Flemming Rose, who wanted to deal with what the newspaper saw as spreading self-censorship with regard to criticizing Islam. Rose named several examples all taken from the Danish context: the Danish comedian Frank Hvam having exclaimed that he was afraid to "bullshit the Koran on live TV"; one Danish artist who in providing illustrations for a book about the life of Muhammad, had wanted to remain anonymous; three contemporary Danish plays criticizing the president of the USA, George W. Bush, while none were taking on Osama bin Laden; and one Danish imam encouraging the Danish Prime Minister, Anders Fogh Rasmussen, to make his influence count in trying to get Danish media to present a more favorable image of Islam. In the name of freedom of speech, Rose had encouraged members of the Association of Danish Cartoonists to provide images of the prophet "as they see him." The result was twelve cartoons, all published by *Jyllands-Posten* (Rose 2006b).

Rose later explained that he published the cartoons to counteract the moral trappings of cultural relativism and political correctness, which were undermining fundamental, liberal values of Europe: freedom of speech, religion, assembly, and movement. He argued that Danish left-wing intellectuals were deceiving themselves in thinking that Denmark and other European countries are inherently racist and Islamophobic and that all Muslim immigrants are victims of oppression. To the contrary, Rose reasoned, it is precisely what he called "the cult of victomology" that breeds exclusion. As long as Muslims are being stigmatized as victims, they cannot become fully integrated in European culture. The act of publishing of the cartoons, in his view, was an act of inclusion, an act of respect and recognition, giving Muslims the democratic treatment of being satirized in cartoons (Rose 2006a).

Despite Rose's intentions, many events followed the publication of the cartoons that did not serve to include Muslims in European values. In particular the one cartoon portraying Muhammad wearing a turban in the shape of a bomb gave rise to fierce reactions from Muslims who believe it is blasphemous to depict or insult the prophet. Having received petitions from Danish imams, eleven ambassadors from Muslim-majority countries asked for a meeting with Danish Prime Minister Anders Fogh Rasmussen on October 12, 2005, in order to discuss what they perceived as an on-going smear campaign in Danish public circles and media against Islam and Muslims. The government responded to the ambassadors' request for a meeting with Rasmussen with a letter only, referring to the freedom of speech as guaranteed in Danish law since 1849. The government letter also pointed out that all Danish newspapers are privately owned and completely autonomous. Even by Western European standards, the freedom of expression in the Danish press is quite far-reaching. For example, the organization Reporters Without Borders ranked Denmark at the top of its Worldwide Press Freedom Index for 2005 (Wikipedia 2010). Led by Prime Minister Rasmussen, the government managed the rest of the Cartoon Crisis in a strictly secular way by trying to keep religion and politics apart. Paradoxically, this led the Prime Minister to adhere to a religious Protestant understanding of the two realms as being independent, which probably hardly contributed to calming down the religious tempers inflamed (Slok 2009).

Several observers described the Cartoon Crisis as the worst foreign policy crisis in Denmark since World War II. Little wonder, then, that the crisis also influenced the Galathea 3 Expedition. In consequence of the Cartoon Crisis, in February 2006 the Danish Expedition Foundation decided to make changes to the expedition's route, cancelling the planned passage through the Mediterranean, the Suez Canal, the Arabian Sea, and the Bay of Bengal. Instead, it was decided to follow the route of the second Galathea Expedition along the Western coast of Africa to the Cape of Good Hope. From Cape Town, the expedition would proceed directly across the Indian Ocean to Broome, Australia. According to the director of the Expedition Foundation, Morten Meldgaard, there were no direct threats to the expedition. The decision was preventive and had been made due to the current tensions in the area, but also because the expedition required long planning. "If we want to pull a cup of water from the ocean, we need permission, and so we may anticipate problems arising. We just have to adjust to the situation," said Meldgaard (Rebensdorff 2006).

The changes did influence a few of the planned research projects. One expert on Islamic culture, Professor Jørgen Bæk Simonsen, applauded the decision, arguing that the expedition planners were being very wise. The planners, he argued, obviously needed to think about the safety of the researchers, the crew, and the journalists, not to mention the few schoolchildren on board. Although nobody really seemed prepared to challenge the wisdom of the decision, a few researchers were still annoyed that their projects were being cut. Associate professor Alf Skovgaard, for example, was planning to study zooplankton in the Mediterranean. He said that the changes made were going to alter his project, although in no significant way. Senior researcher Jakob Toftegaard, who participated in the research project on sound in the oceans, was a little more explicit in commenting on the decision: "It's all very frustrating. We were very excited about the possibility of studying humpbacks in the Arabic Sea, because they haven't been described in detail before" (Grund & Gram 2006).

Other research projects had their status changed from ship-based to land-based expedition projects. For example, this happened to the Tranquebar (or Tranrangambadi) Initiative of The National Museum, which included a number of projects all aiming to strengthen the ongoing Indo-Danish efforts to preserve intercultural heritage and to explore colonial Indo-Danish cultural encounters. The initiative as such was not exclusively a Galathea 3 research project, but an ongoing initiative of the museum and its collaborators. However, the research initiative benefited from its affiliation to the expedition and remained an official part of the expedition even though the expedition vessel did not come to anchor at Tranquebar.

The intervention of the Cartoon Crisis made clear that the Galathea 3 Expedition was tangled up in politics as well as media publicity strategy. From the point of view of the organizers, the expedition needed to take into account developments within Danish foreign policy. The interconnection of policy-making and exploration has historical precedents in Denmark, for example with regards to Danish expeditions in Greenland in the 1930s, when scientific exploration was an important tool of claiming sovereignty (Ries 2002). The Galathea 3 Expedition, however, was not so much an instrument of foreign policy-making; rather it was simply caught up in current foreign affairs. From the point of view of the Danish Government whose primary aim was to manage the crisis in the best way possible, while also staying true to the Protestant ethics of separating religion and politics, changing the route of the Galathea 3 Expedition must have seemed like a sound and pragmatic solution. At the time, the idea of a Danish naval vessel, carrying a machine gun and under the banner of *Jyllands-Posten*, on a cruise through the Suez Canal would not have been politically sound.

Jyllands-Posten also wanted to separate religion from politics and science. As stated above, Flemming Rose explicitly defended the publication of the cartoons with reference to the secular norms of Europe. He based his defense on a distinction between fundamentalist beliefs and democratic freedom. Ten months after the publication of the cartoons, and shortly before the departure of the Galathea 3 Expedition, *Jyllands-Posten* launched its scientific canon with particular reference to the natural sciences as a cultural bulwark against superstition and ignorance. The incentive for launching the canon was partly promoting public understanding of science and generating public attention to the upcoming expedition, partly advancing the sciences as the epitome of rational thought in opposition to religious fundamentalism (K. H. Nielsen 2006). Introduced as the collective achievement of scientists throughout the ages, the canon comprised ten scientific theories or ideals, all of which were presented, somewhat immodestly, as the most significant insights of mankind (Collingnon 2006). The editorial of *Jyllands-Posten* expressed the hope that the canon – in combination with the Galathea 3 Expedition – would help scientific reasoning supersede religious fundamentalism in the public domain and would direct young people's attention to the sciences:

> Even though the Muhammad Crisis almost unexpectedly tripped the Galathea 3 Expedition up, the expedition promises to be a modern voyage of discovery in which the numerous researchers on board the expedition vessel "Vædderen" are anticipating a lot of new knowledge within fields such as biology, geology, climate, environment, culture, and history. Thanks to the latest satellite technology and the internet, the Galathea Expedition will be closely coupled to primary and secondary schools. In a scientific perspective, one should not belittle the value of this historic expedition. If the combination of a scientific canon and the Galathea 3 Expedition can stimulate public debate and contribute to arousing interest in science education, it will unquestionably be of benefit to Danish society (Redaktionen 2006).

Doing Research vs. Doing Research Communication
While the potential conflict between science and religion faded into the background after the changes to the route, another conflict took center stage: the collaboration of media and science. Less than one month after departure, journalist Anders Legarth Schmidt (2006) reported on several minor conflicts between researchers and journalists. Several researchers had responded negatively to news stories about the sewage system of the ship emptying all toilet tanks into the sea just as researchers were taking water samples. A crisis meeting was arranged where researchers vented their frustrations about what they saw as "frivolous journalism," and where the journalists were able to explain that they did not see research coverage as their one and only task. In Schmidt's view, for journalists it all came down to whether they should be acting the role of "media partners" (as the expedition management would have it) or whether they, as independent journalists, were free to write whatever stories they believed served the interests of their readers.

A few weeks later, in a more satirical vein, journalist Anders Lund Madsen criticized the research of the expedition for being too boring to cover:

> According to Helge Sander [minister of science] and the boys from the Danish Expedition Foundation, the Galathea 3 Expedition is essentially about communicating the exciting research on board in a way so that young people in Denmark would want to become researchers instead of being big in the media or in design or turning into fundamentalists or just hanging out on street corners. There are two obvious problems with that plan of action: there is very little interesting research on board. And the communication is a complete farce (Madsen 2006b).

Halfway through the expedition, Anders Lund Madsen put forward a more serious critique of the expedition in which he characterized all research projects as "pre-school research" (a pun on the term "1st class research" used by the Expedition Foundation). He also accused the expedition management of being "a first team of chickens," afraid of making mistakes and thus unable to produce a real success (Madsen 2006a). One marine biologist, who joined the expedition on the same expedition leg as Madsen, retorted that, all along, Madsen had been too shy to approach the researchers simply because he was unable to understand them. Instead of taking an interest in what really went on, she claimed, he was watching videos (Mouritzen 2006). Søren Haslund Christensen, chairman of the board of the Expedition Foundation, simply argued that Madsen was unable to deliver what the expedition management had expected, "namely to capture young people's interest"

(Rønn & Lützen 2006). Minister of education Bertel Haarder basically agreed with Christensen. From his own experiences on the expedition, Haarder had learned that basic research is "laborious" and required "disrespect for work hours," but also that research, even research performed on the Galathea 3 Expedition, would provide solutions to many social and environmental problems. Sadly, he concluded, it is no longer "in" to be researcher, and the criticism raised by Anders Lund Madsen and others would not change this regrettable fact (Haarder 2006). The editor of *Jyllands-Posten* Henrik Thomsen, however, concurred with Madsen, saying that some researchers were treating journalists with contempt and that, contrary to the original agreement, the expedition management were ignoring the needs of the media when making important decisions en route (Thomsen 2006b).

These discussions indicate, as previously mentioned in the introduction, the entanglement of the Galathea 3 Expedition in at least three layers of discourse: politics, science, and news media. From the viewpoint of ministers like Helge Sander and Bertel Haarder, the expedition served political ends (strengthening Danish research while also recruiting young people to the sciences); from the viewpoint of the researchers, the expedition primarily aimed at producing new research-based knowledge and disseminating research results; while, from the viewpoint of journalists, participating in the expedition entailed freedom to produce newsworthy stories (whether they included commentary on the research or not).

Madsen's critique coincided with *Jyllands-Posten*'s internal mid-way assessment in which the Expedition Foundation was criticized for failing to live up to its responsibilities to the media. The newspaper complained that the Foundation put too much emphasis on research and that the journalists on board generally had a hard time trying to get the researchers to communicate their research (with the exception of a few "born" research communicators). Most important decisions on board favored the research of the expedition, not the media. Additional time was being allocated to research projects, while the journalists struggled to find exciting things to write about. It was concluded that since "many researchers complained about the journalists writing too little about research and too much about bingo and bathing, it has to be made clear that Galathea is ALSO an adventure" (*Morgenavisen Jyllands-Posten* 2007).

"The Party Boat"

The mid-way assessment report was an internal document. Besides pointing out some of the differences between the expedition management and the media, the report also made certain claims about the behavior of the crew. It was pointed out that the crew drank heavily during port visits and that, in the ship's canteen, a few crew members had bragged about their visits to prostitutes. This behavior, the report argued, was not only a threat to the reputation of the expedition, but also a problem because of the schoolchildren on board. One of *Jyllands-Posten*'s journalists reported to the ship's management that on her first encounter with the expedition, during which she had been accompanied by two 14 year-old schoolgirls, she had had to confront drunken sailors wearing their uniform. This made her very concerned about the reputation of the expedition. In fact, the rumors of these conditions seemed to have been widespread. For example, the report pointed out that the employees of the American Embassy in Copenhagen had begun referring to the Galathea 3 as "the party boat" (*Morgenavisen Jyllands-Posten* 2007).

Since the mid-way assessment did not result in perceived improvements, upon the return of the expedition in April 2007 *Jyllands-Posten* repeated the same criticisms in a second assessment report. Again, the critique was intended to be internal, but for reasons unknown the Admiral Danish Fleet chose to publish on its homepage the second assessment report of *Jyllands-Posten*, causing intense media attention. The report stated:

> It was and is the assessment of *Jyllands-Posten* that these circumstances constitute an unnecessary threat to the name and the reputation of the expedition. Frequenting prostitutes is being debated in relation to the peacekeeping operations of the United Nations, and it verges on outright stupidity and bad ship management to accept drunkenness to this extent and frequenting of prostitutes on an expedition with massive participation of the media and children. Several expedition members have already – for the time being only jokingly – aired the possibility of writing 'the true story' about the conditions on board the expedition vessel. It is said that there exist several photographs, which, if made public, would only make things worse (*Morgenavisen Jyllands-Posten* 2007).

Even though the said photos were never exposed to the public, the "party boat" story, combining sex, alcohol, and scandal, made headlines in virtually all of Danish news media. Allegations and denials were widespread. Confronted with the observations of *Jyllands-Posten*'s journalist, Søren Haslund Christensen said that all sailors were entitled to a bit of fun when off-duty. He thought *Jyllands-Posten* was being self-righteous (Mortensen 2007). Researchers also got involved in the public debate about "the party boat." Jens Tang Christensen, who took part in all but nine days of the expedition, said that it might have been the case that schoolchildren would have been confronted with drunken people, adding that they could just as well have experienced the same things at home (Tougaard & Olesen 2007). At the initiative of researcher Peter Roepstorff, another 125 researchers, schoolchildren, and expedition leaders signed an open letter declaring that *Jyllands-Posten*'s claims about the sex and alcohol culture on board were "deeply unfair and totally unjustified" (Ejsing 2007a). Carsten Schmidt, captain of the ship on the first part of the expedition, said that the collaboration between *Jyllands-Posten* and the rest of the crew had been strained from the beginning because of *Jyllands-Posten*'s lack of cooperativeness. He thought that the disagreements were really caused by the fact that the newspaper wanted to have special status on the expedition, but never really received this status. He also added that the only known violation of the alcohol rules on board was conducted by one of *Jyllands-Posten*'s journalists who was caught stealing alcohol and consequently was dropped off in Greenland (Astrup 2007).

The Navy quickly produced their own evaluation based on testimony from the ship's two captains. The Navy's report concluded that the allegations were unfounded. The only known abuse of alcohol on board was the case of *Jyllands-Posten*'s own journalist mentioned above. There were no known cases of sailors visiting prostitutes or bragging about such visits in the ship's canteen, although the Navy was unable to deny that incidents such as these could have happened. The Navy's policy was not to interfere in such matters as long as prostitution is legal. When visiting Boston, the crew was made aware that prostitution is illegal in the state of Massachusetts (Nyvang 2007). Refraining from commenting on the allegations of *Jyllands-Posten*, the Board of the Expedition Foundation took note of the Navy's report and declared: "From all sides, there has been expressed praise for the way in which the Navy has carried out the Galathea 3 Expedition and, so, we are pleased with the positive recognition of the naval crew on behalf of the media and researchers" (Ritzau's Bureau 2007).

Final Controversy between Jyllands-Posten and the Rest of the Expedition
The "party boat" controversy exposed differences between *Jyllands-Posten* and the expedition management that had built up during the expedition. The crux of the matter seemed to be the tension between seeing the expedition as a public venture and seeing it as a scientific research project. From the public viewpoint, the one promoted by the media and by a few government politicians trying to legitimize the extraordinary investment to the Danish public, the expedition was to provide live coverage of expedition research and adventure. Importantly, the coverage was to take place on more or less normal media terms. That meant adhering to standard procedures of identifying newsworthy events on behalf of journalists, and that was why the co-operation agreement between the media and the expedition management included a clause regarding the full editorial freedom of the expedition's media

partners. *Jyllands-Posten* clearly felt that the expedition management did not live up to this agreement. There were several disagreements between the newspaper and the expedition management, most of which concerned the prioritization of media interests vs. research projects. On several occasions, the newspaper and its journalists felt that they were being overruled by the expedition management. Moreover, in its mid-way assessment report the newspaper urged the expedition management to encourage or, rather, to force the researchers to live up to their obligation to communicate their research by co-operating with the journalists on board. The critical articles of Anders Lund Madsen, in the employment of *Jyllands-Posten*, seemed to have made things worse. The mid-way report emphasized that the newspaper's journalist had a right and a duty to be critical:

> These principles should be communicated to everyone on board in a way that makes it unmistakably clear that the media alone decides which stories are to be made when and how. The communicators have no obligations in this respect. Nor is the media obliged to participate in particular events during harbor visits – it is the media which will assess the relevance on behalf of our readers. Good ideas and suggestions are welcome, but no obligations can be imposed on the media. The expedition management and others should recognize this (*Morgenavisen Jyllands-Posten* 2007).

A few days after the "party boat" scandal, another public controversy about the Galathea 3 Expedition broke in the media. It was caused by a feature for *Berlingske Tidende* authored by the ship's second captain, Lars H. Hansen, and leader of the research project on the marine carbon cycle (one of the largest projects on board), Katherine Richardson Christensen. Published on 25 April, the title of the feature was "Galathea 3 – two successes and one fiasco" (Hansen & Richardson 2007). While the research of the expedition and the co-operation between researchers and naval officers accounted for the two successes, the two authors had argued, the media coverage was a failure. The two authors noted that the two main Danish TV stations, DR and TV2, had shown almost no interest in the expedition and, also, that there had been very little international media coverage of the expedition. They wondered if collaborating closely with just a few media partners had simply been the wrong media strategy for the expedition:

> The construction is peculiar, because the responsibility of communicating the high-profile and government-funded prestigious project was being handed over to a few selected media whose only purpose was to make money. Besides

responsibility, these privileged Danish media have acquired monopoly, which means that all other media – Danish as well as foreign – have been denied access. Consequently, it was difficult to profile the expedition internationally, and none of the media on board have had an incentive to promote the expedition to the foreign press and the local population during harbor visits. However, it is important to make clear that the collaboration between the researchers and most of the media and journalists affiliated with the expedition has turned out to be positive and instructive. On both sides, prejudices have been overcome, and the basis for much better research communication in the future has been laid (Hansen & Richardson 2007).

The publication of the feature amounted to waving a red rag to a bull. The editor of *Jyllands-Posten* Henrik Thomsen wrote a letter of complaint to the vice-chancellor of the University of Copenhagen, Ralf Hemmingsen, accusing Katherine Richardson of "making false and discrediting claims," of "dishonesty," and of "lying" (Ejsing 2007b). He argued that it was plain wrong to speak of a monopoly since all media partners had to respond to a bidding round. "The problem is that there was very little bidding interest, but it is not the case that anybody was denied access," he explained (Richter 2007). All requests from international media, although admittedly very few such requests were being made, were met (Ejsing 2007b).

On behalf of the university, Ralf Hemmingsen said that he saw no problems in Katherine Richardson taking part in what he saw as a public debate about the evaluation of the Galathea 3 Expedition. Using the word "dishonesty," he added, was misplaced as it usually referred to the scientific work of researchers and had very little to do with their being involved in a media-based case of disagreement. As vice-chancellor, he also saw no reason to pursue the case further. Katherine Richardson agreed, adding by use of some pathos and with explicit reference to the Cartoon Crisis:

> It is interesting that it is the very newspaper who forced the nation to its knees in the name of freedom of expression whose chief editor is now threatening me and making complaints to my employer because I make my opinion known. There is something completely wrong when a chief editor is complaining about my views on the media coverage of the Galathea 3 Expedition. *Jyllands-Posten* wanted to cover an adventure expedition, but the overall purpose was research, and this underlying conflict concerning the way in which the expedition ought to be communicated has impacted on the whole of the trip (Richter 2007).

Although most researchers did not engage in the public evaluation of the expedition, a few did, mostly by agreeing with Katherine Richardson. Marine biologist Bo Barker Jørgensen, who participated in the Galathea 3 research project on oceanic oxygen deficiency zones, referred to the expedition on a Danish radio program as "a romantic, outdated conception of how to plan a spectacular research expedition round the world" (Villumsen 2007). He explained that there were ocean-going research vessels all over the world, and so the very idea that going round the world would turn the expedition into a groundbreaking voyage of discovery was "anachronistic." On the contrary, the ambition of sailing round the world resulted in too little time for the individual projects to explore their topics in depth. The contemporary way of planning an expedition was to begin with important research questions and then plan accordingly. The Galathea 3 Expedition had proceeded the other way around: first planning the route and then finding the important research projects to include. Another problematic issue in terms of getting the optimal research output was the emphasis on Danish research. Although international collaboration was stressed in the planning phases, the reality of most Galathea 3 research projects was not as international as one could have wished for, Jørgensen said. "It was a Danish research expedition, planned and carried out by Danish researchers, which included very few foreign researchers. The degree of internationalization was not as high as one could have expected or wanted," he concluded. Moreover, there was too much emphasis on reporting the things and organisms that were being pulled from the ocean on a daily basis in comparison with the way in which research is actually being conducted and should have been communicated. On the same radio program, Stig Markager, leader of the Galathea 3 research project on dissolved organic matter, added that, although in his opinion the expedition did get the balance right between the media and research, the participation of the media did put pressure on organizers and researchers to rush off before the necessary preparations had been done. For example, he noted, there was not enough time to secure funding for PhD projects, and many necessary preliminary studies had to be skipped. In terms of research output, it would have been better to wait for one more year before launching the expedition (Villumsen 2007).

Concluding Remarks about "Live" entanglements of Science, Media, Navy, and Politics

The Galathea 3 Expedition spanned several institutional boundaries, trying to accommodate many different interests and different narratives about the expedition and its "real" purpose. To simplify matters, we might say that the media, *Jyllands-Posten* in particular, saw the expedition as a means of generating adventure and live coverage, while researchers and to a large extent the expedition management thought of it as yet another opportunity to conduct field research. Although some of the controversies presented in this essay can be described in terms of this simple difference (some of the protagonists such as Richardson and Thomsen clearly did so), this, of course, is not the whole story. Media and science both subscribed to the political vision of strengthening Danish research by providing additional research funding (temporarily) and by promoting young people's interest in science and research. And it appears that not all media partners and not all scientists viewed the difference between research and adventure as so difficult to overcome as Thomsen and Richardson did.

The inclusion of media partners on board the expedition combined with live coverage on the Internet (online news, blogs, etc.), made latent conflicts visible and enabled the participants to use the media as a venue to expose their differences. What served as a means of "making it cool to be a researcher" by showing young people the excitement and adventure of daily field research also became a tool of producing and proliferating different, if not conflicting narratives about the expedition. If we return to the introductory notion of narratives as technologies of travel, we could say that, whereas the adventure, national, and public understanding of science narratives mentioned in the beginning of this essay enabled a lot of interest in the expedition plan and made the expedition possible, the two differing narratives of media vs. research expedition seemed to provide obstacles in terms of producing consent with respect to the internal assessment of the expedition.

To what extent did the narratives about the expedition and the conflicts about defining the expedition presented in this essay reach the general public? Is there any way in which we may speak of the implications of covering the expedition live on the public understanding of expedition life? There is of course no way to answer such questions in a comprehensive way within the framework of this essay, where the aim is to try to get a grasp on the different kinds of understandings at play in the live media coverage of the expedition. Yet what we can say is that *Jyllands-Posten* hired Rambøll Management Consulting to perform a public opinion poll on the impact of the Galathea 3 Expedition among the Danish population. The results are depicted in table 1.

The results indicate that the impact of the Galathea 3 Expedition among the Danish population was quite high (with 82.2 percent having heard about the expedition). Of course, there is no way of knowing from the survey what kind of impression of the expedition all of these people formed from "having heard about the expedition." Most people seem to have heard about the expedition from TV, which indicates that the Hansen and Richardson were right in arguing how important it

Statements	Level of agreement (%)
Have heard about the Galathea 3 Expedition	82.2
Very or predominantly positive about the Galathea 3 Expedition	60.6
Knowledge about the expedition through TV	76.4
Knowledge about the expedition through newspapers	62.5
Have watched the expedition several times a month	23.1
Have watched the expedition several times a week	17.6
Have never watched the expedition	25.2
It is a good or very good idea to allow research and research communication on the same expedition (such as on Galathea)	81.6
Agree or predominantly agree that the expedition will lead more children and young people to take an interest in science	48.9
Agree or predominantly agree that the expedition will lead to more young people seeking science educations	36.6
It is important to strengthen interest in science among young people	57.8

Table 1. Results from interviews conducted with a representative cross section of the Danish population above 17 years of age, performed by Rambøll Management for Jyllands-Posten (Gram 2007).

would have been to have national TV on board – which Henrik Thomsen had also wanted from the beginning, but the second and largest national TV station DR said no, apparently because of the participation of TV2. Most people approved of the special Galathea 3 Expedition construction, combining on one and the same expedition research and media coverage; and most people agreed that the expedition would make a difference to young people's attitudes to science.

Søren Haslund Christensen interpreted the results in this way:

> Some people have described the expedition as a way in which to get young people off the streets. That's a good way of seeing it, I think. I know of many parents who have talked to their kids about the expedition. It is all about getting young people to start working on the climate debate or biological subjects (Gram 2007).

Apart from anecdotal evidence like the one provided by Haslund Christensen, there has been no evaluation of the impact of the live coverage of the Galathea 3 Expedition on young people. Neither has any official evaluation of the research output been performed. In October 2009, Member of Parliament Jesper Langballe wanted to know what came of the 200 million DKK put into the expedition in terms of research results. In his response, minister of science Helge Sander said that, having consulted the board of the Danish Expedition Foundation, he believed that it was too early to assess the research of the expedition. An evaluation should take place five years after the expedition or even later. The data needs to be evaluated, papers need to go through peer review, and articles need to be published. It is a lengthy process. Taking an optimistic view, Sander rejoiced in the fact that at the time the voluntary catalogue of Galathea 3-related publications included 52 primary publications and 29 secondary ones (Rasmussen 2009).

While waiting for the final evaluation of the Galathea 3 Expedition and maybe even the dispatch of a fourth Galathea Expedition from Denmark, there are probably two important conclusions to be drawn from this particular story of the events: the emphasis put on reporting the expedition live, due in part to the expedition's starting point as a newspaper-based expedition and in part to the importance given to promoting public understanding of science, resulted in the expedition being a hybrid venture that combined media interests, government research and educations policies, naval practices, and scientific research. For some researchers, the expedition thus looked more like a scientific expedition of the past, rather than a new and promising way of doing expeditions. However, the live coverage, facilitated by the many more or less critical journalists on board, also meant that controversies and different interpretations of what the expedition was really about were exposed in public. Following the expedition live, people might have felt that they were not only witnessing an animation of the idea of an adventurous expedition round the world as promoted by editors, politicians, and some scientists, but also a clash of the professional cultures of scientists, journalists, government officials, and naval officers. An experiment in carrying out an expedition with live coverage, the Galathea 3 Expedition certainly highlighted the complexities involved in combining science, media, and politics.

References

Astrup, S. 2007. "Kaptajn: Jyllands-Posten er urimelig." *Politiken*, May 13, 3.

Bergsagel, D. 2006. "Galathea: Søslangen er denne gang nok en ål." *Politiken*, August 16, 3.

Boelskifte, E. 2005. "Et hav af spændende projekter." *Jyllands-Posten*, November 30, 4.

Boelskifte, E. 2006. "Galathea 3: Afrejse: Af sted med piber og trommer." *Jyllands-Posten*, August 12, 4.

Boelskifte, E. & J. Grund 2004. "Ny dansk Galathea-ekspedition: Store forventninger til Galathea-ekspedition." *Jyllands-Posten*, August 26, 4.

Bravo, M. & S. Sörlin 2002. "Narrative and Practice – an Introduction." In: M.Barvo & S. Sörlin (eds.) *Narrating the Arctic: A Cultural History of Nordic Scientific Practices*. Canton, MA: Science History Publications, 3-32.

Collingnon, P. 2006. "Menneskehedens 10 største erkendelser." *Jyllands-Posten*, August 6, 1.

Danish Expedition Foundation 2006. "Bestyrelsen." Retrieved August 1, 2011 from http://130.226.56.246/dk/Menu/Om+os/Bestyrelse

Dannemand, H. 2000. "Naturfags-skepsis undersøges." *Berlingske Tidende*, April 27, 2.

DTU Aqua 2009. "The research vessel Dana." Retrieved August 1, 2011 from http://www.aqua.dtu.dk/English/About/Research_vessel.aspx

Editorial 2006a. "Galathea 3." *Jyllands-Posten*, August 11, 10.

Editorial 2006b. "Galathea 3 er stor." *Ingeniøren*, Aurgust 11, 2.

Editorial 2006c. "Videnskab & nostalgia." *Politiken*, August 13, 10.

Ejsing, J. 2007a. "Avis i ordkrig med forskere og sømænd." *Berlingske Tidende*, May 15, 13.

Ejsing, J. 2007b. "Morgenavis og universitet i Galathea-strid." *Berlingske Tidende*, May 22, 6.

Galathea 3 2006. "Galathea 3." Retrieved August 1, 2011 from http://www.galathea3.dk/dk

Gram, S. 2007. "Stort kendskab til Galathea 3-ekspeditionen." *Jyllands-Posten*, June 24, 6.

Grund, J. & Gram, S. 2006. "Galathea-rute bliver ændret." *Jyllands-Posten*, Fabruary 22, 5.

Haarder, B. 2006. "Kronik: Anders Lund Madsen eller Galathea?" *Jyllands-Posten*, December 28, 9.

Hansen, L. H. & K. Richardson 2007. "Galathea 3 – to succeser og én fiasko." *Berlingske Tidende*, April 25, 13.

Haslund-Christensen, S. & M. Meldgaard 2008. "Baggrund og rammer for ekspeditionen." In: L. N. Jørgensen (ed.) *Galathea 3 – 2006-2007.* Copenhagen: Dansk Ekspeditionsfond, 12-35.

JP Explorer 2009. "Eventyr med omtanke." Retrieved August 1, 2011 from http://viden.jp.dk/explorer/om/

Jørgensen, L. N. 2008. *Galathea 3: 2006-2007.* Copenhagen: Dansk Ekspeditionsfond.

Kammeradvokaten 2005. "Samarbejdsaftale om mediedeltagelse i Galathea 3 Ekspeditionen." Retrieved August 1, 2011 from http://www.ft.dk/samling/20051/aktstykke/aktstk.24/pgf/19/bilag/2/228767.pdf

Klein, T. 2000. "Det er handling der giver forvandling." *Berlingske Tidende*, March 5, 2.

Larsen, H. 2005. "Galathea-ekspedition mangler 70 millioner." *Politiken*, November 30, 5.

Latour, B. 1987. *Science in action.* Cambridge, MA: Harvard University Press.

Madsen, A. L. 2006a. "Den sejlende børnefødselsdag." *Jyllands-Posten*, December 13, 13.

Madsen, A. L. 2006b. "Lavtryk over Azorerne." *Jyllands-Posten*, September 20, 24.

Mikkelsen, J. 2004. "Aktuelt protræt: Projektmageren." *Jyllands-Posten*, August 26, 12.

Morgenavisen Jyllands-Posten 2007. "Morgenavisen Jyllands-Postens evaluering af Galathea 3-ekspeditionen Cape Town-København." Retrieved August 1, 2011 from http://www.ft.dk/samling/20061/almdel/uvt/bilag/178/379005/index.htm

Mortensen, M. N. 2007. "Haslund afviser kritik af søfolk." *B.T.*, May 12, 14.

Mouritzen, L. T. 2006. "Debat: Anders Lund gemte sig." *Jyllands-Posten*, December 16, 10.

Nielsen, K. H. 2005. "Between understanding and appreciation: current science communication in Denmark." *Journal of Science Communication*, 4(4). Retrieved July 10, 2011 from http://jcom.sissa.it/archive/04/04/A040402/

Nielsen, K. H. 2006. "In search of the sea monster." *Endeavour*, 30(1): 36-40.

Nielsen, K. H. 2006. "Naturvidenskabernes kanonisering: Forskere, erkendelser eller kulturarv." *Nordisk Museologi* 2: 27-44.

Nielsen, K. H. 2008. "In quest of publicity: the science-media partnership of the Galathea Deep Sea Expedition from 1950 to 1952." *Public Understanding of Science*, 18(4): 464-480.

Nielsen, K. H. & C. Autzen 2011. "Looking Over the Shoulders of Researchers: In Quest of a Research-Media Partnership and Online, News Media–Based Researcher Identities on the Galathea 3 Expedition From 2006 to 2007." *Science Communication*, 33(4): 472-500.

Nielsen, K. H., C. R. Kjaer & J. Dahlgaard 2007. "Scientists and science communication: a Danish survey." *Journal of Science Communication*, 6(1). Retrieved July 10, 2011 from http://jcom.sissa.it/archive/06/01/Jcom0601%282007%29A01/

Nielsen, L. H. 2006. "Eventyret er begyndt." *Berlingske Tidende*, August 11, 6.

Nyvang, M. 2007. "Søværnet: Kun Jyllands-Posten brød reglerne." *Ingeniøren*, May 14. Retrieved July 10, 2011 from http://ing.dk/artikel/76098-soevaernet-kun-jyllands-posten-broed-alkoholreglerne

Rasmussen, S. E. 2009. "Hvad blev der af Galathea-forskningen?" *Universitetsavisen*, October 15. Retrieved July 10, 2011 from http://uniavisen.dk/politik/hvad-blev-der-af-galatheaforskningen

Rebensdorff, J. 2006. "Galathea stikker syd om Afrika." *Berlingske Tidende*, February 22, 6.

Redaktionen 2006. "En nyttig kanon." *Jyllands-Posten*, August 6, 6.

Richter, L. 2007. "Ytringsfrihed: Den er helt Galathea." *Information*, May 23, 12-13.

Ries, C. 2002. "Lauge Koch and the Mapping of North Eat Greenland: Tradition and Modernity in Danish Arctic Research, 1920-1940." In: M. Bravo & S.Sörlin (eds.) *Narrating the Arctic: A Cultural History of Nordic Scientific Practices*. Canton, MA: Science History Publications, 199-231.

Ritzau's Bureau 2007. "Bestyrelsen bag Galathea 3 er tilfreds." *Berlingske Tidende*, May 16, 4.

Rose, F. 2006a. "Why I Published the Mohammad Cartoons." *Spiegel Online*, May 31. Retrieved August 1, 2011 from http://www.spiegel.de/international/spiegel/0,1518,418930,00.html

Rose, F. 2006b. "Ytringsfrihed: Muhammeds ansigt." *Jyllands-Posten*, September 30, 3.

Rønn, L. & M. Lützen 2006. "Kritik af Galathea Ekspeditionen." *Jyllands-Posten*, December 13, 2.

Sander, H. 2004. *Talepapir vedrørende Galathea 3 Ekspeditionen*. Retrieved August 1, 2011 from http://www.ft.dk/samling/20041/aktstykke/aktstk.6/pgf/19/spm/8/svar/122569/118601.pdf.

Schmidt, A. L. 2006. "Galathea 3: Når journalister bliver gjort til mediepartnere." *Politiken*, September 9, 2.

Skovgaard, L. E. 2000. "Erhvervspolitik: Initiativer skal redde velfærden." *Jyllands-Posten*, February 29, 2.

Slok, C. 2009. "Here I Stand: Lutheran Stubbornness in the Danish Prime Minister's Office during the Cartoon Crisis." *European Journal of Social Theory*, 122: 231-248.

Sorrenson, R. 1996. "The Ship as a Scientific Instrument in the Eightteenth Century." In: H. Kucklick & R. E. Kohler (eds.) *Science in the Field. Osiris*, 11: 221-236.

Thomsen, H. 2006a. "Fra JP-idé til virkelighed." *Jyllands-Posten*, August 8, special section on the Galathea 3 Expedition.

Thomsen, H. 2006b. "Galathea halvvejs." *Jyllands-Posten*, December 13, 8.

Thomsen, H. 2008. Personal communication with the author, February 18.

Tougaard, H., & S. Olesen 2007. "Galathea-efterspil: Påstand mod påstand om druk." *Jyllands-Posten*, May 12, 5.

Tænketanken for forståelse for forskning 2004. *Forsk og fortæl!* Copenhagen: Videnskabsministeriet. Retrieved July 10, 2011 from http://vtu.dk/publikationer/2004/forsk-og-fortael

Villumsen, J. H. 2007. "Kritik af Galathea-ekspeditionen." *P1 Orientering*. Retrieved July 10, 2011 from http://www.dr.dk/P1/orientering/indslag/2007/05/02/191522.htm

Wikipedia 2009. "HDMS Vædderen F359." Retrieved July 10, 2011 from http://en.wikipedia.org/wiki/HDMS_V%C3%A6dderen_(F359)

Wikipedia 2010. "Jyllands-Posten Muhammad cartoons controversy." Retrieved July 10, 2011 from http://en.wikipedia.org/wiki/Jyllands-Posten_Muhammad_cartoons_controversy

Jenny Beckmann

The Swedish Taxonomy Initiative
Managing the Boundaries of "Sweden" and "Taxonomy"

"Linnaeus's legacy carries on." Under this impressive heading, the Swedish Taxonomy Initiative was presented in *Science* magazine in February 2005. The article continued: "Following the tradition of its most famous taxonomist, Sweden aims to be the first country to complete an inventory and pictorial guide to its biodiversity." Ultimately, the goal of project was described as taxonomy's "elusive grail" – "a complete catalogue of life on earth" (Miller 2005:1038).

The Swedish Taxonomy Initiative (STI) is an ongoing research project at the intersection of conservation biology, systematics, and popular science. It was funded by the Swedish Government in 2002, and was designed to run for twenty years at an estimated cost of 30 million Swedish Kronor a year – a little more than € 3 million – adding up to about € 60 million over the whole period. Its explicit aim is "to develop keys for the identification of all Swedish multicellular organisms, approximately 50,000 species," and thus, "a considerable improvement of the infrastructure of Swedish taxonomy." It is simultaneously an inventory project, an effort to increase funding, interest, and research in taxonomy, and an exercise in popular education, not least through the publication of a multi-volume *Encyclopedia of Swedish Flora and Fauna* (Svenska artprojektet, n.d.).

In the diversity of its aims and its complicated relationships between participants, funding agencies, audiences, institutions, and geographical dimensions, the STI involves a variety of issues of modern biology and science policy. In this preliminary study, partly based on interviews with project participants and policy-makers, I will focus on two related issues: the status of taxonomy and inventories in environmental and science policy, and the negotiation of national and regional boundaries in inventory projects.

The STI straddles the boundary between academic science and environmental policy. Taxonomy is part of evolutionary biology as well as biodiversity conservation, but funding for environmental monitoring and taxonomic research is, as a rule, handled by different kinds of institutions. While taxonomy depends on academic funding agencies, such as research councils and foundations, practical conservation projects are the responsibility of the Swedish Environmental Protection Agency (SEPA), as well as municipal and county administration. The status of taxonomy depends on successfully navigating the field of scientific research as well as practical conservation.

At the same time, the practice of inventory involves negotiating political and geographical boundaries. The presentation of the STI is emphatically national, not least in its recurring emphasis on Carl Linnaeus. This is a reflection of a political and financial situation in which international agreements on biodiversity are made on the basis of participating nations, and rely on national frameworks for implementation and reporting; and where research funding agencies – whether public or private – often operate in national contexts. The European Union is emerging as an important factor in the borderland between research, environmental monitoring, and infrastructure, but national commitments remain crucial to EU funding, as in the case of LifeWatch, a project to develop a European infrastructure for biodiversity research (LifeWatch: E-science and technology infrastructure for biodiversity data and observatories, n.d.).

On another level, the focus on Linnaeus reflects the idea, promoted in schools, universities, popular education, and the media for almost two centuries, that Swedes, with their Linnaean heritage, are particularly prone to nature study. The national and Linnaean rhetoric is a factor in acquiring funding on the domestic front, as well as in promoting the project internationally. The promoters of the STI hope to set an example to be followed by other nations: if the Swedes can do it, others can as well. In particular, they would like to sell the idea to their closest neighbors, and transform the Swedish Taxonomy Initiative into a Nordic Taxonomy Initiative.

Underpinning the Swedishness is a complex set of regional and international relations. National inventories provide opportunities for local actors to assert themselves; and historical precedent as well as the behavior of plants and animals calls for co-operation across national and other administrative boundaries. As much as inventories are in themselves a way of establishing territories and outlining boundaries, they involve going beyond these boundaries.

Biodiversity and the Status of Taxonomy

Taxonomy in the History of Science

Taxonomy is a scientific specialty with a chequered history. Arguably "top science" in the eighteenth century (certainly so from a Swedish perspective), and hugely successful in education and institution-building in the nineteenth, it lost some of its cachet during the twentieth. Taxonomy is sometimes used interchangeably with "systematics," and it is currently defined by London's Natural History Museum as "the science of identifying and naming species and organizing them into systems of classification" (Natural History Museum 2010, Hine 2008). In a narrower sense, "alpha taxonomy" often refers to the describing, naming, and ordering of living organisms, whereas systematics takes on the evolutionary, genetic, and ecological questions associated with constructing phylogenetic trees and tracing the history of life on earth.

These different definitions reflect long-standing debates about the scientific status of taxonomy. Discussions about a crisis in taxonomy have been going on for at least a century. In Sweden, one expression of this was the attempt, in the 1960s, to turn the Swedish Museum of Natural History into a research center for molecular biology, immunology, and biophysics. This particular threat was averted, but taxonomists repeatedly express feelings of disadvantage in relation to other biological disciplines (Taxonomy for the twenty-first century 2004, Tunlid 2009).

Numerous efforts have been made since the mid-twentieth century to raise the stock of the discipline, and to change its image from an old-fashioned pursuit associated with dusty collections, to a modern theoretically and methodologically advanced discipline, hooked up to its many users through the latest technology.

With the launch of the "evolutionary synthesis" in the 1930s, taxonomy joined with population genetics, forming part of a more comprehensive field of evolutionary biology, using tools from mathematics and ecology to formulate theories of speciation (Smocovitis 1996). In the 1960s and 1970s, the ideas of German entomologist Willi Hennig were employed to give the construction of phylogenetic trees additional rigor, sparking controversy among taxonomists in the process, and establishing the term "cladistics" to describe this very influential approach to taxonomy (Hull 1988, Vernon 1993).

For generations, taxonomists have developed meticulous and craftsmanlike methods for the comparison of morphological features. The advent of new technological tools has been used to promote taxonomy as a modern science, less dependent on individual judgments. In the 1960s, the preferred approach was numerical methods, using computers to create clusters of morphological characters to demarcate taxa, seeking scientific safety in numbers rather than relying on individual scientists' experience to evaluate morphological similarity (Vernon 2001). As DNA sequencing became more accessible (and less expensive), taxonomists began to construct phylogenies on the basis of genetic information rather than only traditional morphological characters. This has had far-reaching consequences, particularly for the self-image of the taxonomic community, which has emerged *as* a community with a set of shared methods and problems, rather than as separate groups of botanists, entomologists, mycologists, or ichthyologists, each with their own specific expertise (Hine 2008). However, some taxonomists argue that molecular methods in themselves do not generate problems, but only serve to shed additional light on questions already raised by morphological approaches (Wheeler 2004).

Finally, taxonomists argue that their science is useful. All biologists need stable nomenclature, reliable catalogues where they can access information required for conservation work, genetic research, population ecology, etc. Moreover, biodiversity inventories conducted by taxonomists may turn up species that can provide material for medicine, agriculture, and industry. In fact, in competition with ecologists, taxonomists claim to possess the key to understanding biodiversity – without "completing the inventory," we won't know what we are losing before it is lost forever (Wilson 2000). And the recent rapid development of a field which is becoming known as "biodiversity informatics" hinges entirely on the notion of a multitude of users outside of the field itself, who are in urgent need of biodiversity information (Stockholm Biodiversity Informatics Symposium 2008). But, as Christine Hine has noted in her study of systematics as "cyberscience," it is not a simple matter to maintain credibility both as a service and as a science (Hine 2008: 214).

Taxonomy and the Convention of Biological Diversity
This ambiguity has become increasingly prominent in the wake of the Convention of Biological Diversity (CBD), opened for signature at the Earth Summit in Rio de Janeiro in 1992. David Takacs has pointed out that the term "biodiversity" itself sits on the boundary between the science and the politics of conservation, as well as being claimed by geneticists and ecologists in addition to taxonomists (Takasc 1996). Taxonomy was singled out as a crucial area for biodiversity work in Rio. The CBD stressed the so-called "taxonomic impediment" to biodiversity work – in their words, "the knowledge gaps in our taxonomic system (including those associated with genetic systems), the shortage of trained taxonomists and curators, and the impact these deficiencies have on our ability to conserve, use and share the benefits of our biological diversity" (CBD 2007). To remove this "impediment," the Global Taxonomy Initiative (GTI) was established in the late 1990s under the aegis of the CBD – essentially a framework for coordinating taxonomic information and expertise to implement the convention (Samper 2004).

However international the CBD and the GTI are, it is the task of national governments, as signatories of the treaties, to implement the conventions and protocols agreed upon. Under the Convention, governments undertake to conserve and sustainably use biodiversity. They are required to develop national biodiversity strategies and action plans, and to integrate these into broader national plans for environment and development (Secretariat of the Convention on Biological Diversity 2000). In Sweden, the University of Agricultural Sciences has been charged with the academic side of environmental analysis, monitoring, and assessment programs, in co-operation with agencies such as SEPA. The university hosts two semi-autonomous institutions with particular responsibility for monitoring biological diversity: the Swedish Biodiversity Center, created in 1993 in the wake of the CBD with a mission to initiate, conduct, and co-ordinate research in biological diversity; and the Swedish Species Information Center (SSIC), established in the mid-1980s as the institutional embodiment of a series of flora and fauna inventory projects (Beckman 2007). The Biodiversity Center initiates and coordinates research, and it describes itself as "the engine of Swedish biodiversity activity," while the Species Information Center monitors endangered species and administers the compilation of the so-called Red Data Lists of endangered species (Centrum för biologisk mångfald, n.d., ArtDatabanken n.d.). The focus of the Swedish Biodiversity Center is on ecology, genetics, and agriculture, and it is primarily research-oriented, whereas the Species Information Center approaches issues of conservation and sustainability from the perspective of the status of individual species, and is more involved with practical monitoring and inventory processes.

The Swedish Taxonomy Initiative and Science Policy

Background and Initiation of the Swedish Taxonomy Initiative

The particular story of the origin of the Swedish Taxonomy Initiative appears to be one of serendipity – especially as told by the participants and beneficiaries. It is presented as a political fairy-tale.

In the late 1990s, several biologists working on systematics and conservation were lamenting the difficulties of making the Rio Protocol and the CBD practical and effective on a Swedish national level. They related these problems to the generally poor status of taxonomy in the scientific community. They also lamented the apparent fact that people were losing the ability to "read" nature: that the interest in field biology – so great in the 1970s – was waning (Gärdenfors 2008, Ronquist 2008). They began working on a proposal to address the shortage of information as well as the shortage of popular knowledge: an inventory of poorly known Swedish species, accompanied by a popular presentation to foster public interest in conservation.

The proposal was presented at an annual conference on flora and fauna management – Flora- och faunavård – in 2000, hosted by the Species Information Center and attended by the minister for the environment. Somewhat surprisingly, the minister picked up the idea immediately, and on the spot invited the directors of the Species Information Center to further discussions. Meanwhile, further political support for the project was solicited in Parliament. In 2000, Sweden was ruled by a minority government of Social Democrats, supported, sometimes precariously, by the Left Party and the Green Party. The Species Information Center approached the supporting parties, who used the project in their negotiations over the 2001 budget proposal. In this way the species project was turned into an asset in the high stakes haggling over political power (Goude 2009).

The result was hugely successful, by taxonomy standards. The newly named Swedish Taxonomy Initiative (Svenska artprojektet) was allocated 30 million Swedish Kronor – about € 3 million – a year for twenty years, with an equal allotment to be distributed to support natural history museums and collections. In addition, the government allocated 440 million Swedish Kronor to the Swedish Research Councils to support research on biological diversity (Regeringen 2002).

Thus, there were two parts to the new, massive biodiversity program. First, the Swedish Taxonomy Initiative, inventorying species, developing identification tools, and raising public awareness and knowledge of species, was hosted by the Species Information Center, and in many ways functioned as an independent funding agency. Second, there was a large infusion of biodiversity money into the conventional research funding system. Some of the money for museums and collections was also channeled through the research councils: Formas, the Swedish Research Council for Environment, Agricultural Sciences and Spatial Planning, and Vetenskapsrådet, the Swedish Research Council.

The research councils were dubious. However happy they were to receive increased funding, they were – as always – reluctant to submit to political restrictions (Nilsson 2008, Sundin 2008). In fact, as the years went by, taxonomists and conservationists were increasingly concerned that the councils did not use the biodiversity money to fund research that could benefit the STI, but concentrated on ecology and other interpretations of biodiversity (Alström 2008, Gärdenfors 2008, Ronquist 2008).

To some extent, this reflects long-standing distrust between ecologists and systematists on the research council committees. For decades, systematists have claimed to be marginalized by ever more powerful ecologists, institutionally as well as financially (Bowker 2000, Gärdenfors 2008, Nilsson 2008, Ronquist 2008). But more fundamentally, it indicates disagreements about the scientific status of taxonomy. The division of the money between two different systems of allocation highlights the ambiguous position of taxonomy in the scientific community.

Taxonomy and the Boundary between environmental and Science Policy
The position of the STI between research and policy was not necessarily an advantage when it came to the funding and the scientific status of the Swedish Taxonomy Initiative. Indeed, the Swedish Research Council for Environment, Agricultural Sciences and Spatial Planning (Formas), charged with administering the museum funding, did not regard this as "research" at all, and eventually managed to get rid of the assignment (Gärdenfors 2008, Nilsson 2008). One particular instance of this problem was the status of collecting and taxonomic inventories as scientific activities. While making an inventory was regarded as a preamble to doing "real science," monitoring, on the other hand, was applied science. The former is too basic, not "hypothesis-driven" enough, for research funding; the latter should be handled by environmental agencies. As expressed by Charles Godfray and Sandra Knapp, editors of a special issue of the *Philosophical transactions of the Royal Society* on "taxonomy for the twenty-first century": "Much of what taxonomists did was perceived as descriptive, at a time when every undergraduate had read a précis of a précis of Popper and equated 'true science' with falsifiability and experimentation" (Godfray & Knapp 2004).

The Research Councils were also suspicious of attempts to align funding with political projects. When a government investigation of knowledge production in biodiversity suggested that current research in the field was not adapted to conservation needs, the Research Councils vigorously defended what might be called a "linear model" of biodiversity research (cf. Grandin, Wormbs & Widmalm 2004). Conservation agencies would, in the long run, be better served through research into fundamental problems in ecology and evolution, than by more applied and politically controlled projects (Utredningen kunskap för biologisk mångfald 2005).[1]

Thus, from the perspective of the research councils, guardians of scientific ideals and standards, taxonomy ended up being at the same time too basic *and* too applied to be considered "research," and thus to be eligible for conventional research funding. At the same time, other conservationists criticized the project as being too expensive, too narrowly focused on Sweden, and diverting funds that would find better use in other conservation efforts, such as habitat protection or climate research (*Upsala Nya Tidning* 2005, 2008).

It is not surprising then that it was only possible to launch the Swedish Taxonomy Initiative by by-passing the funding agencies and applying directly at the top political level. But although the influx of biodiversity funding has strengthened

---

1   See also Vetenskapsrådet 2005. Vetenskapsrådets kommentarer över: *Kunskap för biologisk mångfald*. SOU 2005:94, Dnr 112-2005-8812.

the resources of taxonomy in Sweden, it has not necessarily benefited its scientific status. In fact, some taxonomists are afraid that the existence of separate, political money for the STI makes the research councils even less likely to fund taxonomic research (Alström 2008, Gärdenfors 2008, Sundin 2008).

Inventories in the Swedish Taxonomy Initiative

Most biologists would agree that Sweden is hardly a biodiversity hot spot; moreover, its flora and fauna have been more thoroughly investigated and surveyed than most comparable regions. However, the STI is unique in being one of a very few national ATBIs – All Taxa Biodiversity Inventories – in existence. There are some international parallels, notably the inventory of Costa Rica initiated by David Janzen in 1989, and of the Great Smoky Mountains National Park begun in 1999; but both have shaky support from public agencies and rely primarily on donations (InBio n.d.).

However, there are still new species to be found in Sweden, and the STI was intended to fill such "knowledge gaps." Its administrators have endeavored to direct research funding to organism groups that are poorly known, in terms of distribution and abundance as well as taxonomy. Every year in the applications season, the STI publishes a list of research priorities, in which taxa are allocated a number between 1 and 5, with the lowest numbers found among mammals, flowering plants, and other well-known groups (Sundin 2008, Forskningsprioriteringar 2009). The STI then funds projects according to their scientific quality, the publication plan for the *Encyclopedia*, and the list of research priorities.

In addition to these projects, the STI has funded two general inventories, one on marine fauna and one on two megadiverse orders of insects, Hymenoptera and Diptera. These are rather different projects, in terms of methodology, region, participation, and, not least, expense.

The Malaise Trap Project

The insect inventory was conducted between 2003 and 2006 throughout the country, and used so-called Malaise traps (Karlsson et al. 2005). A Malaise trap is a large tent-like structure designed to capture flying insects in its folds. It is a "passive" trap, in that it does not emit light or sound, or use any other strategy to attract insects (Malaise 1937, Svenska Malaisefälleprojektet 2005). But it does require regular monitoring, to collect the catch and to make sure that it has not been damaged. Malaise traps had recently been used by the entomologist Paul Hanson in conjunction with the highly publicized ATBI run by InBio in Costa Rica, and Hanson was invited to a meeting with Swedish entomologists to discuss the design of the Hymenoptera inventory (Glemhorn & Karlsson 2008). The proposed 148 localities were whittled down to around 50, chosen for biotope significance as well as the availability of reliable trap monitors. All monitors – whether they were academics or local enthusiasts – must be vouched for by someone in the project group, and when the traps had been set up, they were given a crash course in trap maintenance: changing collecting bottles, writing labels, dealing with bats stuck in the folds, nesting dunnocks, or stampeding reindeer. Some of the monitors would bicycle cross-country for dozens of miles every week to look after their traps, and received a diploma, a box of chocolates, and the first volume of the *Encyclopedia of Swedish Flora and Fauna* for their pains.

Forty million specimens were collected during the collecting phase. They were brought to the sorting headquarters, an ecological field station which used to belong to Uppsala University, but had recently been taken over by a local network of nature enthusiasts in the hope of turning it into a center for research, popular science, sustainable development, and tourism – and renamed "Station Linné" (Station Linné & Porten till Alvaret n.d.). The catch was sorted into orders and families by volunteers trained at the station – most of them with a biology background, but some of them secondary school students doing work experience (Glemhorn & Karlsson 2008). The specimens were then sent on to experts on particular genera and species for final identification.

The collecting phase of the Malaise trap project ran for several years, and received some attention in local media, mainly though the activity of the trap monitors. In time for the final collecting round, the project leaders contacted the national media, and ended up on the front page of the country's biggest newspaper: "Scientists find 500 new animal species" (*Aftonbladet* 2006). Visitors to Station Linné were invited to meet the scientists and, on some occasions, witness science in action as the insect sorters went about their business.

The Malaise trap project involved a large number of people, many of whom were volunteers. The traps themselves were fairly simple, in construction as well as maintenance. In many ways, the project resembled recent recording projects organized by local naturalist societies, where plants and butterflies were mapped and reported in elaborately printed volumes (Beckman 2009). The design of the project, however, was not left to the volunteers, but conducted by taxonomists who modeled it on similar academic projects in other countries.

The Marine Inventory

The marine inventory, on the other hand, involved more expensive equipment and personnel, as well as other forms of interaction with local and national environmental administration. Several researchers were interested in a marine recording project, for different reasons. Anna Karlsson, an ecologist employed by the SSIC to compile Red

Lists for Swedish marine fauna, was concerned about deficient data in her field. Matz Berggren and Christoffer Schander, marine biologists specializing in crustaceans and mollusks, were interested in mapping biodiversity change by following up marine inventories that had been conducted along the Swedish west coast in the 1920s and 1930s. Based in Sweden and Norway, respectively, they also wanted to engage several Nordic countries, to connect the project to recent inventories of the North Atlantic benthic fauna, BIOFAR and BIOICE (BIOFAR 2006, BIOICE n.d.). BIOFAR concentrated on the territorial waters of the Faroe Islands, BIOICE on Iceland. Both were supported by their local governments, as well as by the Nordic Council of Ministers and individual research institutions in the Nordic countries. Finally, Rikard Sundin, the newly recruited STI project coordinator, was interested in launching a marine project to fill taxonomic knowledge gaps for the STI (Berggren 2008, Sundin 2008, Karlsson 2009).

A suggestion for a marine inventory was presented to the STI in the early days of the project, but proved difficult to fund. But in 2004, the opportunity arose to join a survey of offshore banks planned by the Swedish Environmental Protection Agency (SEPA); a survey motivated in part by the question of where projected wind power stations should be located (Naturvårdsverket 2007). With a little extra funding, the STI were allowed to join the SEPA team on the boat, although they could not influence the choice of sampling localities. After this initial venture, the STI managed to produce funding for an inventory of their own. After a meeting with marine experts to design the project, and a trial run in the waters close to expedition headquarters at Kristineberg Marine Research Station to evaluate methods and sampling procedure, the boat loaded with taxonomists was launched in the summer of 2006 (*Göteborgs naturhistoriska museums årstryck* 2004, 2005, 2006, 2007, 2008, Karlsson 2009).

The goals of the STI inventory differed from those originally planned by the SEPA in several ways. The SEPA inventory was motivated in part by the question of where projected wind power stations should be located. Although there were more general objectives, the inventory was reminiscent of those carried out in conjunction with development projects, and regulated by law (cf. Sundin 2005). The STI inventory, on the other hand, aimed to revisit the localities studied by marine biologist Leonard Jägerskiöld along the Swedish west coast in the 1920s and 1930s, to re-establish a template for studying change in the marine fauna – although the data could also be used for Red Lists and other projects more closely tied to environmental policy (Karlsson 2009). The intention was also to deposit the material collected for the inventory in a museum. These two aims, in their turn, tied the project to the Göteborg Natural History Museum, host to the Jägerskiöld material and a potential repository for the new marine collections. The STI would strengthen the scientific position of the museum as well as tie the inventory firmly to taxonomic practice, rather than to environmental policy.

At the same time, the project managers were careful to maintain good relations with regional environmental administration. Suggestions for additional sampling localities were solicited and results were reported to the county administration. The STI and the marine inventory would help establish the Species Information Center as an authority in matters of regional habitat protection and nature reserves, as well as encourage high standards of species information and classification in local environmental administration (Karlsson 2009).

Inventories and Geographical Boundaries

The inventory projects also involved complex negotiations of national territories and boundaries. The first problem was with the Swedish Official Secrets Act. Detailed hydrographic surveys can only be conducted by permission of the Swedish military establishment. And even if survey permission is granted, publication of the data remains prohibited. This situation has been criticized by oceanographers, environmental administrators, fishing organizations, representatives of marine transportation, and many others, and over the years several official enquiries have been conducted, so far without finding a practical solution (Riksdagen 2008, Samhav 2008). The problem was especially acute for a benthic marine inventory such as the STI, which focused on the seafloor, looking for potential biodiversity hot spots that had escaped the heavy trawling of the modern fishing industry (Berggren 2008, Sundin 2008, Karlsson 2009). Since almost every fishing boat is equipped with sonar and GPS receivers, detailed information about the seafloor was available in practice, while acknowledging or publishing it was complicated.

Another problem was a familiar one: the fact that plants and animals do not respect national boundaries. In the marine context, this problem took on an additional level of legal complication. The STI managers were interested in exploring deep-sea canyons in Danish territorial waters, since they might serve as breeding grounds for fauna found on the Swedish side of the border. They had to apply to the Swedish Foreign Office, who eventually – after some prodding – contacted their Danish counterparts, who processed the application and, in due course, approved it (Berggren 2008). The process was cumbersome and time-consuming. Even when diplomatic relations were not a problem, funding and research interests had to be co-ordinated across national boundaries. Norwegian waters were of particular interest to the Swedish marine inventory, and the Norwegians were already conducting sea floor mapping, but they focused on different elements of the marine fauna (Berggren 2008).

Thus, a national project such as the STI necessarily involved investigation of neighboring territories. This aspect of species inventories is even more apparent in the related process of compiling so-called "Red Lists" of endangered species.

Red Lists
The first Red Lists – lists detailing the conservation status of endangered species – were published by the Species Survival Commission under IUCN, the International Union for the Conservation of Nature, in the 1960s. Over time, a set of categories and criteria for evaluating the risk of extinction were developed, and adopted in 1994 (IUCN 2008). The IUCN membership is made up of governments and NGOs, and although the Red Lists aim to evaluate the risk of species extinction on a global level, a lot of the data is collected on a national level, and used in national conservation policies. The Species Information Center is responsible for compiling and publishing the Swedish Red Lists. The relation between the global status of a species and its local conditions is not necessarily straightforward. As Ulf Gärdenfors, head of both the Swedish Red Lists and the STI at the Species Information Center, points out: "The dilemma is that much conservation policy is bound within geopolitical borders, including laws, resources available for conservation measures and personal levels of commitment" (Gärdenfors 2001: 511). These circumstances might be possible to reconcile if "entire, totally isolated populations" of endangered species kept themselves within national borders (Gärdenfors 2001: 511). Questions such as these have been crucial in, for instance, the heated debate about the African elephant in the 1980s and 1990s, where representatives from different nations and different organizations disagreed about the status of elephant populations under CITES, the Convention on International Trade of Endangered Species of Fauna and Flora (Thompson 2004). Within the IUCN, a Regional Application Working Group was established in 1996 to try to resolve similar issues, with Gärdenfors as its chairman (IUCN 2010). The problem is particularly difficult in countries such as Sweden, with comparatively small numbers of endemic species, i.e. species that are naturally found in that area and nowhere else. This means that most species found in Sweden occur in other places as well: not least in the neighboring Nordic countries.

"Norden" and "Nordism" in Biodiversity Conservation
"Norden," or the Nordic countries, is a geographical area which includes Sweden, Norway, Denmark, and Finland (and sometimes Iceland, the Faroe Islands, and Greenland as well). But it is also a political, historical, and cultural construct, manifested in a variable set of economic and political institutions, linguistic conventions, traditions, and rhetoric (Sørensen & Stråth 1997).

Almost from the very beginning of the STI, the possibility of involving the other Nordic countries was a priority.[2] There were a number of practical reasons for this. As outlined above, biogeographical boundaries in "Norden" did not coincide with political borders, and in order to assess the status and occurrence of species in Sweden, Nordic collaboration was deemed essential. This was particularly true of the costly marine inventory, and given the expense of going to sea, the STI administrators wanted their Nordic colleagues to either join the project, or launch national inventory projects of their own to cover their share of the marine environment.

There were also historical precedents for treating "Norden" as a biogeographical unit. Floras and distribution maps published during the nineteenth and twentieth centuries often aimed to cover the entire region; one of the most monumental examples being Eric Hultén's atlas of plant distribution from 1950. Botanists, in particular, tended to treat Sweden and Norway as a biogeographical unit (Hartman 1820, Hultén 1950, Hylander 1953-1966, Lid 1963). Biogeographical classifications vary considerably – some lump most of the northern hemisphere together in the "Boreal kingdom," others distinguish between Arctic and Euro-Siberian regions, whereas the European Union has its own system in which "Norden" is a composite of the Alpine, Atlantic, Boreal, and Continental regions. In the 1970s, a classification of Scandinavian biogeographical regions was commissioned by the Nordic Council of Ministers, to facilitate conservation efforts across national boundaries (Nordiska ministerrådet 1977). The biogeographical basis of conservation policies in the Nordic countries was itself worked out in the context of centuries of Nordic co-operation and conflict.

Claire Waterton has studied how the construction of European biotope classifications reflects political cultures in the EU (Waterton 2002, Waterton & Wynne 2004). Similarly, Nordic conventions in biogeographical mapping reflect a history of shifting political alliances over the last 200 years. The biogeographical maps of 1977 were modified in 1984 and are still used in Nordic conservation (Nordiska ministerrådet 1984, Aagaard 2003).

Add to this the very real legal and security problems involved in crossing national boundaries; and the incentive for persuading the Nordic neighbors into forming a joint "Nordic Taxonomy Initiative" was very strong. With a network of Nordic taxonomy projects, research funding from the Nordic Council might

---

2   Svenska artprojektet 2001, 2002, 2003: Ledningsgruppsprotokoll. Minutes of the STI executive group: August 24, 2001; September 18, 2001; March 21, 2002; and June 24, 2003. Archived at the Swedish Species Information Centre, Rikard Sundin.

become available. The countries could form a common research council to handle applications and reduce the bias problems; workshops and joint doctoral programs could bolster the small national taxonomic communities, and reduce planning and expedition costs (Gärdenfors 2008).

There were partial parallels to the STI in other Nordic countries. The Finnish "PUTTE" project, a large-scale inventory of forest species conducted 2003-2007 (Finnish Environment Institute 2010). PUTTE was primarily a conservation project, related to specialized inventories run by the environmental administration, but representatives of the STI did their best to encourage the Finns to extend it into a full-blown taxonomy initiative. The program was extended to 2016 but remained focused on forest species (Finnish Environment Institute 2009). Norway, however, did launch its own Taxonomy Initiative, modeled on the Swedish case, and with similar budget and aims (Miljøværndepartementet 2008). An important factor in the decision was an appeal for a Norwegian Taxonomy Initiative, presented by the Norwegian Association of Natural Scientists and signed by a number of organizations ranging from environmental NGOs to fisheries and energy companies, and delivered to the Norwegian government in February 2008 (Naturviterne 2008). One of the arguments was that Norway needed to catch up with Sweden.

The appeal to Nordic co-operation is tinged with a definite element of competition. Historians Bo Stråth and Øystein Sørensen have pointed out that "Nordism" is, above all, a reinforcing element of *national* identities (Sørensen & Stråth 1997: 22, emphasis added). In the post-war era, "Nordism" has served as a potential alternative to other supra-national conglomerations – NATO, the Warsaw Pact, and the EU. The fact that attempts at realizing this alternative in terms of economic, defense, or customs unions have been half-hearted has not diminished the appeal of a Nordic community on a national and rhetorical level. Although a Nordic Taxonomy Initiative would facilitate the negotiation of national boundaries and funding, it also serves as a reaffirmation of the significance of Sweden.

> National inventories also have the potential of spreading to neighboring countries. Indeed, other Nordic countries are now considering taxonomy initiatives of their own. The completion of the first national biodiversity map will undoubtedly be a significant event. Will the Swedes, inspired by their Linnaean tradition, be first? (Ronquist & Gärdenfors 2003: 270).

Linnaeus is a useful symbol for a project such as the STI. It builds on the national pride invested in Linnaeus, making it attractive to funding agencies and the general public. Linnaeus also serves as a justification for limiting the project to Sweden, with the precedent of *Flora Suecica* and centuries of emphasis on the idea that the Swedes are a particularly nature-loving people (Beckman, forthcoming). At the same time it alludes to the hegemonic ambitions of a man who wanted to name everything in the world, reflecting some of the glory of an imagined "catalogue of everything" onto a project which investigates a tiny corner of the world – and not a particularly diverse one at that.

In the context of national identity versus international community, Nordism and Linnaeanism are ways of having your cake and eating it, too. This approach is also characteristic of the way the STI navigates the troubled waters of research and environmental policy. In addition to national pride and supposed Linnaean traditions of popular engagement with nature, conservation arguments were central to the political success of the STI. However, the proximity to environmental policy has not proved entirely beneficial to taxonomy as a fundamental science. Particularly in the eyes of research councils and university-based academic institutions – as opposed to museums and environmental agencies – taxonomy continues to be problematic, in spite of the influx of funds into the field. However, practical and theoretical experiences from policy-oriented biodiversity work, such as Red Lists, are crucial to the inventory work. Even though the STI inventory is administratively and financially separate from the project of compiling Red Lists of endangered species, the overlap in personnel and planning is substantial.

The STI is simultaneously negotiating the boundaries of taxonomy and the boundaries of Sweden and Scandinavia in a framework of international conventions and global initiatives. In this process, the STI outlines the boundaries of taxonomy while at the same time reaffirming national and regional boundaries – or, in practice, keeping those boundaries fluid. The STI sits on the fence geographically as well as academically, and in the end, neither "Sweden" nor "Taxonomy" is a fixed category.

Acknowledgments

I would like to thank the participants in the workshop "Ways of knowing the field" and in the colloquium at the Department of the History of Science and Technology at the Royal Institute of Technology for helpful comments on earlier versions of this paper. I would also like to thank the Max-Planck-Institut für Wissenschaftsgeschichte and Längmanska kulturfonden for generously supporting my work as a guest researcher at the MPI.

References

Aagaard, K. 2005. "Regional Red Data Lists and invertebrates in the Nordic countries." In: D. Procter & P. T. Harding (eds.) *Proceedings of INCardiff 2003. Red lists for invertebrates: their application at different spatial scales – practical issues, pragmatic approaches.* JNCC Report 367. Peterborough: JNCC, 7-15.

Aftonbladet 2006. "Insektsfest! 500 nya arter i Sverige ger forskare glädjefnatt." *Aftonbladet*, October 11.

Alström, P. 2008. "Interview with Per Alström, head of taxonomy in the Swedish Taxonomy Initiative." Interview conducted by the author on June 30.

ArtDatabanken n.d. "Arbetsuppgifter." Retrieved July 10, 2011 from http://www.artdata.slu.se/arbetsuppgifter.asp

Beckman, J. 2007. "Le projet Linné: Mobiliser et organizer les botanistes amateurs en Suède, 1972-1986." In: F. Charvolin, A. Micoud & L. K. Nyhart (eds.) *Des sciences citoyennes? La question de l'amateur dans les sciences naturalistes.* La Tour d'Aigues: Editions de l'aube, 167-182.

Beckman, J. 2009. "Landskapsfloror: Att inventera över gränser." In: S. Widmalm (ed.) *Vetenskapens sociala strukturer: Sju historiska om konflikt, samverkan och makt.* Lund: Nordic Academic Press, 216-257.

Beckman, J. Forthcoming. "Linneanska traditioner? Hjältedyrkan och praktik i svensk skolbotanik." In: M. Rodell (ed.) *Konstruktionen av den svenska snilleindustrin.*

Berggren, M. 2008. "Interview with Matz Berggren, expedition coordinator." Interview conducted by the author on July 25.

BIOFAR 2006. "A description of a large scale internordic benthic macrofauna/flora project." Updated June 7. Retrieved July 10, 2011 from www.biofar.fo

BIOICE n.d. "Benthic invertebrates of Icelandic waters." Retrieved July 10, 2011 from http://www3.hi.is/pub/smc/bioice.htm

Bowker, G. C. 2000. "Biodiversity datadiversity." *Social studies of science* 30: 643-683.

CBD 2010. "2007: What is the problem? The taxonomic impediment." Updated April 16. Retrieved July, 2011 from http://www.cbd.int/gti/problem.shtml

Centrum för biologisk mångfald n.d. "Centrum för biologisk mångfald." Retrieved July 10, 2011 from http://www.cbm.slu.se/omcbm.htm

Finnish Environment Institute 2009. "Research programme of deficiently known and threatened forest species 2009-2016 PUTTE." Retrieved July 10, 2011 from http://www.environment.fi/default.asp?contentid=228996&lan=EN

Finnish Environment Institute 2010. "Research programme of deficiently known and threatened forest species 2003-2007 PUTTE." Retrieved July 10, 2011 from http://www.environment.fi/default.asp?node=24993&lan=en

Forskningsprioriteringar 2009. "Svenska artprojektet." Retrieved July 10, 2011 from http://www.artdata.slu.se/svenskaartprojektet/filer/ForskningsPrio090511.pdf

Gärdenfors, U. 2001. "Classifying threatened species at national versus global levels." *Trends in Ecology and Evolution*, 16: 511-516.

Gärdenfors, U. 2008. "Interview with Ulf Gärdenfors, research director of the Swedish Taxonomy Initiative." Interview conducted by the author on May 7.

Gärdenfors, U. 2008. "The Swedish Taxonomy Initiative and the PUTTE project: Some viewpoints and thoughts about future cooperation." Presentation given January 15-16, 2008 at Finlands miljöcentral. Retrieved July 10, 2011 from http://www.ymparisto.fi/download.asp?contentid=81012&lan=fi

Gibbons, M., C. Limoges, H. Nowotny, S. Schwartzman, P. Scott & M. Trow 1994. *The New Production of Knowledge: The Dynamics of Science in Contemporary Societies*. London: Sage.

Glemhorn, K. & D. Karlsson 2008. "Interview with Kajsa Glemhorn and Dave Karlsson, administrators of the Hymenoptera/Diptera inventory." Interview conducted by the author on June 10.

Godfray, C. & S. Knapp 2004. "Introduction." *Philosophical Transactions of the Royal Society B*, 359: 559-569.

Göteborgs Naturhistoriska Museum Årstryck 2004: 37-48, 2005: 29-33, 2006: 29-37, 2007: 31-45, 2008: 31-50.

Goude, G. 2009. "Interview with Gunnar Goude, Member of Parliament for the Swedish Green Party 1994-2002." Interview conducted by the author on May 9.

Grandin, K., N. Wormbs & S. Widmalm (eds.) 2004. *The Science-industry Nexus: History, Policy, Implications*. Nobel Symposium 123. Sagamore Beach, MA: Science History Publications.

Hartman, C. J. 1820. *Handbok i Skandinaviens flora, innefattande Sveriges och Norriges Vexter, till och med Mossorna. Med Inledning, afhandlande grunderna för Botanikens studium, samt tvänne planscher*. Stockholm: Zacharias Haeggström.

Hine, C. 2008. *Systematics as Cyberscience: Computers, Change and Continuity in Science*. Cambridge, MA: MIT Press.

Hull, D. 1988. *Science as a process: An Evolutionary Account of the Social and Conceptual Development of Science*. Chicago: University of Chicago Press.

Hultén, E. 1950. *Atlas över växternas utbredning i Norden: Fanerogamer och ormbunksväxter – Atlas of the distribution of vascular plants in NW Europe*. Stockholm: Generalstabens litografiska anstalt.

Hylander, N. 1953-66. *Nordisk kärlväxtflora: omfattande Sveriges, Norges, Danmarks, Östfennoskandias, Islands och Färöarnas kärlkryptogamer och fanerogamer.* Stockholm: Almqvist & Wiksell.

InBio n.d. "What is InBio? Historical background." Retrieved July 10, 2011 from http://www.inbio.ac.cr/en/inbio/inb_antec.htm

IUCN 2008. "Guidelines for application of red list criteria at regional levels." Retrieved July 10, 2011 from http://www.iucn.org/about/work/programmes/species/red_list/resources/technical_documents/guidelines_application/

IUCN 2010. "About the IUCN Red List." Retrieved July 10, 2011 from http://www.iucn.org/about/work/programmes/species/red_list/about_the_red_list/

Karlsson, A. 2009. "Interview with Anna Karlsson, director of the marine inventory project." Interview conducted by the author on April 14.

Karlsson, D., T. Pape, K. A. Johanson, J. Liljeblad & F. Ronquist 2005. "Svenska Malaisefälleprojektet, eller hur många arter steklar, flugor och myggor finns i Sverige?" *Entomologisk tidskrift*, 126: 43-53.

Lid, J. 1963. *Norsk og svensk flora.* Oslo: Det norske samlaget.

LifeWatch n.d. "E-science and technology infrastructure for biodiversity data and observatories." Retrieved July 10, 2011 from www.lifewatch.eu

Malaise, R. 1937. "A new insect-trap." *Entomologisk Tidskrift*, 58: 148-160.

Miljøværndepartementet 2008. "Tidenes budsjettløft for naturmangfoldet i Norge." Press release, October 7. Retrieved July 10, 2011 from http://www.regjeringen.no/nb/dep/md/pressesenter/pressemeldinger/2008/tidenes-budsjettloft-for-naturmangfoldet.html?id=531051

Miller, G. 2005. "Linnaeus's legacy carries on – Taxonomy's elusive grail." *Science*, 307: 1038-1039.

Natural History Museum 2010. "Taxonomy and systematics." Retrieved July 10, 2011 from http://www.nhm.ac.uk/nature-online/science-of-natural-history/taxonomy-systematics/index.html

Naturviterne 2008. "Artsprosjekt i Norge!" Updated October 15. Retrieved July 10, 2011 from http://www.naturviterne.no/?module=Articles;action=Article.publicShow;ID=1843

Naturvårdsverket 2007. *Inventering av marina naturtyper på utsjöbankar.* Rapport 5576. Stockholm: Naturvårdsverket.

Nilsson, L. M. 2008. "Interview with Lars M. Nilsson, director of biodiversity at the Swedish Research Council." Interview conducted by the author on September 26.

Nordiska ministerrådet 1977. *Naturgeografisk regionindelning av Norden.* Arbetsgruppen för naturvårdsfrågor. NU B 1977:34. Stockholm: Gotab.

Nordiska ministerrådet 1984. *Naturgeografisk regionindelning av Norden.* Oslo: Nordiska ministerrådet.

Regeringen 2002. "Skrivelse 2001/02:173 En samlad naturvårdspolitik." Retrieved July 10, 2011 from http://www.riksdagen.se/webbnav/?nid=37&dokid=GP03173

Riksdagen 2008. *Rapport från riksdagen 2008/2009:RFR3 Uppföljning av statens insatser på havsmiljöområdet.* Retrieved July 10, 2011 from http://www.riksdagen.se/webbnav/?nid=3777&doktyp=rfr&rm=2008/09&bet=RFR3&dok_id=GW0WRFR3

Ronquist, F & U. Gärdenfors 2003. "Taxonomy and biodiversity inventories: Time to deliver." *TRENDS in Ecology and Evolution*, 18: 269-270.

Ronquist, F. 2008. "Interview with Fredrik Ronquist, director of the Hymenoptera/Diptera inventory." Interview conducted by the author on April 18.

Samper, C. 2004. "Taxonomy and environmental policy." *Philosophical Transactions of the Royal Society of London B*, 359: 721-728.

SamHav 2008. *Rapport för år 2007 från SamHav, Samordningsgruppen för Havsmiljöfrågor.* Retrieved July 10, 2011 from http://www.naturvardsverket.se/upload/04_arbete_med_naturvard/Havsmilj%C3%B6/N_samhav_rapport_Rk_%2007.pdf

Secretariat of the Convention of Biological Diversity 2000. "Sustaining life on earth: How the Convention of Biological Diversity promotes nature and human well-being." Retrieved July 10, 2011 from http://www.cbd.int/convention/guide.shtml?id=nataction

Smocovitis, V. B. 1996. *Unifying Biology: The Evolutionary Synthesis and Evolutionary Biology.* Princeton: Princeton University Press.

Sørensen, Ø. & B. Stråth 1997. "Introduction: The cultural construction of Norden." In: Ø. Sørensen & B. Stråth (eds.) *The Cultural Construction of Norden.* Oslo: Scandinavian University Press, 1-24.

Station Linné & Porten till Alvaret 2010. "Välkommen till Station Linné!" Retrieved July 10, 2011 from http://www.stationlinne.se/

Stockholm Biodiversity Informatics Symposium 2008. "Så fångas jordens arter på nätet: en introduktion till biodiversitetsinformatik." Retrieved July 10, 2011 from http://artedi.nrm.se/sbis2008/

Sundin, B. 2005. "Nature as cultural heritage." *International Journal of Heritage Studies*, 11: 9-20.

Sundin, R. 2008. "Interview with Rikard Sundin, administrator of the Swedish Taxonomy Initiative." Interview conducted by the author on March 28.

Svenska artprojektet n.d. "Välkommen til Svenska artprojektet!" Retrieved July 10, 2011 from www.artdata.slu.se/svenskaartprojektet.asp

Svenska Malaisefälleprojektet 2005. "Svenska Malaisefälleprojektet Publikationer." Retrieved July 10, 2011 from http://www2.nrm.se/en/malaiseproject/publications.html

Taxonomy for the twenty-first century 2004. Theme issue of *Philosophical Transactions of the Royal Society B*, 359: 559-739.

Takasc, D. 1996. *The Idea of Biodiversity: Philosophies of Paradise*. Baltimore: The Johns Hopkins University Press.

Thompson, C. 2004. "Co-producing CITES and the African elephant." In: S. Jasanoff (ed.) *States of Knowledge: The Co-production of Science and Social Order*. Abingdon: Routledge, 67-86.

Tunlid, A. 2009. "Den nya biologin: Forskning och politik i tidigt 1960-tal." In: S. Widmalm (ed.) *Vetenskapens sociala strukturer: Sju historiska om konflikt, samverkan och makt*. Lund: Nordic Academic Press, 99-136.

Upsala Nya Tidning 2005. June 26 & 29, July 19, August 18 & December 4

Upsala Nya Tidning 2008. June 22, 23 & 24.

Utredningen kunskap för biologisk mångfald 2005. *Kunskap för biologisk mångfald: Inventera mera eller återvinn kunskapen?* SOU 2005:94. Stockholm: Fritzes.

Vernon, K. 1993. "Desperately Seeking Status: Evolutionary Systematics and the Taxonomists' Search for Respectability 1940-60." *The British Journal for the History of Science*, 26: 207- 227.

Vernon, K. 2001. "A Truly Taxonomic Revolution? Numerical Taxonomy 1957–1970." *Studies in the History and Philosophy of Biology and the Biomedical Sciences*, 32: 315–341.

Waterton, C. & B. Wynne 2004. "Knowledge and political order in the European Environmental Agency." In: S. Jasanoff (ed.) *States of Knowledge: The Co-production of Science and Social order*. Abingdon: Routledge, 87-108.

Waterton, C. 2002. "From field to fantasy: Classifying nature, constructing Europe." *Social studies of science*, 32: 177-204;

Wheeler, Q. 2004. "Taxonomic triage and the poverty of phylogeny." *Philosophical Transactions of the Royal Society B*, 359: 571-583.

Wilson, E. O. 1989. "The coming pluralization of biology and the stewardship of systematics." *Bioscience*, 394: 242-245.

Wilson, E. O. 1994. *Naturalist*. Washington, D.C.: Island Press.

Wilson, E. O. 2000. "A global biodiversity map." Editorial. *Science*, 289: 2279.

Mikkel Bunkenborg & Morten Axel Pedersen

The Ethnographic Expedition 2.0
Resurrecting the Expedition as a Social Scientific Research Method

Cruising through the Gobi desert on a pleasant and sunny day in the autumn of 2009, the rear axle of our four wheel drive van suddenly snapped and brought us to a grinding halt. We had undertaken a very long drive through southern and eastern Mongolia as an exploratory part of a larger anthropological project aimed at comparing Chinese infrastructure projects and resource extraction investments in Mongolia and Mozambique, and we were some 150 km from the nearest town when the break-down occurred. Bürnebat, our driver, guide, and research assistant, looked unusually glum as he examined the fractured rear axle, and it dawned upon us that we might have to stay in that particular patch of desert for a while. Despite the fact that we were literally stuck in the middle of nowhere (even the nomads considered the place remote), being stationary felt like an oddly reassuring return to a more conventional form of fieldwork, and our involuntary stop in the desert revealed the novelty, but also the discomfort, of the motile manner in which we were conducting our research. We realized that our seemingly endless journeying through sparsely inhabited grasslands and inhospitable deserts in search of Chinese mines and oil fields had not really conformed to either of the two dominant models of ethnographic fieldwork – the classical community study attributed to Bronislaw Malinowski or the multi-sited ethnography associated with George Marcus. So, as we sat there in mock-colonial style, sipping coffee in front of our tents and embarrassingly incapable of assisting Bürnebat with the daunting mechanical tasks ahead of him, we started wondering why anthropological methods today should be reduced to one of these two forms, and why "the expedition" has been so thoroughly forgotten, if not downright rejected, as a method for generating ethnographic data.

Individual fieldwork and participant observation in single or multiple sites have apparently become so integral to anthropology's auto-narrative that no-one dares tamper with this cornerstone lest the whole disciplinary edifice come tumbling down. Yet before the canonization of the Robin Crusoe ethnographer doing individual fieldwork, and the emergence of his postmodern successor in the guise of a seasoned cosmopolitan traveller (Clifford 1997, Hannerz 2003), anthropology was widely associated with expeditions. Dismissed for failing to provide anything but superficial impressions and later tainted by their association with an aggressive colonialism, expeditions have long been relegated to the margins of anthropology and the social sciences more generally, but the question is whether the expedition – understood as a mobile, collaborative research effort aimed at exploring certain features pertaining to some empirical field – might have something to offer contemporary anthropology, and whether certain kinds of ethnographic objects might actually *call for* the kinds of collaborative and mobile research practices that characterized the classic ethnographic expeditions of the past. As we wish to demonstrate in this chapter, our collaborative research project on Chinese interventions in Mozambique and Mongolia is a case in point.

The Torres Strait and the Strange Death of the Expedition

Before providing some more details on our ongoing research in Asia and Africa, and explaining how it calls for a resurrection of the expedition in an updated, second version, we need to remind ourselves that expeditions were actually quite crucial to the production of ethnographic knowledge until they gradually disappeared in the beginning of the twentieth century. When anthropology emerged as an academic discipline towards the end of the nineteenth century, it largely consisted of what Frazer himself referred to as imaginative speculation about empirical information gathered by others (Willerslev 2010), but Frazerian anthropology was quickly transformed by the idea that social scientists, like many biologists had done since Darwin, might profitably travel to study in the natural surroundings proper to their research objects. The exact nature of this sort of fieldwork was not immediately clear, and for a while it looked as if expeditions might become the accepted means of generating ethnography. In this respect, it seems worthwhile to revisit an earlier "experimental moment in the human sciences": the 1898 Cambridge Anthropological Expedition to Torres Straits.

The leader of the expedition, A.C. Haddon, was originally trained as a natural scientist but he became interested in anthropology when a field trip to study marine biology brought him to the Torres Strait in 1888. On his return to Britain, he was contracted to write an ethnographic book on the Torres Strait but the gaps in his material made him decide that another trip was necessary and thus the ambitious project of an expedition to carry out a comprehensive anthropological study of the islands between Australia and New Guinea was conceived. A multidisciplinary team of seven people including the psychologist W. H. R. Rivers, the specialist in tropical medicine Charles Seligman, and the self-taught linguist Sidney Ray, was assembled and spent seven months in the field. Crucially for our present purposes, the Torres Straits Expedition was from the beginning designed as a collaborative and multidisciplinary effort aimed at anthropological description in the broadest sense, and the team members took charge of specialized research tasks to that end.[1]

Upon its return, the Torres Straits expedition was deemed to be a great success; a large collection of artifacts was brought home along with photographs and even sound recordings on wax cylinders, and the findings of the expedition were published in six volumes over the following decades, contributing crucially to the institutionalization of anthropology in Cambridge. On the strength of the expedition, Haddon was offered a small lectureship and a fellowship that allowed him to pursue his idea of anthropology as a field science and pass it on to a new generation of students, including Radcliffe-Brown and Malinowski (Rouse 1998: 72). At the time of the expedition, the members distinguished between intensive research and survey work, and Haddon explicitly advocated and supported the use of both these methods. Members of the expedition accordingly collected data from separate language groups, and while they spent most of the time on the island of Mer, they also managed to visit many of the other islands in the Torres Strait as well as British New Guinea and Borneo. Curiously, the fact that this fieldwork was conceived as an expedition, a motile and collaborative project rather than an individual sojourn, is strangely absent in the recommendations the participants later passed on to their students. Both Haddon and Seligman, who actually continued to participate in expeditions himself, encouraged Malinowski to pursue the intensive form of research focused on a particular problem in a single location (Urry 1998: 230). And that is what Malinowski did, taking the intensive study of a limited area to a new pitch of perfection and incidentally marginalizing the expedition in the process.

Prior to the institutionalization of anthropology, expeditions were crucial means of procuring ethnographic information and one could make an endless list of journeys that contributed to "Anthropology before anthropology": Columbus' discovery of the West Indies; the voyages of Bougainville, Cook, and Krusenstern who were dispatched by the aspiring empires of France, Britain, and Russia; the Lewis and Clark expedition to the American Northwest; the travels of Alexander

---

[1] In addition to general ethnology and decorative art, Haddon was responsible for physical anthropology, but he later came to doubt the value of racial characteristics and published very little on his actual findings. Rivers experimented on vision and devised a method for recording genealogies. Seligman studied native medicines and pathologies. Sidney Ray took charge of linguistics and even had sufficient foresight to treat the local Pidgin English as a language rather than a degraded form of English. Charles Myers, who was a musician as well as a doctor, worked on hearing and music, while another physician, William McDougall, did research on tactile sensation. Anthony Wilkin, the youngest man on the team, served as photographer (Herle & Rouse 1998).

von Humboldt; and so on (Liebersohn 2003, 2008). As the project in the Torres Strait shows, the expedition was not just a distant legacy, but a research strategy that contributed directly to the establishment of anthropology as a proper academic discipline, only to be abandoned by a younger generation of professionals who may have been eager to distance their cutting edge research methods and monographs from the old-school expeditions and the amateur writings of travellers and explorers that went before. No doubt, part of the reason why expeditions became completely marginalized (if not scandalized) as anthropological methods is the intimate relation between expeditions and colonial domination, an intimacy which is also quite evident in the Torres Strait project.[2] However, it actually seems that expeditions were a lost cause long before anthropology, and expeditionary anthropology in particular was criticized as the handmaiden of colonialism (Asad, Said). One might point to a decreasing interest in material culture as yet another factor, but the object here is not to explain the strange disappearance of the expedition, but rather to present some of the characteristic features of this dead mode of research and to suggest how they might be profitably revived. Etymologically, the word "expedition" is about feet freed of fetters and according to the OED the word has the various meanings of sending forth with martial intentions, a warlike enterprise; a journey, voyage, or excursion with some definite purpose; a body of persons engaged in such a project; and the quality of being expedite. The Torres Strait Expedition had all these characteristics, but for our purposes, we would emphasize the fact that "the ethnographic expedition" denotes a body of research subjects – and research objects – in motion, which stand at right angles, so to speak, from the lonely island ethnographer associated with Malinowski and from the self-assured cosmopolitan anthropologists associated with multi-sited fieldwork (see e.g. Hannerz 2003).

Chinese interventions in Asia and Africa: A Multi-sited Ethnography?

Recent years have seen a massive academic interest in China's new political-economic clout in Africa (Taylor 2006, Alden et al. 2008, Brautigam 2009). However, a similar but largely neglected expansion to satisfy China's thirst for resources is

---

2 The published reports make no secret of the fact that the expedition relied on colonial authorities and missionaries, and it may therefore be referred to as warlike or conquering in its attitude in the sense that it was largely supportive of British imperial domination, even if its members took pains to disown a rumour that Queen Victoria would dispatch a gunboat if the islanders failed to provide truthful answers to their queries (Beckett 1998: 41). Still, the effects of the expedition went beyond this imperial basis. Haddon's interest in pre-colonial culture sparked renewed interest in discontinued ceremonies much to the chagrin of the missionaries who had sought to eradicate them, and the genealogies produced by Rivers were later employed by islanders in land claim disputes with the Australian Government (Beckett 1998: 46).

being played out in Central and Inner Asia (Kleveman 2003, Swanström 2005) and China's political-economic intervention beyond its borders now appear to be truly global. One of the characteristic features of this neo-colonial intervention (if it makes sense to call it that) is a tendency to deliver packaged material solutions. Reminiscent of the modernization theory that inspired Western development assistance in the decades after the second world war, Chinese projects seem to focus on the construction of infrastructure such as roads, power plants, dams, factories, and government buildings, emphasizing the engineering of "material civilization" as a necessary basis for civilized, modern existence. Designed and built by the Chinese with little consideration of local needs and participation, such projects have emerged as "zones of awkward engagement" (Tsing 2005) in which Chinese and (for instance) Mongolian worlds meet, mix, and sometimes clash in the co-construction of complex infrastructural entities, which might be seen as harbingers of an emerging empire.

The journey that left us stranded in the Gobi desert was part of an ongoing anthropological research project entitled *Imperial Potentialities*, in which the two of us along with Morten Nielsen seek to shed light on China's growing involvement in development projects, resource extraction, and trade in Central Asia and Sub-Saharan Africa via a sustained integration of three different ethnographic fieldworks focused on Chinese interventions in infrastructure and resource extraction in Mongolia and Mozambique.[3]

Attempting to study very big things such as mines, oilfields, and highways obviously poses a considerable methodological challenge. Responding to this challenge, our project design involves the implementation of what we call the dual perspective approach: three tightly integrated sub-projects which explore Chinese infrastructure projects from a local and a Chinese perspective, thus defining both internal and external comparative axes. More specifically, the project consists of (1) a study of Chinese interventions in Mozambique from the perspective of African workers and state cadres; (2) a comparable study of similar interventions with respect to the same kinds of agents in Mongolia; and (3) a study carried out in and around the same project sites as the two others but from the perspective of Chinese workers, managers, and government officials. Ideally, this allows us to compare the infrastructural projects at hand as two concrete cases of global friction via their common attainment to a Chinese model of economic intervention in Third World contexts.

---

3   The project is funded by the Danish Council for Independent Research in the Social Sciences (FSE); it commenced in 2009 and will be concluded towards the end of 2012.

Before we had started our exploration of Chinese investments in Mongolia, we had spent some days in the capital of Ulaanbaatar in front of a map lining up a list of potential research objects along a 3,000 km-long route that would take us through some of those regions of southern and eastern Mongolia that were rumored "to be full of Chinese mines" (no-one, including the government officials Pedersen spoke to, seemed to have a full overview of, let alone figures on, the scale of the Chinese intervention). Our planned visits included a Chinese fluorspar mine, a Mongolian-Chinese joint-venture zinc mine and processing plant, a road construction project financed partly by the Chinese government, and an oilfield operated by a subsidiary of PetroChina, a leading star among state-owned Chinese firms and at that point the biggest company in the world. In addition, we came across a number of other projects: a team of Chinese surveyors preparing for the construction of a gigantic power plant also to be sponsored by the government in Beijing, a privately owned Chinese facility for processing iron ore, a sports stadium built by Shanghai workers, a Chinese entrepreneur in agribusiness negotiating access to land, and so on.

Our question now is: did this amount to a conventional multi-sited fieldwork project or was it actually something different? According to George Marcus, who has justifiably emerged as the godfather of multi-sited ethnography within anthropology, this method involves:

> movi[ng] out from single sites and local situations of conventional ethnographic research designs to examine the circulation of cultural meanings, objects, and identities in diffuse time-space. This mode defines for itself an object of study that cannot be accounted for ethnographically by remaining focused on a single site of intensive investigation ... This mobile ethnography takes unexpected trajectories in tracing a cultural formation across and within multiple sites of activity that destabilize the distinction[s] ... by which much anthropology has been conceived (Marcus 1995: 96).

On the face of it, this sounds a lot like what we are trying to do. Still, while our dual-perspective approach bears a certain resemblance to multi-sited fieldwork, it is also, we believe, qualitatively different in at least two senses, which we will now consider in turn.

Arbitrary Locations and Accidental Stoppages

It is evident that our dual perspective design imposes quite different, and in many ways stricter criteria in the demarcation of our fieldwork venues than the multi-sited approach does with its liberal credo of "following the people." Crucially, however, we are not imposing these limitations in the naive belief that we (or anyone else) would actually be able to make a "conventional controlled comparison," which "operates on a linear spatial plane, whether the context is a region, a broader culture area, or the world system" while addressing "homogeneously conceived conceptual units (e.g. peoples, communities, locales)" pertaining to "distinctly bounded periods or separate projects of fieldwork" (Marcus 1995: 102). So, while there is an air of scientific rigor to our dual perspective approach, its promises of fixed axes of comparison and complementary data sets are, as we have always been perfectly aware, fundamentally unrealizable, not only when it comes to rendering data obtained in Portuguese, Mongolian, and Chinese commensurable, but also in terms of identifying cases that could yield the symmetrical data-sets necessary for a "real" controlled comparison.

But if we knew from the start that our impossibly rigid research design would never translate into correspondingly rigid forms of data, why even bother trying to implement it in the first place – why not simply leave for Mongolia and Mozambique and start to follow people, things, metaphors, stories, biographies, and conflicts as Marcus suggests? Here we find inspiration in Matei Candea's (2007) defense of the bounded field site. Via an analogy with two contrasting responses to the virtually boundless technical possibilities available to contemporary film makers, the super hi-tech films directed by Peter Jackson and the ultra low-tech ones directed by Lars Von Trier,[4] Candea concurs that multi-sited ethnography, celebrating the emancipation of the anthropologist from the single and bounded field site, does indeed offer a possible response to contemporary visions of social reality as an unbounded and seamless fabric of relations. But, he suggests, one might more fruitfully respond to this seamless reality with an obverse, Trier-like response – a *dogme* ethnography involving the deliberate imposition of limits to set free creative anthropological experimentation. Importantly, the demarcation of such "arbitrary locations" as Candea provocatively refers to these more or less accidentally defined field sites, does not represent a naïve return to the fetish of the naturally bounded community;

---

4   For Peter Jackson, advances in computer animation made it possible to create *Lord of the Rings: The Fellowship of the Ring*, a film which presented an entirely fictional universe with the detailed accuracy of a realist historical reconstruction. In the same year, however, the same technological possibilities inspired the two Danish directors Lars Von Trier and Thomas Vinterberg to take a "vow of chastity" and set up a manifesto known as Dogme 95 in which they abjured the use of visual effects, artificial lighting, background, music, and make up, and they promised to use only handheld cameras in their attempt to wring the truth from their characters and settings. While both approaches depended on sophisticated technology and shared a concern with realism, Peter Jackson's ready embrace of technological emancipation from all constraints contrasts quite sharply with the self-conscious asceticism of the dogme instructors.

on the contrary, he argues, it constitutes a productive imposition of limits, which is conducive, if not necessary, to the generation of good anthropological knowledge.

> In the above, I have chosen to illustrate the use of self-imposed limitations through a spatial example, revisiting the hackneyed image of the 'village ethnography'. But the wider point about the necessity, both epistemological and practical, of recognizing the value of limitation amidst the calls for freedom could find many other, including non-spatial expressions (Candea 2007: 180).

This is precisely how we would like to think of the formal criteria underwriting the identification of research sites and the calibration of data within our dual perspective approach – not as rules which are to be slavishly followed, but as productive limits that are sufficiently challenging to generate forms of ethnographic experimentation or "methods" and leave the objects under investigation a certain leeway in terms of dictating the course of the project. Such self-imposed limits are playful rather than dogmatic, and perhaps this is better captured by Von Trier's idea of obstructions. In a 2003 film, Trier challenges another Danish director, Jørgen Leth, to remake a movie in five different ways according to the whimsical obstructions or trip-ups imposed by Trier and the creative drive of this film stems from Leth's attempt to work his way around these obstructions. For reasons pertaining to the scale of the infrastructural entities we wanted to study, it made little sense to formulate our obstruction as an "arbitrary location" in spatial terms as Candea does in his study of the Corsican "village" of Crucetta, and so we devised an entirely different kind of trip-up by stipulating that the Chinese interventions in Mongolia and Mozambique had to be sufficiently similar to allow for meaningful comparison (working out exactly what that means is obviously part of the project). Under the guise of contributing to an investigation with overly determined comparative goals, the research design with its built-in comparative obstruction actually came to stimulate "pointless journeys" into the ethnographically unknown.

This, after all, was very much what happened to us during our recent trip to Mongolia. A couple of days into our trip to the Gobi, we heard of another Chinese company drilling for oil in the desert, and that is where we were heading when the rear axle broke, leaving us stuck in the sand for several days, and forcing us to explore whatever was there to be found out. There was no-one in sight, but we had stopped for tea in a nomadic household some kilometers earlier, and, on the horizon, one could glimpse the Chinese oil rigs for which we had originally been heading, so we were, quite literally and entirely accidentally, suspended right in the middle of our ethnographic object. The fact that the nomads' yurts and the oil rigs were separated by some ten kilometers of desert seemed quite in tune with the often rather strained relations between the Mongolians and Chinese, and being forced to set up camp and make a field site out of the no man's land seemed quite apposite to our project. Though the oil rigs to the east were his nearest neighbors, the nomad insisted that he had never visited them, for why would one have anything to do with the Chinese? On their part, the Chinese oil workers complained about the inhospitable environment in which they were working. They lived in containers equipped with bunk beds, electric heating, and air-conditioning, and as their supplies were trucked in, most of them had no interaction with the locals – there's no-one here, they'd claim and point out to the undulating steppe that stretched into the distance without any visible sign of human habitation. Yet what seemed like an inhospitable desert to the Chinese was also full of animal droppings and other signs of human activity, and to the Mongolians the place was apparently so crowded that they could afford to have strained relations with their Chinese neighbors.

Having undertaken similar drives through Mongolia on other occasions, Morten Pedersen was not particularly troubled by the mobile character of our research, but Mikkel Bunkenborg, whose primary fieldwork took place in a Chinese township, found the experience of doing fieldwork on the road oddly discomforting and the nagging problem of motility finally came to a head with the accidental stop in the Gobi. The whole situation reverberated with the distinction between the nomadic and the sedentary. We were driving close to the border separating agrarian China from pastoral Mongolia, the broken rear axle temporarily arrested our own movement, and the Chinese workers ensconced in their containers around the heavy oil rigs and the Mongolian nomads following their flocks seemed entirely different and disconnected. We found ourselves caught mid-way between the two entirely disparate worlds of the stationary and the motile in several ways, but while the whole situation at first seemed like a performance of an absolute distinction between the nomadic and the sedentary in different registers, it turned out that things were rather more complicated. The Mongolians did in fact interact with the Chinese and the oil rigs were actually on the move. On a visit to one of the Chinese encampments, a Mongolian youngster working as a security guard asked us to find out what his Chinese employers were up to. "They never tell me anything," he complained, but he had a hunch that they were disassembling the oil rig and this proved to be true, as the team of Sichuanese workers on site were preparing to move the rig some 500 kilometers to the East. It also turned out that the elderly nomad who denied having ever visited the Chinese was not entirely truthful: we caught him red-handed as he drove his little truck back from the oil rig, and he grudgingly admitted that he was actually their regular provider of fresh meat. These two findings, the moving oil rig and the connected nomad, tended to undermine the impression that motility was a feature of a disconnected Mongolian world surrounding the very big and apparently stationary things we were studying and Mikkel Bunkenborg was thus reconciled to the idea that motility, albeit at different speeds, and connections, though fervently denied, were a basic premise for the research project. The methodological implications of this insight became clearer while Bürnebat dismantled the car, and what we realized as we were sipping coffee in front of our tent entrances was that the "Chinese mine safari" (as Bürnebat jokingly referred to it) had turned into something more than a preliminary search for field sites; it was generating highly diverse and interesting ethnography at a rate we had not anticipated and what was intended as a mere precursor to a "proper" fieldwork in the future had somehow turned into a method in its own right, a method that shared certain characteristics with the object we were investigating.

From Multi-sited Fieldwork to Trans-local Expeditions

This brings us to the second sense in which our collaborative research project differs from a conventional multi-sited ethnography, namely the fact that our object of study cannot really be described as being composed by discrete localities, but rather seems to exist in the form of a single intensively differentiated and extensively distributed spatio-temporal assemblage of imperial scales and relations (Holbraad & Pedersen 2009; Nielsen & Pedersen In Press). In order to illustrate why this is so – and to show why the ethnographic study of such self-scaling entities can be said to represent a return to expeditionary anthropology – it is necessary to take a closer look at the concept of multi-sitedness itself.

Multi-sited ethnography is widely considered an adequate if not sophisticated methodological solution to the epistemological challenges posed by the so-called crisis of representation within anthropology (Falzon 2009). Yet simply adding more sites does not represent a particularly radical departure from the classic concept of the ethnographic field as something comprised by spatially (geographically) discrete settings or localities. As argued by Karen Fog Olwig (2002), migrants and their ethnographers may spend a great deal of time flying from one location to another and still, to all intents and purposes, constitute a single ethnographic field site. Ghassan Hage, who followed Lebanese villagers to Venezuela, Australia, Europe, and the US, makes a similar point and describes how he suffered from jetlag even though he was not doing a multi-sited ethnography: "I was studying one site: the site occupied by the transnational family. It was a globally spread, geographically non-contiguous site, but it was nevertheless one site" (Hage 2005: 466). For both Olwig and Hage, their attempt to heed the multi-sited call to "follow the people" wherever they go thus seems to have led to the paradoxical conclusion that there is only one site. The question is whether something similar happens when the ethnographic object at hand is not a group of transnational migrants, but a global assemblage of infrastructural interventions.

In fact, Marcus himself seems to have foreshadowed the possibility that a so-called multi-sited ethnography is in fact not so much *multi-sited* as *trans-sited*:

> In projects of multi-sited ethnographic research, de facto comparative dimensions develop ... as a function of the fractured, discontinuous plane of movement and discovery among sites as one maps an object of study and needs to posit logics of relationship, translation, and association among these sites. Thus ... comparison emerges from putting questions to an emergent object of study whose contours, sites, and relationships are not known beforehand, but are themselves a contribution of making an account that has different, complexly connected real-world sites of investigation. The object of study is ultimately mobile and multiply situated, so any ethnography of such an object will have a comparative dimension that is integral to it, in the form of juxtapositions of phenomena that conventionally have appeared to be (or conceptually have been kept) 'worlds apart' (Marcus 1995: 102).

In this remarkable passage, it seems to us, Marcus goes far beyond the – in our view rather limited – methodological and epistemological promise of multi-sited fieldwork; he effectively undermines his own and other multi-sited ethnographers' position by relativizing the conventional (modernist) hierarchy of scales and corresponding perspectives on which this approach may implicitly be said to rest (Strathern 1999). Indeed, if one concedes that some entities are so "big" (complex) that they only exist as spatial extensions or, more precisely, as spatio-temporal distributions whose dynamic scales are transversal to the fixed and stale dimensions of the local and the global, it fundamentally challenges the concept of location(s) as such and it would seem to follow that such objects call for a new ethnographic method, one that is neither bound to the merographic fetish of the bounded field site situated *in* a world system nor the corresponding merographic fantasy of a multi-sited fieldwork seeking to follow people and things to achieve an ethnography *of* the world system.

Conclusion

In the introduction to a recent collection of essays on multi-sited ethnography, Mark-Anthony Falzon jokes that anthropologists have developed a standard "road to Damascus" narrative explaining their conversion to multi-sited methodology. Having planned to do conventional single-sited fieldwork, the story goes, the ethnographer suddenly realizes the inadequacy of this single site and proceeds to move about (Falzon 2009: 12). This chapter is also a narrative of methodological epiphany, but it is not one that leads from a single site into the brave but no longer quite so new world of multi-sited ethnography. Instead, it takes us back in time to muse upon the strange death of the expedition as a method in anthropology and then forward to argue that the very big things that constitute the ethnographic object of our study may actually call for the resurrection of the expedition in an updated version.

A parable originating in India tells the story of a ruler who ordered a number of blind men to feel different parts of an elephant and reach a conclusion as to the nature of what they were touching. The story is related in Buddhist scripture (Udana 6.4) but it exists in many different versions, including a poem entitled *The Blind Men and The Elephant* by John Godfrey Saxe (1816-1887). In the poem, the first of the blind men feels the side of the elephant and claims it is very like a wall. The second blind man encounters a tusk and says the elephant is like a spear and so it continues; the trunk being perceived as a snake, the knee as a tree, the ear as a fan, and the tail as a rope.

> And so these men of Indostan
> Disputed loud and long,
> Each in his own opinion
> Exceeding stiff and strong,
> Though each was partly in the right,
> And all were in the wrong!

In most versions of the story, the blind men insist on the correctness of their perception, and they are even said come to blows, but there is also a version where a wise man concludes that all the blind men are right inasmuch as the elephant has all the qualities they have described. The parable may be said to illustrate a sort of perspectivism where reality, in this case an elephant, is imagined to exist unproblematically, but tends to elude accurate description on account of the limited and prejudiced perspectives that are brought to bear upon it.

In an article criticizing the latter-day holism implicit in the multi-sited search for cultural formations, Cook, Laidlaw, and Mair (2009) suggest that multi-sited methodology charges ethnographers to deal with cultural formations in much the same way as the blind men dealt with the elephant. Assuming that a single location is an insufficient basis for reaching conclusions about the whole, the ethnographer is enjoined to feel the object in various places and combine these various insights to reach a more accurate description. However, the parable of the elephant and the blind men is tricky in the sense that it illustrates how all human cognition is inherently partial, how everyone is groping in the dark like the blind men, yet it relies on the elephant being seen in its entirety by men with vision for its effect. We tend to concur with the idea that people have limited cognitive faculties and we are highly sceptical of the assumption that the cosmopolitan anthropologist will be able to see the whole elephant, as it were, merely by zipping between sites. However, unlike Cook, Laidlaw, and Mair, who conclude that anthropologists should simply abandon the idea of studying wider systems – "If there is no elephant, there is no need for us to try to imagine one" (Cook et al. 2009: 69) – we insist on the need to piece together and imagine things of elephantine proportions even if it requires a methodological change which is rather more radical than the shift from single to multiple sites. The problem, in our view, is not an absence of elephants but the fact that they are generally disinclined to stand still for inspection, and this makes it necessary to devise a method for studying very big things on the move.

It seems intuitively true that one can get a better sense of the size and qualities of very big objects by measuring them against one's own movements; circling a stadium, walking through a town, driving through a desert, and so on. In the case of moving objects, it is even more evident that one's own trajectory, speed, and acceleration while walking, cycling, or driving are parameters necessary to approach and judge the characteristics of objects, their movement, and relations. In a sense, that's what the long drive through the Gobi desert was, a journey along the fuzzy contours of very big things in motion, a trip within a widely distributed spatio-temporal assemblage that might just add up to a Chinese empire in the making. If multi-sited ethnography is a question of staying put in multiple locations and then connecting the dots, our "pointless" drive through Mongolia started with the line produced by movement and allowed the dots to emerge as incidental effects. Making movement an integral feature of anthropological research is reminiscent of the traditional ethnographic expedition, the classic version 1.0 of which may be characterized as having four distinct features: 1) mobility, 2) collaboration, 3) colonialism, and 4) speed. In some ways, multi-sited methodology may be said to foreshadow the return of the expedition, but because the issues of mobility and collaboration have not been adequately theorized, there seems to be a real danger of resurrecting only the more unfortunate features of expeditions, viz. the quick condescension of the cosmopolitan ethnographer zipping around the globe. The ethnographic expedition 2.0 that we envision excludes the colonial hauteur of anthropologists making speedy surveys among the natives but it does include the commitment to movement and collaboration that was also a defining characteristic of historical expeditions.

References

Alden, C., D. Large & R. S. de Oliveira 2008. *China Returns to Africa: a Rising Power and a Continent Embrace*. London: Hurst.

Beckett, J. 1998. "Haddon attends a funeral: Fieldwork in Torres Strait, 1888, 1898." In: A. Herle, & S. Rouse (eds.) *Cambridge and the Torres Strait: Centenary Essays on the 1898 Anthropological Expedition*. Cambridge: Cambridge University Press, 23-59.

Brautigam, D. 2009. *The Dragon's Gift: the Real Story of China in Africa*. Oxford: Oxford University Press.

Candea, M. 2007. "Arbitrary locations: in defence of the bounded field-site." *Journal of the Royal Anthropological Institute*, 13(1): 167-184.

Clifford, J. 1997. *Routes: Travel and Translation in the Late Twentieth Century*. Cambridge, MA: Harvard University Press.

Cook, J., J. Laidlaw & J. Mair 2009. "What if There is No Elephant? Towards a Conception of an Un-sited Field." In: M.-A. Falzon (ed.) *Multi-sited Ethnography Theory, Praxis and Locality in Contemporary Research*. Aldershot: Ashgate, 47-72.

Falzon, M.-A. 2009. "Introduction." In: M.-A. Falzon (ed.) *Multi-sited Ethnography Theory, Praxis and Locality in Contemporary Research*. Aldershot: Ashgate, 1-24.

Hage, G. 2005. "A not so multi-sited ethnography of a not so imagined community." *Anthropological Theory*, 5(4): 463-475.

Hannerz, U. 2003. "Being there... and there ... and there!: Reflections on Multi-Site Ethnography." *Ethnography*, 4(2): 201-216.

Herle, A. & S. Rouse (eds.) 1998. *Cambridge and the Torres Strait: Centenary Essays on the 1898 Anthropological Expedition*. Cambridge: Cambridge University Press.

Holbraad, M. & M. A. Pedersen 2009. "Planet M: The intense abstraction of Marilyn Strathern." *Anthropological Theory*, 9(4): 371-394.

Kleveman, L. 2003. *The New Great Game: Blood and Oil in Central Asia*. 1st ed. New York: Atlantic Monthly Press.

Liebersohn, H. 2003. "Scientific Ethnography and Travel, 1750-1850." In: T. M. Porter & D. Ross (eds.) *The Modern Social Sciences*. Cambridge: Cambridge University Press, 100-112.

Liebersohn, H. 2008. "Anthropology Before Anthropology." In: H. Kuklick (ed.) *A New History of Anthropology*. Malden, MA: Blackwell Publishing, 17-31.

Marcus, G. E. 1995. "Ethnography in/of the World System: The Emergence of Multi-sited Ethnography." *Annual Review of Anthropology*, 24: 95-117.

Nielsen, M. & M. A. Pedersen In Press. "Trans-temporal Hinges: Comparing Chinese Investments in Mozambique and Mongolia." *Social Theory*.

Olwig, K. F. 2002. "Det etnografiske feltarbejde: antropologers arbejdsmark eller faglig slagmark?" *Norsk Antropologisk Tidsskrift*, 13(3): 111-123.

Rouse, S. 1998. "Expedition and Institution: A.C. Haddon and anthropology at Cambridge." In: A. Herle & S. Rouse (eds.) *Cambridge and the Torres Strait: Centenary Essays on the 1898 Anthropological Expedition*. Cambridge: Cambridge University Press, 50-76.

Strathern, M. 1999. *Property, Substance, and Effect : Anthropological Essays on Persons and Things*. London: Athlone.

Swanström, N. 2005. "China and Central Asia: a new Great Game or traditional vassal relations?" *Journal of Contemporary China*, 14(45): 569 – 584.

Taylor, I. 2006. *China and Africa. Engagement and Compromise*. London: Routledge.

Tsing, A. L. 2005. *Friction an Ethnography of Global Connection*. Princeton, N.J.: Princeton University Press.

Urry, J. 1998. "Making sense of diversity and complexity: the ethnological context and consequences of the Torres Strait Expedition and the Oceanic phase in British anthropology, 1890-1935." In: A. Herle & S. Rouse (eds.) *Cambridge and the Torres Strait: Centenary Essays on the 1898 Anthropological Expedition*. Cambridge: Cambridge University Press, 201-233.

Willerslev, R. 2010. "Frazer strikes back from the armchair." *The Malinowski Lecture, Department of Anthropology, London School of Economics*. Retrieved October 14, 2010 from http://bit.ly/bGUhJU

Matthew Edney

Field/Map
A Historiographic Review and Reconsideration

To modern eyes, "map" and "field" appear intimately conjoined. After all, what is a map other than an image of the field made to bring the world "out there" to the "in here"? Once "in here," the map can be read in order to comprehend and organize the world "out there." The presumption is that maps are properly grounded in the observation and measurement of the world. They stand as exemplars of objective and truthful representation. Cartography – the activity of making maps – is construed to be a singular, universal activity. It is singular in that it apparently features a specific ontology with one epistemology and (generally) one methodology. It is universal in that this singular practice has been presumed to characterize all mapmaking, at all times and in all cultures; historical and cultural difference is a matter only of technological and archival variation. Map history is thus reducible to a simple narrative of progress in technological ability and archival completeness. From such an unproblematic perspective, cartography seems a fundamental component of the field sciences and it accordingly appears as such in some of the programmatic statements concerning the situatedness of scientific and other cultural practices (most obviously, Latour 1987: 215-32; more recently: Kuklick & Kohler 1996: 1-14, Livingstone 2003: 123-126, 153-163, Shapin 1998: 5-12, Smith & Agar 1998: 1-23, Golinski 2005: 99-101, Withers 2007: 87-111, 201-204). While such statements do recognize the political purposes of cartography,[1] they have not considered the implications of the ongoing critique of the modern cartographic ideal's epistemological and methodological underpinnings.[2]

This last point is ironic, given how closely the modern ideals of Cartography-with-a-capital-c and of Science-with-a-capital-s resemble each other. Both are universalist and placeless; their truth claims depend upon the ultimately naïve criterion of the "direct witnessing" of phenomena. Their larger statements adopt the same rhetorical position of the "view from nowhere." Both have origins in the

---

1   Because they tend to rely upon Revel 1991, see below.

2   For general references throughout this essay, see Edney 2007.

histories told by eighteenth-century practitioners. Both were consolidated over the course of the nineteenth century as progressive endeavors. Both rely on the late-nineteenth century naturalization of objective scientific representation. Both were justified by the creation, in the first half of the twentieth century, of historiographic concepts of past "revolutions" or "reformations." Both rely on the same mythic presumptions of the widespread broadcast of ideas through "print culture." The same rhetorical ideals of public discourse underpin their disinterested "views from nowhere." And so on.[3] Yet, while historians of science now recognize that Science-with-a-capital-s is an unhelpful ideal that obscures much more than it illuminates about the production of scientific knowledge, Cartography-with-a-capital-c persists in infecting and shaping both map and science history.

A key problem is that, despite the centrality of fieldwork to the modern cartographic ideal, map historians have been happy to leave fieldwork and surveying well enough alone, unexamined and untheorized. Even critical map scholars have avoided the issue of the relationship of fieldwork to graphic map. The task of this essay is to unravel this paradox by exploring how the modern cartographic ideal has shaped map history and particularly how it has led the field/map connection to be construed. In the process, it suggests that a newly developing line of research within map history provides an appropriate and useful means to realign our understanding of mapping processes generally and so elucidate their contributions to the field sciences. In its historiographic review, this essay explores the formation of the modern cartographic ideal, the forms of map history it engendered, and the place of the field and of fieldwork in each. In offering perspective, it examines the critical turn in cartographic scholarship with an eye to the integration of fieldwork into map studies.

The Formation of the Modern Cartographic Ideal

A fundamental problem in any discussion of the nature and history of maps is the persistent failure to distinguish specific from generic conceptions of map/*carte*/*karte*/etc. Through the eighteenth century, the understanding of "map" was strictly specific, as a lower-resolution image of regional or world space. For example, Samuel Johnson succinctly defined a map in his *Dictionary* (1755) as "a geographical picture

---

[3] These points were variously made from the history of science perspective, for example, by Daston and Galison 1992, Shapin 1995: 306, 1996, 1998: 7-8, Findlen 1998, Terrall 1998, and Kohler 2002: 191. From the perspective of cartographic history, see especially Edney 1993, 2009, Harley 1989, 2001, Pickles 2004.

on which lands and seas are delineated according to the latitude and longitude."[4] Such geographical maps were understood to be abstractions of the earth's surface that relied for their authority on their spatial correspondence with the terraqueous globe via the graticule of meridians and parallels. They were prepared by combining geographical itineraries, preferably by carefully fitting them to a few known locations determined by latitude and perhaps longitude; they were products of reason and indeed their making was widely held to be an exemplar of rational thought processes (Edney 1994, 1994a, 1997, 1999).

The geographical, regional map was originally defined in opposition to the "plan." Plans were consistently conceptualized as higher-resolution images of the earth's surface. They were understood to be limited and constrained by the "things" they represented, be they buildings, fields, estates, villages, fortresses, or towns. Their truth claims lay in the acts of observation and measurement on which they were based. Thus, when Christopher Packe conceived and implemented his innovative, detailed topographical image of eastern Kent in about 1740 (Figure 1), he made a great fuss about how this work was manifestly *not* a "map" but rather a "landscape" or "portrait" constructed with a perspective "Eye" that was "here, *every where* present by turns." Packe therefore thought it a deprecating calumny when his contemporaries mistook his work for a "map" because of its regional extent; after all, he argued, the abstracted geographical features usually shown on lower-resolution maps could all be interpolated from his detailed plan (Packe 1737: 4 and 15, 1743: 3-4; see Campbell 1949, Charlesworth 1999). In this early argument for the systematic observation of the earth's surface, we see the contours of the future cartographic ideal: the promotion of the epistemological certainty of the surveyed plan; the collapse of the distinction between plan and map; the promotion of the unitary map as the privileged method for representing space, as a view from above; and the formation of "cartography" as a singular endeavor.

Whether intended for military or civil purposes, eighteenth-century surveys were undertaken in a piecemeal fashion; their reach and operations were restricted by the lack of sustained institutional capacity and the fragmentation of political power. Even the great *Carte de France* – usually lauded as the first truly systematic state survey – contained significant technical, institutional, and political fractures. But the consolidation of governmental power over the course of the nineteenth century meant that European states could eventually follow the model presented

---

4   This analysis is based largely on Andrews 1998.

by the *Carte de France* and undertake systematic topographical surveys of extensive territories based on high-quality triangulations.[5]

While systematic surveys continued to suffer from financial, political, and institutional inabilities after 1800 (Edney 2009: 16-17), they nonetheless established what *seemed* to be the ultimate foundation for *all* spatial knowledge for any given area. Detailed surveys became understood as the ultimate data source for all smaller-scale, generalized maps. From this perspective, *all* mapmaking had a common foundation and ethic. This emergent ideal of a unified system of mapmaking was recognized in the 1820s by German scholars, who coined a new term for it: "cartography" (van der Krogt 2006, Edney 2009: 42). The ideal would achieve a general resonance later in the century. The popularization of panoramas and balloon rides promoted the panoptic viewing of map-like landscapes (see Wallach 2005, Byerly 2007, Della Dora 2007). The mid-century reforms of urban infrastructure pushed the systematic surveys into working at the very large scales needed for engineering work. The post-1870 reforms in primary education established new ways to teach geography, in which children began by making maps of their playground or schoolroom – "*as it would appear to a person looking down from the ceiling*" – by a process of observation and measurement; as one educationalist declared in 1874, "geography is commenced, as it should be, with TOPOGRAPHY."[6] Finally, the increasing popularity of bicycling after 1880 and of automobile travel after 1900 dramatically expanded map consumption and reified the conviction that the fundamental purpose of maps is to aid travel; conversely, the repeated use of maps for travel enforced the belief that *all* maps mirrored the physical world (Akerman 2006, Dando 2007). The new ideal received a further imprimature when mathematicians took to calling a structure-preserving transformation between two sets – originally termed *Abbildung* (picture, representation) by Richard Dedekind in the 1870s (Sieg & Schlimm 2005) – a "mapping"; a usage that seems in the twentieth century to have recursively supported the conviction that cartography is necessarily grounded in measurement and science.

Implicit in the singular concept of cartography is that there exists just one kind of cartographic product: a "map," comprising a "picture" of the world as if seen from above, constructed from direct observation and measurement. This generic, idealized "map" encompasses all the different specific kinds of charts, plans,

---

5   Kretschmer et al. 1986 remains the best summary of the nineteenth-century surveys. The French surveys have a large literature, including: Konvitz 1987, Pelletier 1990, Mukerji 2002, Blond 2008. On failings and fractures see, e.g., Licoppe 2002, and, more generally, Edney 1997.

6   Sullivan 1883: 8 (original emphasis); Andrews 1998 noted that this pedagogic plan first appeared in Sullivan's 1874 edition. Generally, see Keltie 1885.

Figure 1. Detail of Christopher Packe, A New Philosophico-Chorographical Chart of East-Kent (London, 1743). By permission of the British Library (maps 3065(1)).

globes, relief models, and (yes) smaller-scale geographical maps; it is what we refer to whenever we blithely say "maps *are* X" or "maps *do* Y." It is what lexicographers – as well as cartographers and historians of cartography – bend themselves into pretzels to define. While some philosophers criticized the ideal on logical grounds (and early aviators on practical grounds), such that Alfred Korzybski could famously proclaim in 1933 that "the map is not the territory," the manifest efficacy of maps as navigational and engineering tools sustained the ideal in popular culture.[7] And it is

---

7   Korzybski 1933: 58. Among many early aviators, Antoine de Saint-Exupéry made pragmatic observations that maps are never the territory in his *Vol de Nuit* (Paris, 1931). Earlier, see Carroll 1893: 169.

this idealized "map" that serves as a crucial component of the modern commitment to representation as a properly mimetic process. Indeed, the idealized "map" has commonly served as a metaphor for communication, representation, and scientific theories and models (Toulmin 1953: 105-39; see also Robinson & Petchenik 1976, Sismondo & Chrisman 2001: 40-41). Yet despite the massive cultural investment in the cartographic ideal, it remains simply that: an ideal. It has little relevance for any history that seeks to understand maps according to their contemporary circumstances. A hindrance to understanding, the generic "map" should never be mistaken for anything other than an idealized concept.

The Practice of Map History: Traditional, Internal, and Socio-Cultural

Elucidating Map Content: Traditional Map History

Two interrelated approaches to old maps developed in the eighteenth century. Some geographers presented narratives of intermittent progress according to which critical scholars, like Gerard Mercator and Guillaume Delisle, periodically made significant advances in the quality and density of the small-scale geographical archive (constituted as large, coherent atlases); but most maps were made by hacks who uncritically copied and inevitably degraded the critical archive, until its next critical reform. At the same time, historians began to read geographical maps – and world maps in particular – as literal "world views" that enshrined, for a given era, European culture's curiosity about the world and its ability to indulge that curiosity; geographical maps became convenient devices for tracing the rise and progress of civilization. The basic methodology of map history was therefore the cartobibliographic listing of geographical maps by regions, in chronological order, to track the growth of geographical knowledge and civilization (Withers 2010).

The evolving cartographic ideal led in the later nineteenth century to the formation of a coherent traditional (a.k.a., modern or empiricist) approach to map history. This approach did perpetuate the basic purpose of Enlightenment map history, to study maps as repositories of knowledge and as markers of the progress achieved by human civilization. As R. A. Skelton would state in 1966, "any map is a precipitation of the spirit and practice of its time" (Skelton 1972: 5). Map history's basic methodology remained the regional cartobibliography. No attempts were made to explore *why* maps were made; their existence was alone sufficient to explain the efforts of untold fieldworkers to collect geographical data. Little attention was paid even to the written texts accompanying and surrounding maps (van den Broecke 2009).

The crux of the traditional approach – distinguishing it from Enlightenment practices of map history – was its construal of a new metanarrative. In line with the new ideal, historians began in the 1870s to narrate the "history of cartography" as a singular endeavor advanced by each dominant culture in turn: Egyptians, Greeks, Romans, Arabs, and then Latin Europeans (beginning with the Germans, then Italians, Dutch, French, and British – e.g., Daly 1879). From this enlarged historical perspective, cartography's progress became a line of continuous advance. In 1905 Christian Sandler crystallized the narrative of progress with a single moment – cartography's "reformation" – when circa 1700 the French academicians apparently imbued cartography with a scientific spirit. Sandler's emblem of this reformation was Jean Picard and Philippe de La Hire's 1693 map depicting the French coastline before and after its correction by new astronomical observations for latitude and longitude (Figure 2, Sandler 1905; see Picard & de La Hire 1693: pt.3, 91-92). Map historians subsequently emulated Sandler and have used the Picard-La Hire map to introduce chapters on the development of modern scientific mapping and the triangulation-based systematic surveys of countries. Not that the map historians who paid attention to these official surveys actually considered field techniques and practices; rather, they remained concerned only with broad institutional histories and the overall progress of the surveys.[8]

The awareness of modern cartography's achievements ossified the history of cartography as a narrative of scientific progress and of maps as repositories of spatial data. As Gerald Crone declared in 1953, "the history of cartography is largely that of the increase in the accuracy with which these elements of distance and direction are determined and in the comprehensiveness of the map content" (Crone 1953: xi). Historical explanation was channeled into a predefined sequence of normal progress (see esp. Buczek 1982: 7, who also drew on Sandler 1905). The cartographic canon was accordingly reconfigured to reinforce this utterly presentist history of a universal, data-driven cartography (Raisz 1937, 1937a). But, as indicated by the fixation on the Picard-La Hire map, evaluations of map accuracy were strictly impressionistic and rooted in visual comparisons with modern maps. (The reduction of map history to the history of decontextualized spatial data is the reason why some commentators have labeled traditional map history as "empiricist.")

Furthermore, influenced by the cartographic ideal, after 1890 historians began to study larger-scale maps for the information they provided about the past. Both

---

[8] Brown 1949: 241-279, Crone 1953: 128-140, Kish 1980: 53-58. See also Revel 1991: 155-156, Thrower 1996: 105-106, Stockhammer 2001: 295-297, and "Map", Encyclopædia Britannica Online, retrieved July 11, 2011 from http://www.search.eb.com/eb/article-51772.

August Meitzen and Frederic Maitland examined early Ordnance Survey maps for their evidence of past landscapes, and in particular of Celtic and Saxon settlement patterns; other scholars interested in large-scale plans included Frederick Jackson Turner in the USA, Marc Bloch in France, and Karol Buczek in Poland in the 1920s. Such sporadic interest intensified after 1950 when historical geographers began to use archival evidence to augment their landscape studies; in particular, Brian Harley sought to codify and regularize the study of larger-scale maps for the benefit of other historians. Yet the character of such historical analysis remained focused on the map *image* and on the landscapes and regions symbolized. That is to say, map historians remained minimally interested in the actual fieldwork of the surveyor and explorer, and they continued to construe larger-scale maps as icons of observation, measurement, and unproblematic representation (Edney 2005a: 19-31, 135-138; one work not discussed there is Buczek 1982).

Elucidating Map Form: Internal Map History
The twentieth-century development of cartographic design as an academic field of study influenced the study of map history in two ways. First, it produced a new approach to map history – the internal (a.k.a., cartographic) approach – centered on the elucidation of the look and artifactual nature of maps. Internal map historians were concerned with map form and so, to some degree, with mapping practices; some attention was accordingly paid to fieldwork. Second, academic cartography's intellectual and institutional needs generated a certain degree of skepticism about the modern cartographic ideal among academic cartographers, thereby laying the groundwork for the development of a socio-cultural approach to map history.

The core of the new, internal map history was the historical vision and justification for academic cartography projected by Max Eckert in the 1920s and by Arthur Robinson after 1945. Both men argued that the nineteenth-century development of the social sciences, and more especially the mapping of social statistics, created a new form of mapping called "thematic" or "special-purpose" distinct from locational "base" or "general-purpose" mapping. While engineers and geodesists had perfected base mapping, the design of smaller-scale maps apparently lagged and was susceptible to abuse. Social scientists had indeed created new symbolic strategies to represent socio-spatial statistics, but their work remained unsystematic and naïve. What was needed, according to both Eckert and Robinson, was a new discipline geared to the scientific study and codification of effective map design and a new professionalization to regulate the production of objective maps (Eckert 1921-1925, esp. vol 1: 24-25, also Arnberger 1966; see Wright 1942, Scharfe 1986, Herb 1997: 34-48; on Robinson, see Edney 2005). The result was the creation of an intellectual

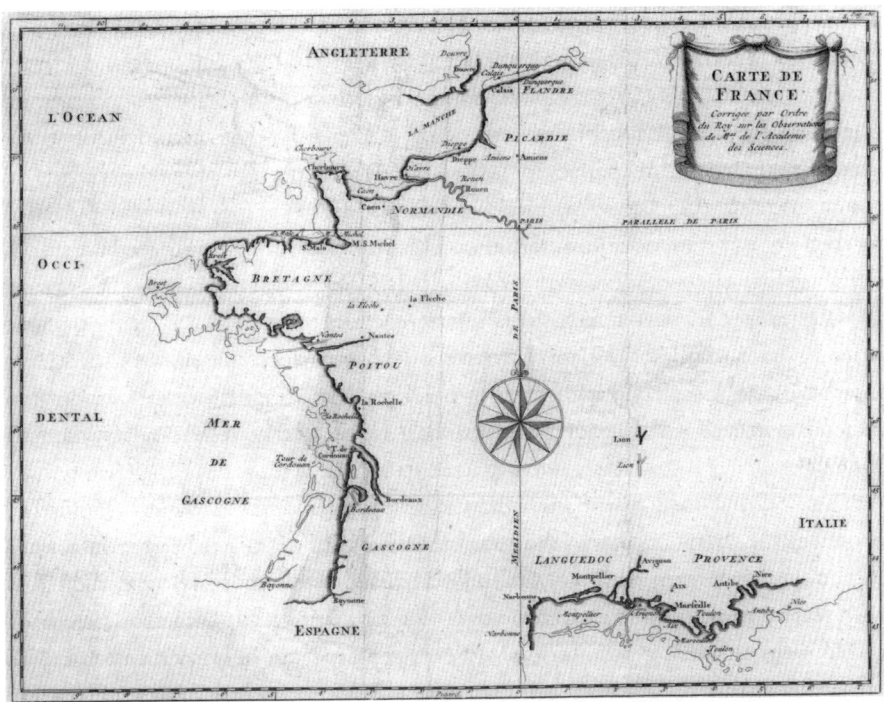

Figure 2. 1729 edition of Jean Picard and Philip de La Hire's Carte de France corrigee par ordre du Roi sur les observations de Mrs. de l'Academie des sciences (Paris, 1693), showing the coastlines of France before and after their correction by astronomical observations (by permission of the Osher Map Library and Smith Center for Cartographic Education, University of Southern Maine).

space within which academic cartographers narrated internalist histories of the development of cartography as a profession: Robinson himself studied thematic mapping; David Woodward, the influence of printing technologies on cartographic aesthetics; Günter Schilder, commercial practices; Ingrid Kretschmer, color map printing; Eduard Imhof, relief depiction; and so on.

Internal map history did not mesh well with the traditional approach. It posited multiple past technological revolutions in cartographic practice, not just the one (Robinson 1982: 12-15, Monmonier 1985). It also relied extensively on archival research to define the contexts within which maps were produced, abandoning the traditional emphasis on studying the maps themselves. Advocates exercised plenty of emotion in the 1970s in arguing for the relative merits of their respective approaches to map history; in particular, the tensions generated by this "form *versus* content" debate pervade Michael Blakemore and Harley's key 1980 monograph, *Concepts in the History of Cartography* (Blakemore & Harley 1980).

The models of cartographic communication promulgated by academic cartographers after 1960, in order to define and justify their new field of study, explicitly included the collection of spatial information within the circuit of communication (Guelke 1976, Edney 2005a: 36-41). Woodward, in the principal methodological statement for internal map history, accordingly embraced field work as part of his matrix of processes, personnel, technologies, and products (Woodward 1974). However, internal map historians had generally been trained as cartographers, working with small-scale maps, rather than as engineers and surveyors with an interest in large-scale surveying, so in practice they remained interested only in issues of map design and continued to take fieldwork for granted (see, e.g., Godlewska 1988).

Yet internal map historians, when they considered the products (if not the processes) of "base" mapping, did share the commitment of traditional historians to the modern cartographic ideal. Indeed, the internal approach contributed to the ideal by providing new methodologies for evaluating the geometrical accuracy of old maps through cartometric analyses. Such analyses began with the graphic technique of the "deformation grid," in which the course of modern meridians and parallels are traced across the features of an old map to give a visual statement of the pattern of geometric accuracy; they developed into a variety of statistical methods (Blakemore & Harley 1980: 60-75, Maling 1989). But such analyses did *not* consider any underlying surveying techniques and were intended only to distinguish accurate from less accurate maps (as made explicit by Lloyd & Gilmartin 1987).

At the same time, academic cartographers' focus on thematic mapping and their formulation of communication models implicitly challenged the modern cartographic ideal. In particular, they held that smaller-scale maps are in fact *interested* and so constitute mediated rather than unproblematic representation. The communication models opened up map studies to new questions about both field work and map use. While they tended to construe map readers as passive absorbers of information, some map scholars did begin to understand that map readers work actively and selectively with maps, further casting doubt on the unproblematic nature even of base maps (Petchenik 1977: 117-128; also Robinson & Petchenik 1976: 36-37).

Elucidating Map Function: Socio-Cultural Map History
By the later 1970s, tensions within both academic cartography and internal approaches to map history prompted some map historians to develop a more culturally open approach to maps. The internal commitment to archival scholarship

demonstrated the existence of many more historically significant maps than had ever fitted within the traditional historical canon; indeed, such scholarship increasingly drew attention to the inadequacies, if not outright flaws, of the ideal, as contextual analyses inevitably led to new considerations of why and how maps were made and read (Woodward 1980; generally: Edney 2005a: 51-56). Academic cartography's fundamental concern for the mediated nature of maps underscored a growing distrust of the modern cartographic ideal, such that it became common after 1980 to suggest that if maps cannot be objective, they must therefore be subjective. That is, each map must be the product of the personal knowledge and biases of its maker; it must *per force* be a "mental map" or a truth-distorting work of propaganda (Axelsen & Jones 1987, Black 1997).

The result was the development of socio-cultural map history. This new approach holds that an effective map history must be culturally and socially minded, that *all* maps are made for specific reasons, and that they are *all* mediated by those reasons. The new approach's underlying concern is therefore with neither map content nor map form, but with map function. This is readily apparent in the new definition of "map" that Harley and Woodward inserted into the first volume of *The History of Cartography* in 1987, that "maps are graphic representations that facilitate a spatial understanding of things, concepts, conditions, processes, or events in the human world" (Harley & Woodward 1987: xvi).

At the same time, academia's "postmodern turn" in the 1970s and 1980s engendered an interest in maps and mapping across a broad intellectual spectrum. A crucial aspect of this intellectual move was to make the basic concern of twentieth-century philosophy – the nature of representation – of general interest across the humanities and social sciences. Just as Korzybski had drawn on the modern cartographic ideal to express the logical and semiotic failings of representation, so too did postmodern critics (Borges 1964: 90, Eco 1985, 1994). In the process, those critics took Korzybski's pragmatic position – that while representation can never make a perfect copy, it can make a functional one – and turned it into a crisis.[9] They have curtailed Korzybski's modernist formula to make the blanket assertion that "the map is not the territory"; some have inverted it so that, within a particular discursive space, the map *is* the territory and *precedes* the territory in the reader's cognition (Baudrillard 1983: 2, Turnbull 1993, see also: King 1996: 1-17). From the postmodern perspective, the relationship of map to territory can never be necessary but only ever arbitrary, just as Saussurian semiotics holds that the relationship of signifier to signified within the linguistic sign is arbitrary. The map is thus construed

---

9   Korzybski's modernism is widely misinterpreted as postmodernist; e.g., Mitchell 2008: 3.

as a semantic construct, its meaning independent of the territory; it is a vehicle for imposing complex meanings on the concept of a specific territory. Of course, the postmodern turn has generally deployed "map" and "territory" as broad metaphors for representation, but it has led some scholars from beyond traditional and internal map history to study actual maps and mapping practices.

Exchange between the two intellectual movements has been considerable. Dedicated map historians have drawn on work in several other fields to inform and support their own analyses. For example, in creating the initial manifesto for a new, culturally minded approach to map history, Blakemore and Harley were very much influenced by the application by some art historians of iconography to urban views and maps to elicit their socio-cultural rather than their spatially factual significance (Blakemore & Harley 1980: 76-86, Edney 2005a: 72-78). Harley subsequently went on, in a series of provocative essays, to explore some of the political contexts within which maps were made and to engage in a series of readings of maps as socio-cultural texts; in this he drew on the work of Michel Foucault, Jacques Derrida, and several other philosophers and social theorists (Edney 2005a: 85-111). His work has in turn proven particularly important in shaping a wider intellectual interest in maps and their histories. The result has been the development of a wide ranging, active, and exciting field of scholarly inquiry that has followed three main trends, each with a different attitude towards elucidating the relationship of field to map.

In the first scholarly trend, the emphasis of traditional map historians on content has been reconfigured as an emphasis on reading maps as texts, in which a map is carefully read and put into the precise context of its production to define its purpose and its meaning to its contemporaries. This approach has meshed well with the concerns of literary and art historical scholarship and has led to a proliferation of textual readings. Suffice to say, this line of scholarship has not been at all interested in fieldwork.

In the second, the interest of internal map historians in the precise contexts within which maps have been made and used has broadened into studies of the political and social contexts of map production. Detailed studies have examined specific instances of official cartography, such as the detailed surveys of France, India, or Guyana in the eighteenth and nineteenth centuries, or of state-sponsored thematic mapping in the nineteenth and twentieth. Of special note is Roger Kain and Elizabeth Baigent's comparative study of modern cadastral map making (Kain & Baigent 1992). But only a very few historians have perpetuated the internal approach's secondary interest in the processes of mapmaking so as to examine the practices and conditions of fieldwork (Edney 1997, and Burnett 2000, to some degree; more specifically see, e.g., Collier & Inkpen 2002, 2003).

In the third, an ethnographic interest in maps made in non-modern cultures – whether traditional Eurasian societies or indigenous societies elsewhere – has demonstrated that such maps are neither "primitive" nor "simple." Rather they are culturally complex and functionally effective texts that are rooted in culturally distinct ways of conceptualizing spatial relationships. Indeed, such studies have tended to break down the conceptual barriers around the graphic, material map by recognizing that maps can comprise the spoken word, gesture, and performance (Woodward & Lewis 1998: 1-10). This line of scholarship has served profitably to destabilize the modern commitment to maps as the necessary products of observation and measurement (See, esp., Wood 1992, Turnbull 1993).

Unfortunately, comparisons of modern with non-modern maps have underpinned two inadequate theorizations of the relationship of field to map. First, a political critique of modern cartography perpetuates the transcendentalist critique of the 1970s, which took the modern ideal of mimetic representation at face value and so condemned modern maps as sterile and culturally impoverished (Roszak 1972: 404-12, Tuan 1977). The political critique holds cartography to be a technology of modern capital and the centralizing state that obscures and negates the individual, local, personal, sensuous, and ultimately *authentic* human experience of space. (This critique accordingly emphasizes the direct experience of space, as through Michel de Certeau's "itinerary" or Charles Baudelaire's "flâneur.") By contrast, authentic spatial experience is more than evident in non-modern maps, which are therefore to be celebrated for their genuine representation (still mimetic!) of their makers' world (e.g., Harley 1989, 1989: 241-59, Revel 1991, Curry 2005, Olsson 2007). Scholars less antagonistic to modern cartography have traced the opposite argument. Following Denis Wood, they have divided cartography into two practices: mapping, or the creation of a personal "cognitive map" (which can only ever be a metaphor), and mapmaking, or the making of an artifactual map, whether ephemeral or durable (Wood 1992: 28-43, 1993). Mapping is an activity common to all cognitively developed, adult humans; mapmaking is not. However, scholars have been seduced by the universality of mapping to presume that the individual's externalization of the cognitive map as some form of artifact – an act typified by the many encounters between Western voyagers and indigenous peoples[10] – defines the character of *all* acts of mapmaking. In doing so, they have only reconfirmed the modern ideal's conviction that cartography is an ineluctably individual act, and as such a personal and "subjective" statement. We can thus read, for example, that "*every* act of

---

10   Such an encounter is, of course, key to Latour's understanding of cartography as a field science: Latour 1987: 215-32. See: Lewis 1998, Feinberg et al. 2003, Brown & Laurier 2005, Turnbull 2007, etc.

mapping embodies elements of autobiography" (Balm 2009: 102 – emphasis added). The result is an emphasis on local knowledge that has little concern either for the semiotic conditions of its expression or for how it might have been incorporated into standardized structures for knowledge.

There are significant problems with these constructions of the relationship of field to map. Clearly, we still need to discard the modern cartographic ideal, in all its aspects, and turn our attention more to tracing the actual histories of maps as part of the ways in which humans have sought to inscribe an order to the world. In this respect, maps should neither be privileged nor critiqued as the only spatial representations that humans make. "Mapmaking" does not comprise a singular and universal process, but encompasses a wide and varied array of practices intertwined with technologies, institutions, and discursive conventions. Despite lip service to Foucauldian "discourse," map scholars have yet to develop effective analyses of specific cartographic discourses. Beyond discourses about maps, "cartographic discourses" do not exist; rather, there are spatial discourses featuring a mix of oral, written, graphic, numerate, and cartographic strategies. "Mapping" entails both personal, direct knowledge of the world and also the comprehension of the world through other inscriptions and practices. That is, we still need to break away from cartography as a singular practice, and to consider it as a multitude of processes (variously embodied, technological, intellectual, and discursive). In this respect, we need to develop a new, processual map history, within which we might be able to come to terms with the relationship of field to map.

Elucidating Mapping Processes: Processual Map History
The challenge therefore is both to reassert the founding principle of socio-cultural map history – that the modern cartographic ideal is simply inadequate for modeling map history – and to give it firm foundations by truly questioning how maps have been variously produced, circulated, and consumed. This argument is not new: I made a preliminary statement in 1993, drawing on a more explicit earlier statement by Bob Rundstrom; more recently, it is implicated in several conceptual discussions, including that by Sismondo and Chrisman which revisits the use of maps as metaphors for scientific theories (Rundstrom 1991, Edney 1993, Sismondo & Chrisman 2000, Edwards 2006: 5-7, Del Casino & Hanna 2006, Kitchin & Dodge 2007). It also meshes very well with the recent trend in book history and bibliography for a "sociology of texts" (Darnton 1982, Warner 1990, Johns 1998, McKenzie 1999), and with Latourean actor network theory (Latour 1987, 1990, 2005, Bijker 1995, Bijker et al. 1987).

Generally, map historians need to engage properly with spatial discourses and the mapping processes that they engender. The nature of discourse remains, for some, a controversial topic. I prefer Michel Foucault's approach to discourse as a regulated network of representation (Foucault 1972: 80; see, Mills 2004: 1-25). Each discourse comprises a particular network of people and institutions that produce a coherent set of texts including maps, circulate those texts, and consume them. The semiotic conventions underpinning the texts, participation within the network, and the practices of producing, circulating, and consuming the texts are all communally regulated. In this respect, each discourse can be legitimately understood as a structure that at once enables and constrains representation. Foucault might have focused more on the constraints and restrictions of discourse, but he did not ignore the manner in which discourses also set the conditions for the possibility of knowledge (e.g., Foucault 1970: xxi-xxii). Moreover, the structures of discourse are neither fixed nor static, but mutable. Their semiotic conventions and institutional rules are altered by their participants as they create, circulate, and consume texts, all in accordance with their participants' various needs and desires. In this respect, specific discourses should more properly be called "discursive formations."

In this sense, discourse is an analytical device. Discourses can be identified and studied at a variety of conceptual scales, depending upon the character of the network that can be discerned and delineated through tracing the circulation of texts. We can identify certain discourses that are concerned to various degrees with representing spatial aspects of the world. Each spatial discourse supports specific mapping practices and so generates specific kinds of map texts. The interpersonal discourses associated with wayfinding, for example, construe space in terms of the individual's place in the landscape; they produce verbal and gestural maps through which spatial complexity – one's bewildering immediate vicinity – is simplified into key landmarks and routes expressed in terms relative to one's own body. By contrast, the modern discourse of shipboard navigation construes location in terms of a global co-ordinate system of latitude and longitude, measured by repeatedly and carefully deploying specialized instruments to make astronomical and chronometric observations, and represented through equally specialized numeric and graphic inscriptive sign systems. While both might be said to be concerned with getting from A to B, they are neither readily combined nor comparable.

Ben Orlove has provided a telling example of the differences between bureaucratic and indigenous maps of reed beds in Lake Titicaca, Peru. In addition to distinguishing the formal characteristics of bureaucratic and indigenous maps – based on consistency of scale, respective emphasis on towns or villages, consistent and uniform orientation, emphasis of boundaries, numbers of copies produced of each map, and so on – Orlove undertook an ethnography of how the two communities produced, read, and interacted with their maps. The state officials placed maps in their reports and on their walls and treated them as intellectual infrastructure, to be consulted as necessary but otherwise part of the common structure of knowledge; they refused to call the indigenous images "maps" and instead referred derisively to them as "sketches." The local inhabitants made their unique maps communally and placed them on the table in front of them in negotiations and discussions, referring to them constantly; unable to access the official reports, they did not normally see the official maps (Orlove 1993). We can identify in this example two networks within which circulated maps of place, all made of the same portion of the earth's surface, and broadly for the same purpose, but which entailed quite distinct spatial discourses.

In summary, we can identify groupings of spatial discourses by the kind of scale-dependent spatial conception in which they are interested: global, regional, place, property, territorial, marine, urban, and so on. Each grouping brings together a broadly similar set of mapping practices in a mode of cartography (Edney 1993; Edney 2007 is a more refined statement). The mode of property mapping, for example, deals with very-high resolution representations of real property, usually assuming a flat earth and plane geometry, with each text circulating in small numbers, often in manuscript, within legal institutions. By contrast, the mode of geographical mapping deals with low-resolution representations of the suprapersonal space of regions or the whole world – areas that no one person could ever hope to see and map in detail by themselves – based on a curved earth and spherical geometry, with texts circulating amongst educational and governmental institutions generally in large numbers, and often printed. These are technologically distinct modes, supporting distinct sets of spatial discourse. Regardless of the modern ideal's insistence that smaller-scale maps are inevitably based on larger-scale ones, we can only accept that connections exist if they can be traced through intervening discourses.

Much can be said about this processual and discursive approach to map history. Four points in particular need to be highlighted. First, we need to distinguish precise discourses by following the flow and circulation of maps. For example, in tracing the trajectories of several cartographic works as they passed to and fro across the eighteenth-century Atlantic, between London and North America, we can distinguish three sets of geographical discourse active in both the colonies and the metropol. Governmental discourses featured the commissioning and tight circulation of manuscript maps; some of these leaked via certain individuals into private circulation among those gentlemen interested in geographical matters, who also made and distributed their own compilations; some maps were eventually drawn from these private archives and combined to make maps for public circulation according to the dictates of the market for printed, public discourse (Edney 2008: 67-70).

Second, as studies of indigenous mappings again remind us, the consideration of spatial discourses means that we can no longer privilege maps as *the* primary, or even only, form of spatial representation. Even in our own map-obsessed society, there are plenty of other representational strategies through which to organize and comprehend space. Mariners have traditionally used a combination of oral lore, written directions, and views of headlands, as well as charts to move through the seas; even with the mid-nineteenth century perfection of the determination of position at sea, modern mariners continue to deploy the same array of representational forms. Different representational strategies are, of course, geared to different functions. In property mapping, the graphic plan serves as a conceptual device to aid interpretation of written metes-and-bounds descriptions and the monuments set in the landscape itself. Geographical texts can comprise written maps. The mapping of discrete places is intertwined with landscape imagery more generally. Indeed, spatial discourses do not require production and consumption of cartographic representations: whether or not graphic maps are deployed within a discourse, and how they are deployed in conjunction with other representational strategies, depends upon that discourse's particular conventions.

Third, the broad sweep of map history can be reconfigured as the history of spatial discourses, their internal development, and their intersections with other discourses. One example of such an intersection is provided by an 1837 map prepared by leaders of the Iowa to outline their territorial claims against neighboring Native American groups. The Iowa had already sought to adapt to the material culture of the encroaching United States and they now tried to make a map according to Western spatial conventions. The Iowa strategy nonetheless failed because the U.S. officials did not appreciate the Western aspects of this hybrid map and they granted large portions of Iowa territory to the Iowas' rivals (Bernstein 2007). Or we might consider how, before the conventions for printed portraits and geographical maps became codified in seventeenth century England, Simon de Passe was able to playfully merge both nascent discourses to create a unique cartographic portrait of Captain John Smith and his putative colony of New England. This famous, complex, and iconic image has been seriously misunderstood by historians of all stripes; we can only make proper sense of it by considering the still-flexible discursive practices that engendered it (Edney 2010).

Fourth, and most important, a discursive approach permits us to understand that the relationship of map to territory is neither determined, as the modern cartographic ideal would have it, nor arbitrary and ineluctably fictitious, as postmodernists would argue. Rather, the relationship is contingent. It is in this contingency that the history of the field sciences can be properly integrated with map history. We can study those particular discourses in which people go out into the field, interact with local inhabitants, and observe the landscape, and consider how these discursive formations define and delimit the precise practices of fieldwork – how they condition a specific survey technique, say, or construe the approach to be taken to local informants – and we can also explore how the conditions in the field affect the discursive formation. Or we might trace the moments when ideas, data, and technologies translate from one discourse to another, perhaps conceptually distant from the field itself, and in doing so are given new meanings. We should answer Bob Kohler's question, "are boundary objects *mutable* mobiles?" with a resounding "yes" (Kohler 2002a: 13). And as we do so, we will find that the complexity of the divide between the field and more physically contained and static sites – laboratories, museums, and libraries – can be effectively conceived through the diversity of particular networks.

That is, it is simply inappropriate to conceptualize a singular field/map relationship, as if there were just one kind of map and one kind of field. The conviction that "map" legitimately describes a single, coherent category of texts underpins the conviction that "field" is similarly a singular and coherent category of situation. So, just as historians of science have come to appreciate the multiplicity of field situations that are defined contextually rather than through any presumably inherent characteristics, so too we must appreciate the multiplicity of map forms and mapping practices. The role of fieldwork within mapping is thus not a given but a question to be explored historically; it is, moreover, a question that lies at the core of a processual map history.

Acknowledgments

Too many people have contributed to this work, through discussions, their writing, and their aid to be thanked here. I must however specifically thank Karl Longstreth for his infrastructural support, Carla Lois for helping me clarify the final task, and Charlie Withers for his kind comments.

References

Akerman, J. R. 2006. "Twentieth-Century American Road Maps and the Making of a National Motorized Space." In: J. R. Akerman (ed.) *Cartographies of Travel and Navigation*. Chicago: University of Chicago Press, 151-206.

Andrews, J. H. 1998. "Definitions of the Word 'Map,' 1649-199." *MapHist Discussion Papers*, retrieved July 11, 2011 from http://www.maphist.nl/papers/199801.html

Arnberger, E. 1966. *Handbuch der Thematischen Kartographie*. Vienna: Franz Deuticke.

Axelsen, B. & M. Jones 1987. "Are all Maps Mental Maps?" *GeoJournal*, 14: 447-464.

Balm, R. 2009. "Discovery as Autobiography: The Machu Pichu Case." *Terrae Incognitae*, 40: 102-113.

Baudrillard, J. 1983. *Simulations*. Trans. P. Foss, P. Patton & P. Beitchman. New York: Semiotext(e).

Bernstein, D. 2007. "'We are not now as we once were': Iowa Indians' Political and Economic Adaptations during U.S. Incorporation." *Ethnohistory*, 54(4): 605-637.

Bijker, W. E. 1995. *Of Bicycles, Bakelites, and Bulbs: Toward a Theory of Sociotechnical Change*. Cambridge, MA: MIT Press.

Bijker, W. E., T. P. Hughes & T. J. Pinch (eds.) 1987. *The Social Construction of Technological Systems: New Directions in the Sociology and History of Technology*. Cambridge, MA: MIT Press.

Black, J. 1997. *Maps and Politics*. London: Reaktion Books.

Blakemore, M. J. & J. B. Harley 1980. *Concepts in the History of Cartography: A Review and Perspective*. Cartographica Monograph 26, *Cartographica* 17, no. 4. Toronto: University of Toronto Press.

Blond, S. 2008. "L'atlas de Trudaine: Pouvoirs, administrations et savoirs techniques (vers 1730-vers 1780)." PhD dissertation, École des Hautes Études en Sciences Sociales.

Borges, J. L. 1964. *Dreamtigers*. Trans. M. Boyer & H. Morland. Austin, TX: University of Texas Press.

Brown, B. & E. Laurier. 2005. "Maps and Journeys: An Ethno-Methodological Investigation." *Cartographica*, 40(3): 17-33.

Brown, L. A. 1949. *The Story of Maps*. Boston: Little, Brown & Co.

Buczek, K. 1982. *The History of Polish Cartography from the 15th to the 18th Century*. Trans. A. Potocki. 2nded. Amsterdam: Meridian Publishing Co.

Burnett, D. G. 2000. *Masters of All They Surveyed: Exploration, Geography, and a British El Dorado*. Chicago: University of Chicago Press.

Byerly, A. 2007. "'A Prodigious Map beneath his Feet': Virtual Travel and the Panoramic Perspective." *Nineteenth-Century Contexts*, 29(2-3): 151-168.

Campbell, E. M. J. 1949. "An English Philosophico-Chorographical Chart." *Imago Mundi*, 6: 79-84.

Carroll, L. 1893. *Sylvie and Bruno Concluded*. London: Macmillan.

Charlesworth, M. 1999. "Mapping, the Body and Desire: Christopher Packe's Chorography of Kent." In: D. Cosgrove (ed.) *Mappings*. London: Reaktion Books, 109-124.

Collier, P. & R. Inkpen 2002. "The RGS, Exploration and Empire and the Contested Nature of Surveying." *Area*, 34(3): 273-283.

Collier, P. & R. Inkpen 2003. "The Royal Geographical Society and the Development of Surveying, 1870-1914." *Journal of Historical Geography*, 29: 93-108.

Crone, G. R. 1953. *Maps and Their Makers: An Introduction to the History of Cartography* London: Hutchinson.

Curry, M. R. 2005. "Toward a Geography of a World without Maps: Lessons from Ptolemy and Postal Codes." *Annals of the Association of American Geographers*, 95(3): 680-691.

Daly, C. P. 1879. "The Early History of Cartography, or What We Know of Maps and Map-Making before the Time of Mercator." *Journal of the American Geographical Society of New York*, 11: 1-40.

Dando, C. E. 2007. "Riding the Wheel: Selling American Women Mobility and Geographic Knowledge." *ACME: An International E-Journal for Critical Geographies*, 6(2): 174-210.

Darnton, R. 1982. "What is the History of Books?" *Daedalus: Proceedings of the American Academy of Arts and Sciences*, 111(3): 65-83.

Daston, L. & P. Galison. 1992. "The Image of Objectivity." *Representations*, 40: 81-128.

Del Casino Jr., V. J. & S. P. Hanna 2006. "Beyond the 'Binaries': A Methodological Intervention for Interrogating Maps as Representational Practices." *ACME: An International E-Journal for Critical Geographies*, 4(1): 34-56.

Della Dora, V. 2007. "Putting the World into a Box: A Geography of Nineteenth-Century 'Travelling Landscapes'." *Geografiska Annaler Series B-Human Geography*, 89B(4): 287-306.

Eckert, M. 1921-25. *Die Kartenwissenschaft: Forschungen und Grundlagen zu einer Kartographie als Wissenschaft*. 2 vols. Berlin: Walter de Gruyter.

Eco, U. 1985. "Map of the Empire." Trans. S. E. Scalia. *The Literary Review*, 28(2): 233-238.

Eco, U. 1994. "On the Impossibility of Drawing a Map of the Empire on a Scale of 1 to 1." In: U. Eco. *How to Travel with a Salmon & Other Essays*. New York: Harvest, 95-106.

Edney, M. H. 1993. "Cartography without 'Progress': Reinterpreting the Nature and Historical Development of Mapmaking." *Cartographica*, 30(2&3): 54-68.

Edney, M. H. 1994. "Cartographic Culture and Nationalism in the Early United States: Benjamin Vaughan and the Choice for a Prime Meridian, 1811." *Journal of Historical Geography*, 20(4): 384-395.

Edney, M. H. 1994a. "Mathematical Cosmography and the Social Ideology of British Cartography, 1780-1820." *Imago Mundi*, 46: 101-116.

Edney, M. H. 1997. *Mapping an Empire: The Geographical Construction of British India, 1765–1843*. Chicago: University of Chicago Press.

Edney, M. H. 1999. "Reconsidering Enlightenment Geography and Map-Making: Reconnaissance, Mapping, Archive." In: D. N. Livingstone & C. W. J. Withers (eds.) *Geography and Enlightenment*. Chicago: University of Chicago Press, 165-198.

Edney, M. H. 2005. "Putting 'Cartography' into the History of Cartography: Arthur H. Robinson, David Woodward, and the Creation of a Discipline." *Cartographic Perspectives*, no. 51: 14-29.

Edney, M. H. 2005a. *The Origins and Development of J. B. Harley's Cartographic Theories*. Cartographica Monograph 54, *Cartographica* 40(1/2). Toronto: University of Toronto Press, 2005, 19-31 and 135-138.

Edney, M. H. 2007. "Mapping Parts of the World." In: J. R. Akerman & R. W. Karrow Jr. (eds.) *Maps: Finding Our Place in the World*. Chicago: University of Chicago Press, 117-157.

Edney, M. H. 2007a. "Recent Trends in the History of Cartography: A Selective, Annotated Bibliography to the English-Language Literature." Ver. 2.1, *Coordinates: Online Journal of the Map and Geography Round Table, American Library Association*, Series B, no. 6. Retrieved July 11, 2011 from http://www.stonybrook.edu/libmap/coordinates/seriesb/no6/b6.pdf

Edney, M. H. 2008. "John Mitchell's Map of North America (1755): A Study of the Use and Publication of Official Maps in Eighteenth-Century Britain." *Imago Mundi*, 60(1): 63-85, 67-70.

Edney, M. H. 2009. "The Irony of Imperial Mapping." In: J. R. Akerman (ed.) *The Imperial Map: Cartography and the Mastery of Empire*. Chicago: University of Chicago Press, 11-45.

Edney, M. H. 2010. "Simon de Passe's Cartographic Portrait of Captain John Smith and a New England (1617)." *Word & Image*, 26(2): 186-213.

Edwards, J. 2006. *Writing, Geometry and Space in Seventeenth-Century England and America: Circles in the Sand*. London: Routledge.

Feinberg, R., U. Dymon, P. Paiaki, P. Rangituteki, P. Nukuriaki & M. Rollins 2003. "'Drawing the Coral Heads': Mental Mapping and its Physical Representation in a Polynesian Community." *Cartographic Journal*, 40(3): 243-253.

Findlen, P. 1998. "Between Carnival and Lent: The Scientific Revolution at the Margins of Culture." *Configurations*, 6(2): 243-267.

Foucault, M. 1970. *The Order of Things: An Archaeology of the Human Sciences*. New York: Random House.

Foucault, M. 1972. *The Archaeology of Knowledge and the Discourse on Language*: Trans. A. M. Sheridan Smith. New York: Pantheon Books.

Godlewska, A.1988. *The Napoleonic Survey of Egypt: A Masterpiece of Cartographic Compilation and Early Nineteenth-Century Fieldwork*. Cartographica Monograph 38-39, *Cartographica* 25, nos. 1 and 2. Toronto: University of Toronto Press.

Golinski, J. 2005. *Making Natural Knowledge: Constructivism and the History of Science*. 2nd ed. Chicago: University of Chicago Press.

Guelke, L. 1976. "Cartographic Communication and Geographic Understanding." *The Canadian Cartographer*, 13(2): 107-122.

Harley, J. B. & D. Woodward (eds.) 1987. *Cartography in Prehistoric, Ancient, and Medieval Europe and the Mediterranean*. Vol. 1 of *The History of Cartography*. Chicago: University of Chicago Press.

Harley, J. B. 1989. "Deconstructing the Map." *Cartographica*, 26(2): 1-20.

Harley, J. B. 1989a. "Historical Geography and the Cartographic Illusion." *Journal of Historical Geography*, 15: 80-91.

Harley, J. B. 2001. *The New Nature of Maps: Essays in the History of Cartography*. Ed. B. Laxton. Baltimore: Johns Hopkins University Press.

Harvey, D. 1989. *The Condition of Postmodernity: An Enquiry into the Origins of Cultural Change*. Oxford: Basil Blackwell.

Herb, G. H. 1997. *Under the Map of Germany: Nationalism and Propaganda, 1918–1945*. London: Routledge.

Johns, A. 1998. *The Nature of the Book: Print and Knowledge in the Making*. Chicago: University of Chicago Press.

Kain, R. J. P. & E. Baigent 1992. *The Cadastral Map in the Service of the State: A History of Property Mapping*. Chicago: University of Chicago Press.

Keltie, J. S. 1885. "Geographical Education: Report to the Council of the Royal Geographical Society." *Supplementary Papers of the Royal Geographical Society*, 1(4): 439-594.

King, G. 1996. *Mapping Reality: An Exploration of Cultural Cartographies*. New York: St. Martin's Press.

Kish, G. 1980. *La Carte: Image des civilisations*. Paris: Éditions du Seuil.

Kitchin, R. & M. Dodge 2007. "Rethinking Maps." *Progress in Human Geography*, 31(3): 331-344.

Kohler, R. E. 2002. "Place and Practice in Field Biology." *History of Science*, 40(128): 189-210.

Kohler, R. E. 2002a. *Landscapes & Labscapes: Exploring the Lab-Field Border in Biology*. Chicago: University of Chicago Press.

Konvitz, J. W. 1987. *Cartography in France, 1660-1848: Science, Engineering, and Statecraft* Chicago: University of Chicago Press.

Korzybski, A. 1933. *Science and Sanity: An Introduction to Non-Aristotelian Systems and General Semantics*. Lakeville, CT.: The International Non-Aristotelian Library Publishing Company, 1933.

Kretschmer, I., J. Dörflinger & F. Wawrik (eds.) 1986. *Lexicon zur Geschichte der Kartographie von den Anfängen bis zum ersten Weltkrieg*. Vol. C of *Die Kartographie und ihre Randgebiete: Enzyklopädie*, 2 vols. Vienna: Franz Deuticke.

Krogt, P. van. 2006. "Kartografie of Cartografie?" *Caert-Thresoor*, 25(1): 11-12.

Kuklick, H. & R. E. Kohler 1996. "Sciences in the Field: Introduction." *Osiris* 11: 1-14.

Latour, B. 1987. *Science in Action: How to Follow Scientists and Engineers Through Society*. Cambridge, MA: Harvard University Press.

Latour, B. 1990. "Drawing Things Together." In: M. Lynch & S. Woolgar (eds.) *Representation in Scientific Practice*. Cambridge, MA: MIT Press, 19-68.

Latour, B. 2005. *Reassembling the Social: An Introduction to Actor-Network-Theory*. Oxford: Oxford University Press.

Lewis, M. G. (ed.) 1998. *Cartographic Encounters: Perspectives on Native American Mapmaking and Map Use*. Chicago: University of Chicago Press.

Licoppe, C. 2002. "The Project for a Map of Languedoc in Eighteenth-Century France at the Contested Intersection between Astronomy and Geography: The Problem of Co-ordination between Philosophers, Instruments and Observations as a Keystone of Modernity." In: M.-N. Bourguet, C. Licoppe & H. O. Sibum (eds.) *Instruments, Travel, and Science: Itineraries of Precision from the Seventeenth to the Twentieth Century*. London: Routledge, 51-74.

Livingstone, D. N. 2003. *Putting Science in its Place: Geographies of Scientific Knowledge*. Chicago: University of Chicago Press.

Lloyd, R. & P. Gilmartin 1987. "The South Carolina Coastline on Historical Maps: A Cartometric Analysis." *The Cartographic Journal*, 24(1): 19-26.

Maling, D. H. 1989. *Measurements from Maps: Principles and Methods of Cartometry*. Oxford: Pergamon Press.

McKenzie, D. F. 1999. *Bibliography and the Sociology of Texts*. 2nd ed. Cambridge: Cambridge University Press.

Mills, S. 2004. *Discourse*. 2nd ed. London: Routledge.

Mitchell, P. 2008 *Cartographic Strategies of Postmodernity: The Figure of the Map in Contemporary Theory and Fiction*. London: Routledge.

Monmonier, M. 1985. *Technological Transition in Cartography*. Madison: University of Wisconsin Press.

Mukerji, C. 2002. "Cartography, Entrepreneurialism, and Power in the Reign of Louis XIV: The Case of the Canal du Midi." In: P. H. Smith & P. Findlen (eds.) *Merchants & Marvels: Commerce, Science, and Art in Early Modern Europe*. New York: Routledge, 248-276.

Olsson, G. 2007. *Abysmal: A Critique of Cartographic Reason*. Chicago: University of Chicago Press.

Orlove, B. 1993. "The Ethnography of Maps: The Cultural and Social Contexts of Cartographic Representation in Peru." *Cartographica*, 30(1): 29-46.

Packe, C. 1737. *A Dissertation upon the Surface of the Earth, As Delineated in a Specimen of a Philosophico-Chorographical Chart of East-Kent*. London: J. Roberts.

Packe, C. 1743. Ανκογραφια, *sive Convallium Descriptio. In which are Briefly but Fully Expounded The Origine, Course and Insertion; Extent, Elevation and Congruity Of all the Valleys and Hills, Brooks and Rivers, (as an Explanation of a New Philosophico-Chorographical Chart) of East-Kent*. Canterbury, Kent: J. Abree for the Author.

Pelletier, M. 1990. *La Carte de Cassini: L'Extraordinaire Aventure de la carte de France*. Paris: Presses de l'École nationale des Ponts et Chaussées.

Petchenik, B. B. 1997. "Cognition in Cartography." In L. Guelke (ed.) *The Nature of Cartographic Communication*. Cartographica Monograph 19. Toronto: B. V. Gutsell, 117-128.

Picard, J. & P. de La Hire 1693. "Pour la carte de France corigée sur les observations de MM Picard & de la Hire." *Recueil d'observations faites en plusieurs voyages par ordre de sa Majesté, pour perfectionner l'astronomie et la géographie*. Paris.

Pickles, J. 2004. *A History of Spaces: Cartographic Reason, Mapping and the Geo-Coded World*. London: Routledge.

Raisz, E. 1937. "Outline of the History of American Cartography." *Isis*, 26(2): 373-391.

Raisz, E. 1937a. "Time Charts of Historical Cartography." *Imago Mundi*, 2: 9-16.

Revel, J. 1991. "Knowledge of the Territory." *Science in Context*, 4(1): 133-161.

Robinson, A. H. & B. B. Petchenik. 1976. *The Nature of Maps: Essays Toward Understanding Maps and Mapping*. Chicago: University of Chicago Press.

Robinson, A. H. 1982. *Early Thematic Mapping in the History of Cartography.* Chicago: University of Chicago Press.

Roszak, T. 1972. *Where the Wasteland Ends: Politics and Transcendence in Postindustrial Society.* Garden City, N.Y.: Doubleday.

Rundstrom, R. A. 1991. "Mapping, Postmodernism, Indigenous People and the Changing Direction of North American Cartography." *Cartographica*, 28(2): 1-12.

Saint-Exupéry, A. 1931. *Vol de Nuit.* Paris.

Sandler, C. 1905. *Die Reformation der Kartographie um 1700.* Munich: R. Oldenbourg.

Scharfe, W. 1986. "Max Eckert's *Kartenwissenschaft*: The Turning Point in German Cartography." *Imago Mundi*, 38: 61-66.

Shapin, S. 1995. "Here and Everywhere: Sociology of Scientific Knowledge." *Annual Review of Sociology*, 21: 289-321.

Shapin, S. 1996. *The Scientific Revolution.* Chicago: University of Chicago Press.

Shapin, S. 1998. "Placing the View from Nowhere: Historical and Sociological Problems in the Location of Science." *Transactions of the Institute of British Geographers*, 23(1): 5-12.

Sieg, W. & D. Schlimm 2005. "Dedekind's Analysis of Number: Systems and Axioms." *Synthese*, 147(1): 121-170.

Sismondo, S. & N. Chrisman 2001. "Deflationary Metaphysics and the Natures of Maps." *Philosophy of Science*, 68(3 suppl., Proceedings of the 2000 Biennial Meeting of the Philosophy of Science Association): S38-S49.

Skelton, R. A. 1972. *Maps: A Historical Survey of their Study and Collecting.* Ed. D. Woodward. Chicago: University of Chicago Press.

Smith, C. & J. Agar 1998. "Introduction: Making Space for Science." In C. Smith & J. Agar (eds.) *Making Space for Science: Territorial Themes in the Shaping of Knowledge.* London: Macmillan in association with the Center for the History of Science, Technology and Medicine, University of Manchester, 1-23.

Stockhammer, R. 2001. "'An dieser Stelle': Kartographie und die Literatur der Moderne." *Poetica: Zeitschrift für Sprach- und Literaturwissenschaft*, 33(3-4): 273-306.

Sullivan, R. 1883. *Geography Generalised: Or, an Introduction to the Study of Geography on the Principles of Classification and Comparison.* Ed. S. Haughton. 65th ed. Dublin: Sullivan Brothers.

Terrall, M. 1998. "Heroic Narratives of Quest and Discovery." *Configurations*, 6(2): 223-242.

Thrower, N. J. W. 1996. *Maps and Civilization: Cartography in Culture and Society.* 2nd ed. Chicago: University of Chicago Press.

Toulmin, S. T. 1953. *The Philosophy of Science*. London: Hutchinson.

Tuan, Y.-F. 1977. *Space and Place: The Perspective of Experience*. Minneapolis, MN: University of Minnesota Press.

Turnbull, D. 1993. *Maps are Territories: Science is an Atlas: A Portfolio of Exhibits*. Chicago: University of Chicago Press.

Turnbull, D. 2007. "Maps, Narratives and Trails: Performativity, Hodology and Distributed Knowledges in Complex Adaptive Systems – an Approach to Emergent Mapping." *Geographical Research*, 45(2): 140-149.

van den Broecke, M. P. R. 2009. *Ortelius' Theatrum Orbis Terrarum (1570-1641): Characteristics and Development of a Sample of 'on verso' Map Texts*. Utrecht: Koninklijk Nederlands Aardrijksundig Genootschap and Faculteit Geowetenschappen Universiteit Utrecht.

Wallach, A. 2005. "Some Further Thoughts on the Panoramic Mode." In: P. Earenfight & N. Siegal (eds.) *Within the Landscape: Essays on Nineteenth-Century American Art and Culture*. Carlisle, Penn.: The Trout Gallery, Dickinson College, 99-128.

Warner, M. 1990. *The Letters of the Republic: Publication and the Public Sphere in Eighteenth-Century America*. Cambridge, MA: Harvard University Press.

Withers, C. W. J. 2007. *Placing the Enlightenment: Thinking Geographically about the Age of Reason*. Chicago: University of Chicago Press.

Withers, C. W. J. 2010. "Histories of Cartography." In: M. H. Edney & M. S. Pedley (eds.) *Cartography in the European Enlightenment*. Chicago: University of Chicago Press, in prep.

Wood, D. 1992. *The Power of Maps*. New York: Guilford Press.

Wood, D. 1993. "The Fine Line between Mapping and Mapmaking." *Cartographica*, 30(4): 50-60.

Woodward, D. 1980. "The Study of the Italian Map Trade in the Sixteenth Century: Needs and Opportunities." In: C. Koeman (ed.) *Land- und Seekarten im Mittelalter und in der frühen Neuzeit*. Wolfenbütteler-Forschungen, 7. Munich: Kraus International, 137-146.

Woodward, D. & G. M. Lewis (eds.) 1998. *Cartography in the Traditional African, American, Arctic, Australian, and Pacific Societies*, vol. 2.3 of *The History of Cartography*. Chicago: University of Chicago Press.

Woodward, D. 1974. "The Study of the History of Cartography: A Suggested Framework." *The American Cartographer*, 1: 101-115.

Wright, J. K. 1942. "Map Makers are Human." *The Geographical Review*, 32: 527-544.

Author Information

Casper Andersen is an assistant professor of history of ideas at Aarhus University. He has published on British Engineers and the British Empire and his current research pursues this theme into the period of decolonisation.
See http://pure.au.dk/portal/en/persons/ideca@hum.au.dk.

Jenny Beckman is senior lecturer in History of Science and Ideas at Uppsala University. Her research interests include the history of biology, science pedagogy, and amateurs in science.

Mikkel Bunkenborg is a postdoc at the Department of Anthropology, University of Copenhagen, and he is presently engaged in a collaborative research project on Chinese interventions in Mongolia and Mozambique funded by the Danish Research Council for Independent Research in the Social Sciences.
See http://anthropology.ku.dk/staff/beskrivelse/?id=76464.

Palle Ove Christiansen is a researcher at The Royal Library, Copenhagen, Denmark. His research includes a scientific biography of folklorist Evald Tang Kristensen (1843-1929) and the history of cultural knowledge in the last part of the nineteenth century.

Daniel E. Clinkman is a doctoral candidate in the School of History, Classics and Archaeology at the University of Edinburgh. His current research focuses on constitutional and legal thought in the British Enlightenment.

Matthew H. Edney holds the Osher Chair in the History of Cartography, University of Southern Maine, USA, and directs the History of Cartography Project, University of Wisconsin–Madison, USA.
See http://www.usm.maine.edu/~edney/.

Kasper Risbjerg Eskildsen is associate professor of history of science at Roskilde University, Denmark. His research focuses on German intellectual history and the history of human sciences.
See http://forskning.ruc.dk/site/person/eskild.

Esther Fihl is dr.phil and professor. She is head of Center for Comparative Cultural Studies, Department of Cross-Cultural and Regional Studies at University of Copenhagen. She is trained an anthropologist and her current research focuses on alternative spaces relatively independent of well-established political institutions of the nation state.

Michael Harbsmeier, associate professor at Roskilde University, studied anthropology and teaches history. He has written about early modern European travel writing and expeditions, and is currently also working on how travellers from other parts of the world have described Europe and the West.
See http://forskning.ruc.dk/site/person/miha.

Anke Fischer-Kattner is a research and teaching associate at the University of the Bundeswehr, Munich. She has recently completed her doctoral dissertation on European travelogues about Africa, c. 1760-1860, examining narrative and performative dimensions of the processes of knowledge production.
See http://www.unibw.de/sowi2/fnz/personen/Fischer-Kattner.

Kristian H. Nielsen is an associate professor of history of science and science communication at the Center of Science Studies, Aarhus University. His current research focuses on science communication during the Cold War.
See http://pure.au.dk/portal/en/khn@ivs.au.dk.

Morten Axel Pedersen, who is associate professor in the Department of Anthropology, University of Copenhagen, is the principal investigator of the research project on Chinese investments in Sub-Saharan Africa and Inner Asia and the author of *Not Quite Shamans: Spirit Worlds and Political Lives in Northern Mongolia* (Cornell University Press, 2011).
See http://antropologi.ku.dk/ansatte/beskrivelse/?id=255694.

Serge Reubi is lecturer in history at the Institut d'histoire, University of Neuchâtel, Switzerland. His current research focuses on the process of professionalization in the social/human sciences in Europe, on the historicity of concepts in social sciences, and on the frontiers between science, art and trade.
See http://www2.unine.ch/cms/site/histoire/op/edit/serge.reubi.

Christopher J. Ries is a postdoc researcher at the Center for Science Studies, Aarhus University. He is currently working on geological investigations in Greenland during the Cold War.

Rengenier C. Rittersma is a postdoc researcher, specialised in the field of history of imagination and of historical myths. He is currently writing a book on the cultural history of the truffle in Europe. Since autumn 2007, he serves as the secretary of the editorial board of the journal *Food & History*.

Jeppe Strandsbjerg is an associate professor in International Relations at the Department for Business and Politics, Copenhagen Business School. He has previously published on the relationship between space, territory and cartography, and his current research investigates the relationship between cartography and geopolitics in the Arctic Region.

Jeremy Vetter is an assistant professor of history at the University of Arizona in Tucson. He is the editor of *Knowing Global Environments: New Historical Perspectives on the Field Sciences* (Rutgers University Press, 2010). He works at the intersection of the history of science and technology, environmental history, and the history of the American West.

Index

A
d'Abbadie, Antoine 13-4, 147-9, 152-166
d'Abbadie, Arnauld 13-4, 147-9, 152-166
Abyssinia 13, 153, 156, 160, 162, 164
Académie Royale des Sciences 31, 153
Academy of Sciences of St. Petersburg 39
Acta Eruditorum 103
Adanson, Michel 151, 167
Afghanistan 268-9, 285, 287-91, 298
Africa 13, 132, 147, 151, 153, 158, 160, 165, 169-86, 283, 294, 377, 416, 418-9
African Society 183-4
Agricultural experiment station 232, 237, 239, 242. See also Indiana Agricultural Experiment Station
Agriculture 40, 238-9, 398-9
Alabama Expedition 335
Alberti, Lodewyk 44
All Taxa Biodiversity Inventory (ATBI) 19, 402
Amateur 12, 25-7, 30, 137, 249, 286, 288, 309, 346, 349, 356, 418
American Indian 26, 31, 44, 145, 231, 233-4, 249-50

American Museum of Natural History 233, 311
Anthropo-geography 292
Anthropology 9, 11, 134, 223, 26-7, 29-30, 424, 51, 150, 160, 165, 192, 219, 233-4, 309-12, 315, 319-21, 324-6, 415-18, 420, 424, 426. *See also* History of anthropology
Anthropometry 312, 314
d'Anville, Jean Baptiste Bourguignon 151
Archaeological Survey of India (ASI) 21, 260-1, 267-8, 270-1, 274-79
Archaeology 21, 25, 248-9, 259-61, 264-5, 267-8, 270, 275-79, 297, 309, 321, 344
Arctic 19, 283, 285, 287, 330, 333-6, 338, 341-2, 344, 347-52, 356-7, 407
Aristotle 80
Armchair 9, 14, 24, 29-31, 42, 191, 221, 233, 291, 309, 329-3, 345-7, 352
Arnold, Gottfried 106-9
Artifact 11, 233-4, 248-9, 264, 272-3, 275, 301, 312, 314-5, 317, 322-3, 417, 443
Asbjørnsen, P. Chr. 197
Ashburner, Charles 236
Asia 18, 266, 273 283-89, 291, 293, 296, 298, 301, 304-5, 416, 418-20,

Astronomy 31, 34, 40, 153
Atlas 57, 59-60, 65, 269, 407, 436
Australia 29, 44, 376, 378, 416, 418, 424

B
Bachelard, Gaston 51, 326
Bacon, Francis 93
Bahnson, Kristian 301, 304
Bailey, Vernon 230, 236
Baird, Spencer 230
Ballad 23-4, 191-2, 194, 196, 202, 206-210, 212, 217, 219
Baluchistan 264, 266-8, 274
Banerji, Rakhal Das 261
Banks, Joseph 124-6, 128, 130-1, 136-7, 144-5
Barbour, Erwin H. 231
Bartholomew, Elam 239
Bartolucci, Francesco 83, 85, 89
Basel Ethnographical Museum 23, 311, 323
Basnage, Jacques 102
Bayle, Pierre 102, 106, 110
Beeler, Henry C. 231
Benthem, Henrich Ludolff 103-12
Berggren, Matz 404
Bering, Vitus 39
Berlin Ethnographical Museum 311
Berlingske Tidende 369, 374-5, 384
Bertelsen, Alfred 334
Bessey, Charles 230, 232, 243
Biodiversity 15, 395-6, 398-401, 404-6, 408,
Biogeography 5, 77
Biology 11, 242, 363, 379, 395, 397, 399, 403
Blagden, Charles 124, 134, 145
Blakemore, Michael 440, 442

Bloch, Marc 438
Board of Longitude 137
Board of Ordnance. *See also* Ordnance Survey 20, 124, 128-9, 137, 142
Boas, Franz 27, 29, 196, 219, 233, 250
Böhme, Jacob 108-10
Bondeson, August 197, 208
Book 13, 16, 22, 27-30, 32-4, 36-8, 44-5, 57, 68, 81, 85, 102-9, 113-4, 125, 135, 148, 150, 164, 170, 172, 175-7, 184, 193-4, 196-8, 200-1, 204-5, 208, 211, 213-4, 216, 219, 285, 289, 296, 301-2, 336-9, 345, 348, 357-8, 366, 376, 416, 444
Bookshop 106
Botanical garden 101, 242
Botanical Garden (New York) 233
Botanical Museum (Copenhagen) 294-5
Botany 31, 32, 34, 42, 79, 82, 123, 232, 283, 286, 305, 324, 340, 344
Boyle, Robert 31, 234
Brahe, Tycho 61-3
Breckling, Friedrich 106-8
British Association for the Advancement of Science 235
British Museum 267-8
Bruce, James 153, 157, 160
Bruner, Lawrence 243
Brønlund, Jørgen 334
Buczek, Karol 438
Buddhism/buddhist 40, 261-4, 270, 276, 426
Bulletin of the Washburn College Laboratory of Natural History 232
Bureau des Longitudes 164
Bureau of American Ethnology (BAE) 229, 249, 250

Bureau of Biological Survey 234, 236
Bureus, Andreas 70
Burge, Charles O. 173
Byron, John 132, 140
Bøggild, Ove Balthasar 339-40, 345-7, 351

C
Cairo 153, 156, 158, 164
Cannabis 34
Cape of Good Hope 36, 127, 140, 141, 377-8
Cape Town 35, 378
Carlsberg Foundation 288, 340, 373
Carnegie Museum 246-9
Carte de France 433-4, 439
Cartographer 59, 62, 63, 70, 349
Cartography 12, 16-7, 27, 52-6, 59, 61-2, 65-8, 70-1, 431-5, 437-44, 446
Cartoon Crisis 376-8, 385
Cassius, Andreas 102-4, 108, 110
Ceccarelli, Alfonso 85, 86, 92
Center of calculation 54, 63, 68, 69
Ceylon 33-4, 38. *See also* Sri Lanka
China 286-9, 297-8, 418-9, 423
Christensen, Jens Tang 382
Christensen, Katherine Richardson 384
Christensen, Søren Haslund 380, 382, 388
Churchill, Winston 172
Civilization 126, 171-2, 184, 259, 260, 262, 268, 270-1, 273-9, 318, 334, 419, 436
Clarke, Charles 132, 140
Claus, Jacob 106
Claus, Jan 106, 107
Clements, Edith 244
Clements, Frederic 244
Clifford, James 27, 150
Climate 25, 38, 39, 90, 239, 260, 262, 263, 289, 297, 363, 365, 379, 388, 401
Climatology 123, 286
Cockerell, T. D. A. 233, 246
Collection 9, 11, 13, 16, 24, 26-7, 30, 32, 51, 101, 109, 162, 191-2, 194-5, 197, 232, 235, 285, 288, 294, 297, 300-1, 309, 312, 314-7, 323, 332-3, 336, 345-7, 397, 400, 405, 417, 440
Collector 24, 192-4, 197, 199, 202, 215, 217, 226-7, 232, 234, 237, 245-8, 294, 312
Collins, Harry 239
Colony/colonial 21, 22, 30, 35, 56, 65, 127, 153, 158, 169, 173-4, 179-81, 183-5, 187, 264-5, 277-79, 284-5, 289, 291, 293-4, 312-7, 322, 329, 331-2, 334-6, 350, 378, 415-6, 418-9, 427, 446-7. *See also* Empire and Imperial/imperialism
Colorado 225, 228, 231-6, 238, 241-6, 248-50
Colorado Biological Survey 233
Colorado Museum of Natural History 234, 251, 254
Colorado School of Mines 241, 246, 254
Colorado Scientific Society 228
Commission for Geological and Geographical Investigation in Greenland 285, 329, 331, 333, 336, 342, 346, 356
Commission for Scientific Investigations in Greenland 344-5, 356

Compass 124, 133, 141, 144, 159, 296
Condra, George E. 231
Conservation 261, 265, 395-6, 398-401, 406-9
Consulting engineer 174, 179, 181, 183, 185. See also Engineer
Convention of Biological Diversity (CBD) 398-9
Coode, Sir John 185
Cook, James 43, 124, 128, 130, 136, 141-3, 151, 419
Cope, E. D. 245
Copenhagen 12, 38, 59, 68, 114-5, 204, 209, 284-8, 291, 294-5, 301, 303, 332-4, 336, 338, 347, 353-4, 358, 363, 367, 382
Costeau, Jacques 368
Cragin, Francis W. 232
Craigie, W. A. 204
Credibility 16, 329-33, 335, 349, 398
Croese, Gerard 107
Crone, Gerald 437
Curtis, Roger 133, 143
Curzon, George 261-2, 296
Cutler, Ira E. 242

D
Dahlberg, Erik 62
Danckwerth, Caspar 68
Danish Geographical Society 288
Danish Geological Society 331, 338-9, 348, 354, 356.
Danmark Expedition (1906-08) 335, 346
Dedekind, Richard 434
Deichsel, Johann Gottlieb 102, 109
Delisle, Guillaume 436
Della Porta, Giambattista 81-2

Denmark 12, 15-6, 22, 24, 26, 45, 59-62, 64-5, 67-9, 191-4, 202, 208, 221, 284-8, 297-9, 329-30, 332-3, 335-6, 338-9, 343, 345, 350, 358, 363, 366, 368, 370, 372, 377-8, 380, 389, 406
Derrida, Jacques 442
Cartesianism 80-1, 108
Diary 24, 134, 176, 198, 216, 291, 293-5, 299-300
Dikshit, Kasinath Narain 271, 275
Dinosaur 246-8
Dinosaur National Momument (Utah) 249
Discipline 9-14, 16, 25, 27, 29-31, 35, 51, 116, 147, 150-2, 154, 158-9, 161, 163, 165, 191, 207, 234, 248, 251, 259-60, 267, 277-79, 296-7, 304-5, 309-10, 320-2, 324-6, 341, 344, 365, 369, 373, 397, 416, 418, 438. *See also* Interdisciplinarity, Pre-disciplinary
Discourse 17, 19, 158, 206, 292, 372, 381, 432, 444-8
Dixon, Jeremiah 127-8, 140
Dorsey, George 234

E
East India Company 33, 124-5, 130
Eckert, Max 438
Egede, Hans 38-9
Ehrenberg, Bonifacius Heinrich 102, 104
Emmons, Samuel F. 228
Empire 19, 20, 21, 55, 5873, 123, 126, 153-4, 156, 169, 172, 179-80, 183, 186, 235, 254, 262, 267 270, 279, 285-6, 329, 417, 419, 427. See also Colony/colonial

Engineer 19-20, 24-5, 125, 129-31, 169-90. See also Consulting engineer and Explorer-engineer

England 44, 102, 106-7, 109, 113, 116, 124, 128, 132, 144, 297, 333, 347, 447. *See also* Great Britain

Enlightenment 13, 29, 102-3, 113, 123, 125, 135-6, 151-2, 160, 164, 436, 437

Entomology 321

Environment 19-20, 25-6, 67-8, 70, 78, 94, 163, 174-5, 179, 194, 209, 212, 225-7, 238-9, 242, 244, 336, 363, 379, 399, 401, 407, 423

Erasmus of Rotterdam 104, 115, 117

Ethiopia 13, 147, 153-66

Ethnicity 262, 264, 272

Ethnographic Museum in Basel 311

Ethnography 11, 24, 34-6, 38-9, 150, 191-2, 196, 218, 221, 233, 249-50, 283, 286, 301, 304-5, 311-2, 325, 415-6, 418, 420-2, 424-7

Ethnological Society of London 235

Ethnology 23, 30, 43, 165, 325, 344, 417

Europe 13, 23, 30, 35-6, 51-2, 55, 57-61, 68, 71, 101, 104, 111, 115-7, 126, 136, 147-8, 151, 153-8, 160-5, 193, 208, 261-2, 285, 293-4, 301, 303-5, 311-2, 315, 317, 319, 324, 337, 347, 352, 377, 379, 396, 407, 424, 433, 436-7

European Republic of Letters 13, 101, 103, 109, 112, 114-7

Excavation 21, 246, 248, 259, 261, 268, 271, 273, 275-8, 321

Expedition 9-15, 22, 24-5, 28, 31, 39, 42, 44, 68, 82-3, 124, 126-8, 132, 153, 155, 157, 181-2, 184, 195, 201, 211, 213, 219, 226, 233, 235, 247-8, 283-305, 310-2, 314, 317, 322, 329, 331, 332-8, 340-4, 346-50, 353-8, 404, 408, 415-18, 424, 426-7. *See also* Scientific expedition

Experiment 9, 26, 33-4, 81-4, 86, 92, 94, 133, 143-5, 149, 237-45, 316, 389

Expert 26, 92, 123-4, 129, 169, 182-5, 192, 195, 230, 235, 284, 288, 293, 332-3, 336, 339, 345, 347, 349-51, 352, 373, 378, 403-4

Expertise 26, 78, 124, 128-9, 134-7, 303, 335, 349, 352, 357-8, 398

Exploration 9, 19, 37, 51, 71, 124, 126, 128, 151, 155, 162, 165, 169, 171-3, 179, 184, 186, 227, 246, 260-1, 274-5, 283, 286, 289, 296-8, 305, 311, 322, 329-33, 335-6, 339-41, 344, 347, 348-50, 352, 356, 357, 358, 363-4, 366, 368-9, 378, 420. See also Scientific exploration

Explorer-engineer 19, 24-5, 169, 171-5, 179, 181, 184, 186. See also Engineer

Extensive fieldwork 10, 23, 312, 316-18, 320, 323-5, 352

F

Fabricius, Johann Andreas 116

Faroe Islands 202, 211, 333, 339, 404, 406

Fauna 19, 244, 288, 298, 300, 399, 402, 404, 405

Federal University in Zürich 311

Fernow, Bernard 243
Field Station 18, 26, 226, 232-3, 237-45, 283-4, 291-2, 294-6, 298-305, 306, 307, 353, 403. *See also* Rolling field station
Field Surveying Office in Denver 226
Fieldwork 9-15, 17-20, 22-32, 34-39, 41-5, 51-2, 55, 58, 70-1, 77, 107, 109, 110, 112, 114, 152, 163, 181, 191-3, 195-7, 200-1, 209, 211, 216-21, 225-6, 229-35, 237, 249-50, 261, 271, 283-4, 291, 304-5, 309-12, 316-18, 320-5, 331-3, 339, 347, 352, 354, 356, 415-8, 420-1, 423-6, 432, 438, 440, 442, 448. *See also* Extensive fieldwork, Fieldworker, Historiography of fieldwork, Intensive Fieldwork, Multi-sited, Scientific fieldwork
Fieldworker 14, 18, 26, 102, 163, 211, 215, 346
Figgins, J. D. 234, 250
Fischer, Johann Eberhard 39
Flora 86, 90, 288, 294, 300, 399, 402, 407
Folk culture 26, 191, 194, 218, 220
Folklore 11, 23-4, 26, 191-3, 197, 200, 202, 207, 209, 212, 215, 217-20, 286
Forster, Johann Reinhold 43
Fortis, Alberto 43, 46
Foucault, Michel 442, 444
Fox, Douglas, 181
Fox, Francis 183-4
Fox, George 106
France 44, 70, 115-6, 123, 126, 130, 132, 134, 140, 147, 152, 154, 158, 159, 163-5, 309, 417, 438, 442
Francis, A. 271

Frank, Albert Bernhard 84
Frankfurt am Main 109
Franklin, Benjamin 128-9, 142
Frazer, James 29, 416
Frederick, Charles 128, 129, 132, 142
Freuchen, Peter 335, 348-9, 351, 357

G
Gadd, Cyril 270
Galathea 3 Expedition (2006-07) 6, 15, 21, 363-89
Ganges 267, 273
Gannett, Henry 235
Gärdenfors, Ulf 406, 411
Gascoigne, John 125
Geertz, Clifford 45, 150
General Land Office 226, 227
Geography 11, 13, 19, 31, 43, 55, 59, 123-4, 132-3, 136, 155, 159, 160, 164, 171, 234, 244, 283, 285-6, 297-8, 304-5, 344, 349, 434
Geological Museum in Copenhagen 347, 353
Geology 16, 19, 31, 42, 153, 227, 232, 286, 330-2, 338-40, 344-7, 354-5, 363, 379
Geophysics 344
George, Russell D. 231
de Gérando, Joseph-Marie 44-5
Germany 101, 103, 108, 109, 114, 130, 152, 309, 347, 353
Gibelli, Giuseppe 84
Girouard, Percy 173-4
Glenie, James 134, 143
Global Taxonomy Initiative (GTI) 398
Gobi desert 415, 419, 422-3, 427
Gojam 155-8
Great Britain 49. *See also* England

Greenland 16, 19, 38-9, 65, 285, 287, 329-58, 364, 378, 383, 406
Greenland Geological Survey (GGU) 330, 356, 358
Greenwich observatories 130. *See also* Royal Observatory
Greenwich survey 123, 131, 135
Gronovius, Jacob 102
Grundtvig, Svend 23, 26, 192-4, 196-7, 201-4, 206, 209-10, 212, 214-9, 221
Guha, B. S. 273-4

H
Haarder, Bertel 381
Haddon, Alfred Cort 27, 29, 416-8
Hage, Ghassan 424
Hall, Edwin 231
Hall, Marie Boas 124-5, 135
Handbook 105, 345
Hansen, Lars H. 384, 387
Harappa 21, 260, 261, 263-6, 268, 270, 274-5, 278
Hargreaves, Harold 266, 268, 275
Harley, Brian 438, 441-2
Hartlib, Samuel 234
Hawkshaw, John 172
Haworth, Erasmus 231
Hayden, Ferdinand V. 227, 246-7
Hazard, Paul 30, 31
Hemmingsen, Ralf 385
Henderson, Junius 233
Hennig, Willi 397
Hero/heroic/heroism 19, 24-5, 170, 173-5, 186, 333-6, 347-52, 356
Heumann, Christoph August 13, 101-4, 106-12, 115
Hilton, H. R. 239
Hindu/Hinduism 262, 264, 273

Hindu Kush 285-6, 296-7
de La Hire, Philippe 437, 439
History of anthropology 22, 26, 29, 289, 309, 325-6. See also Anthropology
History/historiography of cartography 16, 53, 55, 65, 437. See also Map history
History/historiography of fieldwork/ expeditions 9, 11-2, 14, 26-8, 29-30, 44. See also Fieldwork, Expeditions
History of ideas 326
History of science 25, 239, 251, 336, 396, 432
Hjuler, Anthon 288, 293-4, 299
Holbek, Bengt 207
Holberg, Ludvig 115-7
Holland 101-7, 109-11, 113-7. *See also* Netherlands, The
Holmes, Arthur 339-40
Hooke, Robert 33-4, 38, 80-1
Hultén, Eric 407
Hume, David 135
Huntington Southwest Survey 233
Hvam, Frank 376
Hybridity 158
Hydrography 136, 228
Hydrology 228

I
Iceland 65, 333, 339, 404, 406
Imhof, Eduard 439
Imperial/imperialism 19, 21-2, 24, 45, 52, 56, 58, 78, 85, 126, 154, 158, 164-5, 169-70, 173, 180-1, 183, 185-7, 259-65, 267, 269, 272, 274, 276-7, 278-9, 283, 288, 291, 293, 329, 366, 418-9, 424. *See also* Empire, Colony

Imperial Gazetteer Atlas 269
India 21-2, 36, 65, 173, 176-8, 259-79, 285, 289, 297, 442
Indian Ocean 378
Indiana Agricultural Experiment Station 229. See also Agricultural Experiment station
Indian Museum 265
Indigenous 21, 24, 154, 156-7, 159, 173, 260, 266, 278, 324, 366, 443, 445-7
Indo-Sumerian civilization 21, 260, 266, 268, 270, 278
Indus 21, 259-60, 263-4, 266-79
Indus River 259-60, 269, 274
Informant 14, 22, 24-6, 31-2, 34-5, 41, 192, 194-200, 205-6, 208-21, 283, 301-2, 304, 317, 448
Intensive fieldwork 22, 23, 192, 195, 200, 311, 316, 317, 318, 323, 324, 325. See also Fieldwork
Interdisciplinarity 12. See also Discipline
Inuit 38, 335-6, 350
Inventory 15, 19, 251, 321, 395-6, 398-405, 407-9
Iraq 22, 266, 270-1, 273, 278
Italy 25, 77, 85-7, 89, 101, 112, 114, 116

J
Jägerskiöld, Leonard 404
Jamison, Claude E. 231
Jansen, Albert 108
Japan 34-5, 48
Jena 102-4, 108, 116
Jensdatter, Maren 203, 213
Johnstrup, Frederik 331
Jordan, Marcus 58-60

Journal 13, 16, 40, 101-6, 108-5, 116, 147, 153, 162, 172, 175, 177, 181-2, 184, 265, 267-8, 332, 336, 348, 354, 369, 375. See also Travel journal, Scientific journal
Jubilee Expedition (1921-23) 334-6, 348-9
Jurieu, Pierre 106
Justamond, J. O. 134, 140
Jutland 23, 191, 193-4, 199-200, 202-3, 209, 211, 219, 220
Jyllands-Posten 365, 367, 369-72, 374-6, 379, 381-88
Jørgensen, Bo Barker 386

K
Kaempfer, Engelbert 34-5, 38, 44
Kafiristan 287, 298, 302
Kamchatka 39, 42
Kansas 225-6, 230-2, 237-9, 241-3, 246, 250
Karlsson, Anna 403
Kater, Henry 137
Khoikhoi 36-8, 45
King, Clarence 227
Kipling, Rudyard 172
Knowledge production 18-20, 26, 51-4, 67, 70, 149, 152, 225-6, 233, 242, 251, 258, 357, 401. See also Scientific knowledge
Knox, Robert 33-6, 38
Koch, Lauge 16, 19, 329-31, 334-58
Kohler, Robert 30, 284, 321, 325-6
Kolb, Peter 35-8, 45
Korzybski, Alfred 435, 441
Krafft, Hans Ulrich 32
Kretschmer, Ingrid 439

Kristensen, Evald Tang 23-4, 26, 191-21
Kuklick, Henrika 29-30, 284

L
Laboratory 11-2, 17, 53-4, 57-8, 65, 71, 78, 82, 85, 149, 225-6, 239, 241, 271, 291, 295-7, 321, 345, 347, 366
Lafitau, Joseph-François 30-1
Lämmermann, August 102, 113-4
Lancisi, Giovanni Maria 82, 85, 89
Langdon, Stephen 273
Latour, Bruno 16, 52-5, 58, 70-1, 149-50, 364, 443-4
Lauremberg, Hans 62-4
Lay network 226, 236
Leth, Jørgen 422
Library 62, 77, 85, 104, 217, 298, 453, 457, 473
Linguistics 11, 13, 164, 288, 417
Linnaeus, Carl 15, 44-5, 151, 395-6, 408
Linton, Edwin 230
Livingstone, David 165, 171, 189
London 33, 35, 43, 81, 123, 130-1, 136, 170-1, 173, 177-82, 185-6, 275-6, 285, 446
Lorda, Domingo 162-3
von Luschan, Felix 311
Lutheran 103, 105-6, 110, 112
Løffler, Ernst 286, 304

M
Mackay, Ernest John Henry 271-3, 275
Madsen, Anders Lund 380-1, 384
Malaise trap 402-3

Malinowski, Bronislaw 22-3, 27, 29, 31-2, 195, 216, 309-10, 321, 325, 415, 417-8
Malpighi, Marcello 78, 81, 84-5
Map 16-7, 22, 52, 55-68, 70-1, 151, 154, 227-9, 236, 240, 269, 271, 288, 348-9, 354, 369, 408, 420, 431-48
Map history 17, 431-2, 436-42, 444, 446-8. *See also* Historiography of Cartography
Mapping 11, 16-7, 19, 22, 53, 55, 58-61, 65, 68, 71, 132, 137, 171, 226-9, 288, 304, 329, 331, 336, 343, 348-9, 355-6, 404-5, 407, 432, 434, 437-48. *See also* Cartography
Marchant fils, Jean 81-2
Marcus, George E. 150, 415, 420, 425
Marind-anim 311, 317-20
Markager, Stig 386
Marshall, John H. 21, 260-1, 263-6, 268-76, 278-9
Marsh, O. C. 245
Marsili, Luigi Ferdinando 25, 77-9, 81-94
Maskelyne, Nevil 133, 142, 144
Mason, Charles 127-8, 140
Massawa 153-4, 156, 158-9
Material culture 18, 270, 272, 275, 310, 312, 314-5, 317, 322-4, 418, 447
Mathematics 20, 59, 67, 126, 344, 397
Meddelelser om Grønland 332, 336, 341, 344, 354, 359-61
Media 15, 182, 335, 363-8, 370-1, 373-8, 380-9, 391, 397, 403
Medicine 17, 32, 40, 44, 156, 298, 344, 398, 416-7
Mediterranean, the 266, 270, 376-8, 452

Meitzen, August 438
Mejer, Johannes 22, 62-5, 67-9
Meldgaard, Morten 364-5, 374, 378
Mencke, Johann Burkhard 102-3
Mennonites 106, 109
Mercator, Gerard 58, 60, 436
Merriam, C. Hart 229
Merv oasis 291-3
Mesopotamia 260, 266
Meteorology 11, 344
Method 11, 14, 22-3, 26-7, 43, 61, 79, 85, 89, 91, 93, 110, 112, 126, 160, 184, 191, 197, 211, 234, 237, 243-4, 272, 275-9, 289, 300, 304-5, 309-11, 314, 319-21, 323, 325-6, 397-8, 404, 415, 417-8, 420, 422, 424-27, 433, 441
Methodology/methodological 11, 14, 23-4, 26, 150-3, 191, 196-7, 218, 283-4, 312, 322, 331, 365, 397, 402, 419, 424-7, 431, 436, 440
Michaelis, Johann David 44-5
Micheli, Pier Antonio 82, 84
Mielche, Hakon 368
Mikkelsen, Einar 335
Mineralogy 31, 42, 153
Mining 171-2, 226, 228, 236, 241-2, 250
Moe, Jørgen 197
Moe, Moltke 197
Mohenjodaro 21, 260-6, 268-76, 278-9
Molesworth, Sir Guildford 178
Moltke, Harald 334
Mombasa 175
Mongolia 13, 415-6, 419-24, 427, 428
Monro, Donald 124, 134, 141
Monument 249, 264-5, 272, 350
Mooney, James 250
Moore, Raymond C. 231
Morgan, Glenn B. 231

Motility 423-4
Mozambique 415-6, 419, 421-2
Mudge, B. F. 231
Müller, Gerhard Friedrich 39-43
Multi-sited 13, 14, 27, 415, 418, 420-1, 424-7
Mycology 79, 81, 82
Mylius-Erichsen, Ludvig 334

N

Narrative 15-9, 21-2, 24-7, 33-4, 65, 147-50, 156, 163, 173, 175, 208, 364, 366-7, 372, 386-7, 415, 425, 426, 431, 437
National Co-operative Soil Survey 230
National Museum of Denmark 284, 286-290
Native 13, 25, 35-6, 42, 116, 136, 141, 152, 154-6, 158, 161-2, 166, 174, 182, 229, 298, 302, 310, 313-5, 323, 350, 417, 447
Natural history 29, 31, 34-5, 38, 42-3, 82, 151, 225-8, 232, 309, 321, 400
Natural History Museum (Gothenburg) 404
Natural History Museum (London) 397
Natural History Museum (New York) 311
Navigation 55, 71, 124, 435, 445
Nebraska 225-6, 230-2, 238, 242-5, 247-8, 250, 252, 254, 256-8
Nebraska's Botanical Seminar ("Sem Bot") 232
Nelson, Aven 233
Netherlands, The 44, 105. *See also* Holland
New Guinea 23, 311, 317, 320, 416-7

New Hebrides 311-6, 323-4
Newton, Isaac 86, 93, 135
de Nicolay, Nicolas 70
Niebuhr, Carsten 45
Nile, The 154-5, 157, 164
Nomad/nomadic 10, 156, 268, 284, 288-9, 292-4, 415, 422-4
Norden/Nordism 406-7
Nordenskiöld, Adolf Erik 345
Nordmann, Valdemar 358
Norway 12, 16, 65, 69, 141, 197, 329, 333, 338, 356, 404, 406-8
Notebook 105, 196-8, 208, 292-3, 300, 302
Numismatics 262

O
Objects 9, 11, 13, 23, 29- 31, 219, 232, 246, 250, 268, 273, 284-5, 301, 304, 314-6, 318-20, 416, 418, 420, 422, 425, 427, 448
Observation 16, 18, 22, 30-4, 36, 40-1, 43-4, 55-6, 63, 67, 77, 81-2, 84, 87, 92-4, 105, 112, 124, 126-7, 133-4, 136, 140-5, 151, 154, 158-9, 192, 197, 206, 234-7, 291, 296-7, 301, 303, 305, 319, 322-3, 331-3, 336, 346, 348-9, 352, 354-6, 382, 415, 431, 433-5, 437-39, 443, 445
Observatory 26, 35, 55, 127, 137
Observer 20, 49, 70, 81, 133, 161, 202, 226, 234-7, 245, 377
Oceanography 78
O'Harra, C. C. 231
Oklahoma 250
Olrik, Axel 207-8, 212-3
Olsen, Gabriel 336
Olufsen, Ole 18-9, 283, 285-98, 301-2
Olwig, Karen Fog 426
Ordnance Survey 130, 132, 137, 438. *See also* Board of Ordnance
Orlove, Ben 447
Ortelius, Abraham 57-8, 60, 70
Osborn, Henry Fairfield 247
Oxford University 268, 273

P
Packard, Alpheus Spring 227, 252
Packe, Christopher 433, 435, 450
Pack, Frederick J. 231
Pakistan 21-2, 259, 276-8
Paleontology 247-9, 252-6, 354
Pallas, Peter Simon 42, 46
Pamir 283, 286-90, 293, 296-302
Pandsh River 287-8, 290, 296, 298, 301
Paradigm 23, 77, 305, 310-1, 320-1
"Party boat" 381-4
Pasteur, Louis 84, 149
Patriotism 21, 135, 137
Patterson, John Henry 175-9, 188
Paulsen, Ove 18, 288, 293-5, 299-303, 305
Peary, Robert E. 335, 350
Peary Channel 335, 347-9, 352
Peary Land 335-6, 346, 348-9
Penn, William 106
Perisho, Ellwood C. 231, 233
Perizonius, Jacob 102
Persia 33-4, 43, 266-267, 270, 274, 292, 298
Peterson, O. A. 247
Philipsen, Victor 287, 291-5, 305
Philology 192
Philosophical Transactions of the Royal Society 21, 31, 95, 123-5, 128-35, 401
Philosophy 109-11, 113, 160, 170, 441

Philosophy, natural 126, 135-6, 151
Philosophy of science 93
Physics 153, 344
Picard, Jean 437, 439
Pickersgill, Richard 133, 144
Pickstone, John 251
Pliny 79, 93
Plumb, C. S. 229
Political science 51
Politiken 364, 369, 374-5
Pool, R. J. 243-4
Popular/popularity 15, 21, 55, 135, 150, 164-5, 170, 172-3, 175, 177, 191, 228, 249, 260-1, 264-5, 267-8, 273, 279, 309, 348, 360, 367, 395-6, 399, 403, 409, 434-5
Portugal/Portuguese 55-6, 313, 325, 421
Possehl, Gregory 259, 278
Post-colonial 22. *See also* Colony, Empire
Powell, John Wesley 227-9
Prairie 231, 252
Pre-disciplinary 13, 14, 26. *See also* Discipline, Interdisciplinarity
Ptolemy 55, 57, 59
Public discourse 19, 186, 432, 447
Public understanding of science 364, 371, 379, 387, 389. *See also* Research communication, Science communication
Punjab 260, 264-5, 267, 274

Q
Quakers 106-7, 109
Quarry, mode of fieldwork 26, 226, 245-51
Questionnaire 25, 45, 90-1, 108

R
Radcliffe-Brown, Alfred Reginald 29, 310, 320-1, 325, 417
Railway 24, 171-3, 175-84, 198, 245, 271, 283, 289-95. *See also* Transcaspian railway
Ramaley, Francis 233, 244
Rantzau, Heinrich 60, 71
Rasmussen, Anders Fogh 370, 376-7
Rasmussen, Knud 334-6, 346, 348-51
Rauwolf, Leonhard 32, 34
Ravn, Niels Frederik 332
Ray, Sidney 416-7
Religion 11, 34-5, 40, 42, 104, 106, 111-2, 160, 264, 273-4, 310, 322, 377, 379-80
Renaissance 32, 48, 55, 73, 75-6
Representation 16-7, 22-4, 52, 54-6, 62, 65-71, 148, 150, 152, 169, 182, 284, 310, 424, 431-2, 434, 436, 438, 440-7
Research communication 363, 367-8, 370-1, 380, 385, 388. *See also* Public understanding of science, Science communication
Riggs, Elmer S. 248
Riis-Carstensen, Eigil 340-4
Rink, Hinrich 332
Rivers, W. H. R. 27, 29, 249-50, 416-8
Robertson, George Scott 285
Robinson, Arthur 438-40
Rocky Mountain Herbarium 233
Rocky Mountains 228, 231, 233, 241-6
Roepstorff, Peter 383
Rolling field station 18, 283, 291-2, 294-5, 305. See also Field station
Rose, Alexander 134, 140
Rose, Flemming 376, 379

Royal Geographical Society 177, 235, 253, 297
Royal Navy 20, 123-8, 132-3, 136-7, 140, 142, 178
Royal Observatory 126, 137, 139, 141. *See also* Greenwich observatories
Royal Society of Arts 135, 177, 179, 275
Royal Society of London 20-1, 33-5, 39, 86, 123-37, 140-2, 234
Royal Society of Northern Antiquaries 213
Roy, William 123-4, 128, 130, 135-8, 143-5
Russia 39, 42-3, 174, 283-94, 297-8, 304, 417
Russian Imperial Geographical Society 297
Rycault, Sir Paul 31
Rydberg, Per Axl 233

S
Sahni, Daya Ram 261, 263, 275
Said, Edward 186, 418
Salmon, Samuel 237
Samarkand 283, 290-1
Sander, Helge 371-4, 380-1, 389
Sandler, Christian 437
Sarasin, Fritz 311-2
Sarasin, Paul 311-2
Säve, P. A. 197
Sayce, Archibald 268
Schander, Christoffer 404
Schilder, Günter 439
Schlaginhaufen, Otto 311
Schleswig 60, 63, 68
Schmidt, Carsten 383

Scholar 9, 11-3, 22, 26-30, 39, 43, 53, 55, 77, 79-80, 82-5, 90, 92, 94, 101-5, 109-10, 112-7, 173, 192-4, 204, 207, 217, 221, 260, 262, 270, 273, 278, 309-10, 312, 321-2, 344, 432, 434, 436, 438, 440, 442-4. *See also* Scientist
Scholar-scientist 42
Science communication 364, 370-2. *See also* Public understanding of science, Research communication
Scientific expedition 12, 19, 287, 342, 363, 389. See also Expedition
Scientific exploration 286, 329, 344, 363-4, 366, 368, 378. See also Exploration
Scientific fieldwork 24, 221, 225. See also Fieldwork
Scientific journal 332, 348. See also Journal
Scientific knowledge 20, 53, 148-9, 260, 316, 346, 349, 357, 370, 432. See also Knowledge production
Scientific traveller 286, 288, 305
Scientist 9-13, 15, 18-19, 22, 25-30, 42, 54, 60, 127-8, 137, 149, 153, 164-5, 171, 192, 225, 227-8, 230, 235, 237, 239, 247, 251, 283, 288, 295, 297, 302, 305, 321, 329-30, 333-4, 340-6, 363-4, 370-1, 373, 375, 379, 387, 389, 397, 403, 416, 438. *See also* *Scholar*
Scudder, Samuel H. 246
Seidenfaden, Gunnar 337, 340-1, 350, 357
Seligman, Charles 29, 416-7
Settler 238-9, 314, 317
Seybold, David Christoph 114

Sewell, R. B. Seymour 273
Shantz, H. L. 244
Shelford, Frederic 174, 179, 180-85
Ship 66, 126-8, 133, 140-4, 154, 330-1, 333, 336, 340-3, 363, 365, 374-6, 380-4. See also "Party boat"
Siberia 39, 40, 42
Smiles, Samuel 170
Smith, J. Alden 235
Smith, Sidney 268-9, 273
Société de Géographie 154, 160-1, 164
Society 23, 33-5, 51, 59, 70-1, 123, 131, 176, 180, 259, 277, 279, 301, 325, 350, 368, 370, 372, 374, 379, 447
Society, ancient 260, 270, 274, 278
Society, learned 20, 128, 135, 137
Sociology 12, 51, 444
Soil 38, 77, 79, 83, 85-6, 89-90, 93, 97-9, 159, 229-30, 238-40, 292, 300-1
Soil survey 230
Source 9, 24, 91, 94, 103, 116, 154, 192, 206, 217, 219, 231, 245, 270, 353, 434
Source of the Nile 154, 157
South Dakota 225, 230-1, 233, 237, 245, 247-8, 250, 257
Sovereignty 22, 52, 59, 65, 67-8, 71, 260, 278, 287, 289-90, 331, 335, 337, 340, 359, 366, 380
Spain 55-6, 7-5, 112, 134, 142, 144, 482
Spallanzani, Lazzaro 84
Sparrmann, Anders 45
Specimens 42, 83, 124, 163, 232-4, 236, 245-6, 248, 268, 294, 300, 403
Speiser, Felix 22-3, 310-26
Spinoza 108
Sri Lanka 33. *See also* Ceylon

St. Petersburg 39-40, 42, 48, 286, 289
Stanley, Henry Morton 165, 171-2, 179
Statistics 236, 438
Steenstrup, K. J. V. 332, 345
Stein, Marc Aurel 261, 274, 298
Steinmann, Alfred 311
Stocking, George W. 30, 309, 321
Stolle, Gottlieb 102-3, 106, 108-9, 112-14
Stupa 261, 263, 271-2
Surveyors 52, 54, 57-8, 65, 67-8, 70-1, 230, 261-2, 271, 420, 440
Swallow, G. C. 231
Swart, Claudius Claussøn 59
Sweden 15, 19, 44, 60, 68, 197, 202, 333, 347, 395, 397, 399-401, 404, 406-9
Swedish Biodiversity Center 399
Swedish Environmental Protection Agency (SEPA) 396, 399, 404
Swedish Museum of Natural History 397
Swedish Species Information Center (SSIC) 399, 403
Swedish Taxonomy Initiative (STI) 15, 18-9, 395-6, 399-402, 404-9
Switzerland 309, 311, 347

T

Talbot, Frederick A. 172
Tales 23-4, 191, 193-4, 197, 207, 209-10, 212, 219
Taubman, George Goldie 179
Taxonomy 15, 19, 225, 251, 395-402, 407-9
Technology 17-9, 63, 212, 347, 350, 367-8, 372, 379, 396-7, 421, 443
Teichert, Curt 337, 352-6

Territory 16-7, 22, 51-3, 58-71, 132, 154, 156, 181, 227, 234, 292, 298, 302, 435, 441-2, 447
"The Eleven" 338-40, 345-6, 348-54, 357-8
Theology 81, 102, 106, 110-1
Theologians 81, 107, 109, 110
Thomasius, Christian 103, 106, 108-115
Thomasius circle 102-3, 108, 110, 112
Thomsen, Henrik 367-8, 370, 372, 381, 385, 388
Three Year Expedition (1931-34) 19, 329, 337-8, 347, 353-4, 356
Thule Expeditions (1912-13, 1916-18) 335-6, 346, 348-9, 351
de Thury, Cassini 130
van Til, Salomon 102
Todd, J. E. 231
Topography 16, 227, 304, 434
Torres Straits Expedition (1898) 27, 29, 416-8
de Tournefort, Joseph Pitton 81-2
Transcaspian railway 283, 290, 295, 305. See also Railway
Transportation 18, 20, 40, 126, 128, 251, 300, 312, 369, 405
Travel journal 102, 106, 110
von Trier, Lars 421
Trionfetti, Lelio 84-5
Truffle 25, 77-94
Trumbull, Loyal W. 231
Turner, Frederick Jackson 438
Twenhofel, W. H. 231
Tygesdatter, Johanne 205-6, 210

U
von Uffenbach, Zacharias Conrad 102-6, 109, 113-5
Uganda 175-9
Ulloa, Don Antonio 134, 144
University of Agricultural Sciences 399
University of Basel 311
University of Colorado Mountain Laboratory 244
University of Colorado Museum 233
University of Copenhagen 59, 194, 285-6, 332, 385
University of Denver 242
University of Göttingen 44, 101
University of Halle 113
University of Minnesota 244
University of Nebraska 230, 232, 244, 256
University of Pennsylvania 259
University of Utah 241
Uppsala University 403
Ur 21, 266-9, 271, 273, 275
Urry, James 309
USA (United States) 44, 229, 235-6, 250, 339, 358, 376, 278, 438
U.S. Biological Survey 229-30
U.S. Bureau of Mines 241
U.S. Commission on Fish and Fisheries 230
U.S. Department of Agriculture (USDA) 229-30, 236-7, 240, 242-44
U.S. Geological Survey (USGS) 227-9, 235-6, 248
U.S. National Museum (Smithsonian) 247-8
Utah 225, 227, 231, 233, 241, 243, 247-9

V

Varelaz, Joseph 134, 142
Venus 123-4, 126, 128, 130, 132, 140
Vespucci, Amerigo 56
Victorian culture 169-72, 186
de Volney, Constantin François 44

W

Walker, James R. 237
Ward, Freeman 231
Weather Bureau 236, 243
Wheeler, George M. 227
Wheeler, Robert Eric Mortimer 275-6
Whitehouse, George 176-9, 185
Whitney Milton 229-30
Wichell, George 133, 142
Wilson, Benjamin 128-9, 142
Wirz, Paul 22-3, 310-2, 317-23, 325-6
Wissler, Clark Wissler 233, 237

Woodward, David 439-41
Woodward, John 31
Woolley, C. Leonard 266
World War I 241-2, 260, 262, 266, 267, 309-10, 312, 325
World War II 262, 274, 278, 317, 343, 377
Wortman, Jacob L. 247
Writing 14, 24-5, 32-3, 36, 39-40, 61-2, 111, 116, 131, 148-9, 150, 161, 164-6, 203, 208, 217, 264, 286, 291, 295, 298, 301, 343, 381-2
Wyoming 225, 231, 233, 238, 243, 245, 247-8, 258

Z

Zoology 9, 31, 42, 153, 232, 283, 286, 300, 305, 324